Chiral Liquid Chromatography

Edited by
W.J. LOUGH, BSc, PhD, MRSC, C. Chem
School of Pharmaceutical and Chemical Sciences
Sunderland Polytechnic
Sunderland, UK

formerly of

Beecham Pharmaceuticals Research Division
Medicinal Research Centre
Essex, UK

Blackie

Glasgow and London

Published in the USA by
Chapman and Hall
New York

Blackie and Son Ltd
Bishopbriggs, Glasgow G64 2NZ
7 Leicester Place, London WC2H 7BP

Published in the USA by
Chapman and Hall
a division of Routledge, Chapman and Hall, Inc.
29 West 35th Street, New York, NY 10001-2291

First published 1989

British Library Cataloguing in Publication Data
Chiral liquid chromatography.
1. Enantiomers. Resolution. Liquid
chromatography
I. Lough, W.J.
541.2'23

ISBN 0-216-92499-5

For the USA, International Standard Book Number is
0-412-01741-5

Phototypesetting by Thomson Press (India) Ltd., New Delhi
Printed in Great Britain by Bell and Bain Ltd., Glasgow

Preface

While working as a chromatographer in the pharmaceutical industry, it became apparent to the editor that there was a pressing need for a comprehensive reference text for analysts working on the resolution of enantiomers by liquid chromatography (LC). This need arises from the fact that, whereas previously it was very difficult to determine enantiomers by direct means, there is now a wide choice of direct LC methods. At the same time, regulatory authorities have been changing their attitudes towards the administration of pharmaceuticals as racemates, partly because it is now possible to study the individual enantiomers. Clearly this abundance of new information needs to be rationalized. More importantly, the chiral LC systems which are commercially available or readily accessible to the practising chromatographer needed to be reviewed and, to a much greater extent than in existing reviews or books, discussed in terms of their practical application.

Accordingly this book is very much orientated towards the practical aspects of these commercially available and readily accessible chiral LC systems. To this end, it is written for practising chromatographers by a team of practising, experienced chromatographers who have spent many years tackling the problems presented by resolving enantiomers by LC.

The practical aspects of common chiral LC systems cannot be fully understood if discussed in isolation. Thus great care has been taken to ensure that this major theme of the book is supported by extensive background materials on the nature of chirality, on the very important reasons for wanting to resolve enantiomers in the first place, and on the historical development of chiral LC. For those relatively experienced in the use of chiral LC, there is a wealth of information on past, present and likely future research work. Also, as befits a practically orientated book, an appendix listing commercial suppliers is provided.

In preparing this book it was recognized that in such a fast-moving field the subject material might date rapidly. However, it is becoming clear that the rate of development of radically novel chiral LC systems is slowing down, and therefore that the logical approach adopted here of rationalizing chiral LC systems by grouping them on the basis of their chiral recognition mechanisms, would automatically have a knock-on 'future-proofing' effect.

It is my sincere hope that, both now and for some years to come, this text will provide valuable assistance to analysts faced with the task of developing a method for the resolution of enantiomers.

WJL

Shortly before this book went into production, one of the contributors, Dr Andrew Pryde, died suddenly. Andrew was a committed and enthusiastic member of the team of contributors, and will be missed both as a friend and as a widely respected scientist who combined dedication to his work with a keen sense of humour.

Contributors

M. Ahnoff A.B. Hässle, Kärragatan 5, 431 83 Mölndal, Sweden.

W.H. de Camp Food and Drug Administration, 5600 Fishers Lane, Rockville, Maryland 20857, USA.

L. Coventry Severn Analytical Ltd., Unit 2B, St. Francis' Way, Shefford Industrial Park, Shefford, Bedfordshire SG17 5DZ, UK.

T.D. Doyle Food and Drug Administration, 5600 Fishers Lane, Rockville, Maryland 20857, USA.

S. Einarsson Department of Analytical and Marine Chemistry, University of Göteborg, S-412 96 Göteborg, Sweden.

A.F. Fell School of Studies in Pharmaceutical Chemistry, University of Bradford, Bradford West Yorkshire BD7 1DP, UK.

D.M. Johns Technicol Ltd., Brook Street, Higher Hillgate, Stockport, Cheshire, SK1 3HS, UK.

B. Kaye School of Studies in Pharmaceutical Chemistry, University of Bradford, Bradford, West Yorkshire, BD7 1DP, UK.

S. Lam Albert Einstein College of Medicine, 1300 Morris Park Avenue, Bronx, NY 10461, USA.

W.J. Lough School of Pharmaceutical and Chemical Sciences, Sunderland Polytechnic, Galen Building, Green Terrace, Sunderland SR2 7EE, UK.

S.A. Matlin Chemistry Department, The City University, Northampton Square, London, EC1V OHB, UK.

T. Noctor Pharmaceutical Chemistry, University of Bradford, Bradford, BD7 1DP, UK.

A. Pryde Maag, Dielsdorf, Switzerland.

G. Szepesi Institute for Drug Research, Budapest, PO Box 82, Szabadsá-gharcosok U47-49, Hungary.

D.R. Taylor Department of Chemistry, UMIST, PO Box 88, Sackville Street, Manchester, M60 1QD, UK.

I.W. Wainer Pharmaceutical Division, Pharmacokinetics and Pharmacodynamics Section, St Jude Children's Research Hospital, 332 North Lauderdale, PO Box 318, Memphis, Tenesee 38101, USA.

Contents

Part 1 Introduction

1 Molecular asymmetry

S.A. MATLIN

1.1 Chirality

Molecular isomerism is a subject of fundamental interest to organic chemists. This is particularly true of the subject of stereoisomerism, which has been extensively covered in the literature. Texts by Eliel (1962), Herz (1964), Hallas (1967), Kagan (1979) and Ramsay (1981) deal with all aspects of organic stereochemistry and are recommended for further reading. In this context lack of molecular symmetry is important in that it allows the classification of stereoisomers as either diastereomers or enantiomers. A molecule has the property of chirality (dissymmetry) when its mirror image is non-superimposable on itself. The non-superimposable mirror image isomers are called enantiomers, but also may be referred to as enantiomorphs, optical isomers, or optical antipodes. The enantiomers of a compound have different configurations: they have the same bond connectivities but differ in three-dimensional structure due to different arrangements of the bonds in space. In order to interconvert two configurations, it is necessary to break at least one

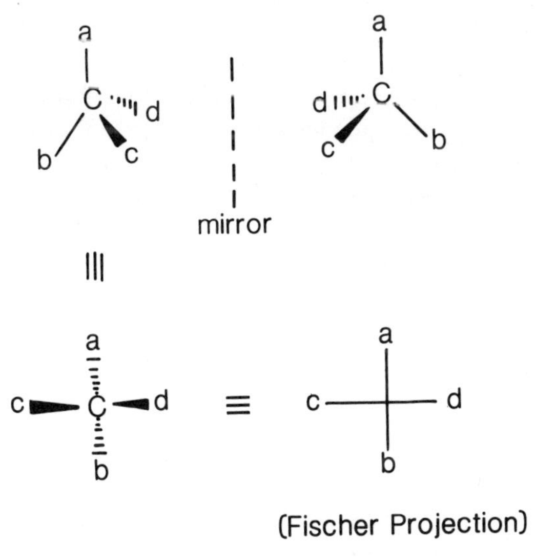

Figure 1.1 Representations of absolute configuration.

pair of bonds in the molecule and exchange the positions of the substituents, or to carry out some other high-energy twisting process (see below) which would not normally occur in the non-excited molecule. Configurations can therefore be distinguished from conformations, which are also differing three-dimensional shapes of the same compound, but shapes which are inter-convertible at ambient temperatures by means of low energy barrier, single-bond rotations.

Absolute stereochemistries of molecules can be depicted in perspective drawings by using uniform lines to represent bonds in the plane of the paper and solid and dashed wedges to represent bonds projecting in front of, or behind, the plane, respectively. Alternatively, Fischer projection formulae may be used, adopting the convention that a plain cross represents a carbon atom with four bonds, the vertical bonds always pointing away from the observer and the horizontal bonds always towards the observer (Figure 1.1).

1.2 Chiroptical properties

Enantiomers of a given compound have identical chemical properties in relation to their reactions with non-chiral reagents, although they will give products of different configurations. In addition, they may show differences in behaviour (both in reaction rates and in product stereochemistries) in their interactions with one enantiomer of a chiral reagent. Their physical properties (e.g. melting point, boiling point, solubilities and partition coefficients in different phases) are all identical. However, they can be distinguished from each other by observation of their effects on plane polarized light: equimolar solutions of two enantiomers will rotate the plane of polarization of plane polarized light by equal amounts, but in opposite directions. The enantiomer causing a right-handed (clockwise) rotation is the dextrorotatory (d) or (+) isomer; that causing a left-handed (anticlockwise) rotation is the laevorotatory (l) or (−) isomer. An equal mixture of two enantiomers is called a racemic form (racemic modification). The term racemic mixture is often used as if it were synonymous with racemic form, although strictly it refers to a solid mixture containing equal quantities of individual crystals of the (+) and (−) isomers. Biot's Law states that the optical rotation α observed for a given solution of an enantiomer, at a wavelength λ and temperature T, is proportional to its concentration c in g per 100 ml and to the cell path length l in dm, allowing calculation of the specific rotation which is characteristic of a given enantiomer (Equation 1).

$$[\alpha]_{\lambda}^{T} = 100\frac{\alpha}{lc} \qquad (1)$$

There is no direct method by which the absolute configuration of an enantiomer of a new compound can be deduced from its sign of rotation. For example, when lactic acid is esterified with diazomethane, the resulting methyl

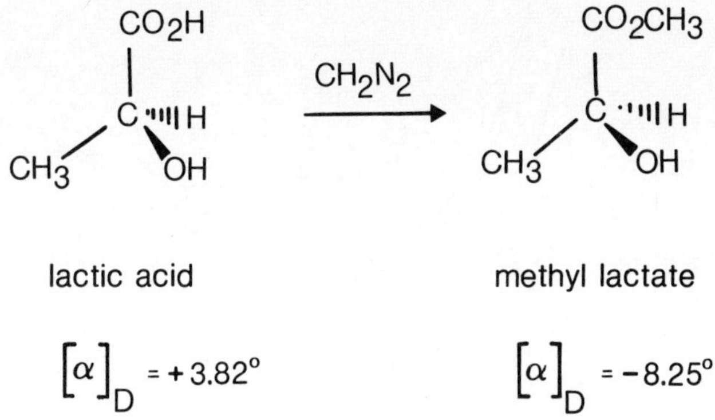

Figure 1.2 Non-correlation of sign of rotation with absolute configuration.

lactate is clearly formed without any alteration to the absolute configuration of the molecule. However, when this reaction is performed on (+)-lactic acid, the product is (−)-methyl lactate (Figure 1.2).

The optical rotation of a given solution of an enantiomer will vary with the wavelength of plane-polarized light used, The plot of rotation against wavelength is called the optical rotatory dispersion (ORD) curve. If the molecule does not contain a chromophore absorbing in the region being scanned, the resulting ORD plot will be a simple, smooth curve either increasing (plain positive) or decreasing (plain negative) towards shorter wavelength, and in some cases the curve will cross the rotation axis, i.e. changing from (+) to (−) rotation at a particular wavelength (Figure 1.3). If the molecule contains a chromophore absorbing within the region scanned, the result is an anomalous dispersion curve, containing a peak and a trough. This anomalous appearance is known as the Cotton effect. In the direction of decreasing wavelength, if the peak appears before the trough the curve is said to show a positive Cotton effect (Figure 1.3); the enantiomer will have a mirror-image ORD curve exhibiting a negative Cotton effect.

If the molar extinction coefficients of a solution of an enantiomer are measured at different wavelength using right- and left-circularly polarized beams, they will be found to differ if the enantiomer contains a chromophore absorbing within the region scanned. A positive difference ($\varepsilon_R - \varepsilon_L$) corresponds to a positive Cotton effect (Figure 1.4).

Whereas absolute configuration cannot be deduced by inspection of the sign of rotation of an enantiomer at one wavelength, it transpires that empirical correlations can be made between the sign of the Cotton effect (or the equivalent sign of the CD curve) and absolute configuration. Empirical rules have been deduced for many simple classes of compound (e.g. the octant rule for substituted cyclohexanones) and the reader is referred to other texts for details of these (Crabbe, 1972; Ciardelli and Salvadori, 1973). For new classes of compounds, correlations between members can often be quickly established

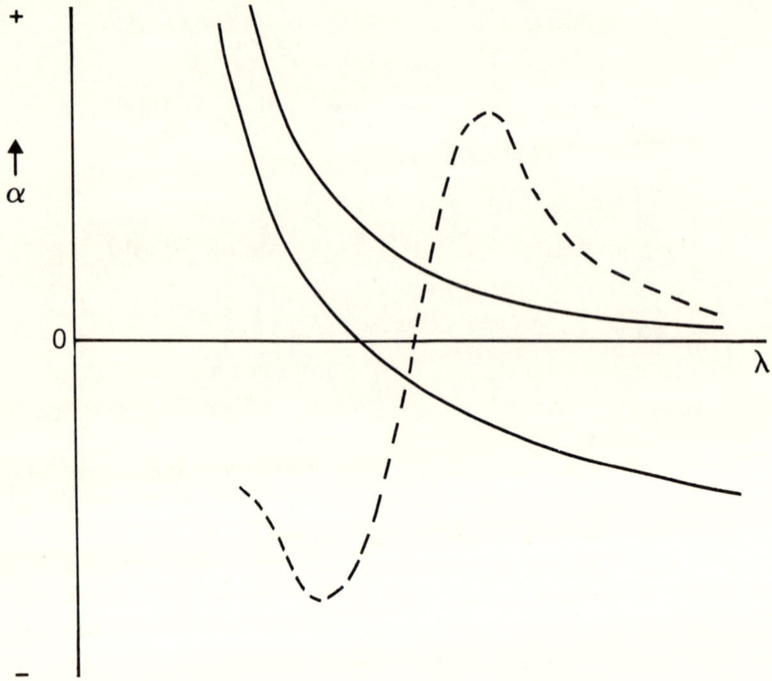

Figure 1.3 Optical rotatory dispersion (ORD) curves.
—Plain positive curves;---anomalous dispersion curve with single positive Cotton effect.

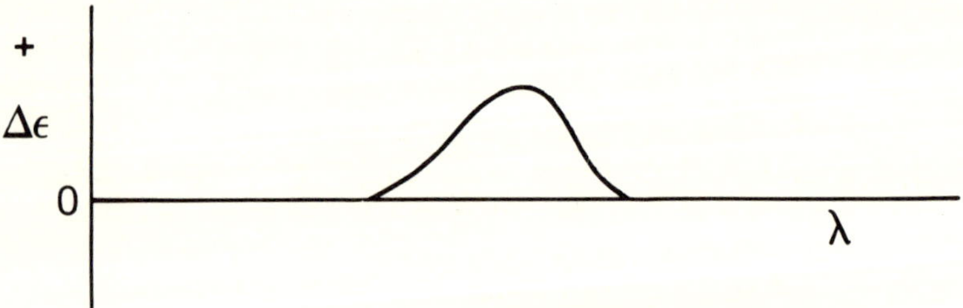

Figure 1.4 Positive circular dichroism (CD) curve.

by means of the Cotton effect, once the absolute configuration of a typical member of the class has been determined.

1.3 Asymmetric centres

Chirality in a molecule may be due to the presence of an asymmetric carbon atom, i.e. a carbon atom carrying four different substituents. An example is the

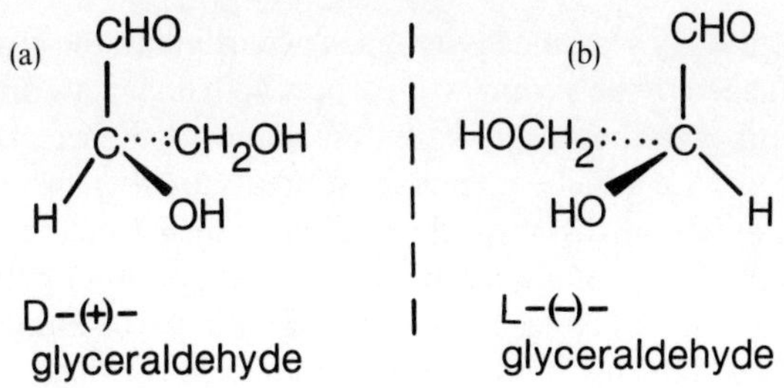

Figure 1.5 Absolute stereochemistry and designation of glyceraldehyde enantiomers.

case of glyceraldehyde, whose dextrorotatory (d) and laevorotatory (l) isomers are shown in Figure 1.5. Before the absolute configurations of organic compounds could be determined, Fischer chose glyceraldehyde as a standard to which other molecules, such as sugars and amino acids, could be compared. He arbitrarily assigned the absolute configuration (Figure 1.5a) to the d isomer and gave this the designation D to describe its configuration: any carbon atom having an arrangement of functional groups which could be chemically related to those in D-glyceraldehyde was similarly designated a D centre. Fortunately, Fischer's arbitrary choice of the absolute configuration of (d)-glyceraldehyde later turned out to be correct, otherwise all early texts in the field would have had to be corrected.

With the later development of interest in the stereochemistries of many other, diverse classes of compounds which could not be easily correlated with glyceraldehyde, a more universal system for designating the absolute configurations of chiral centres was devised by Cahn, Ingold and Prelog. A priority sequence $1 > 2 > 3 > 4$ is assigned to the four substituents on an asymmetric carbon atom, and the molecule then viewed so that the group of lowest priority is pointing away from the observer. If the remaining three groups then follow a clockwise progression 1-2-3, the asymmetric centre is designated R (rectus); if the sequence 1-2-3 is anticlockwise, the designation is S (sinister), as shown in

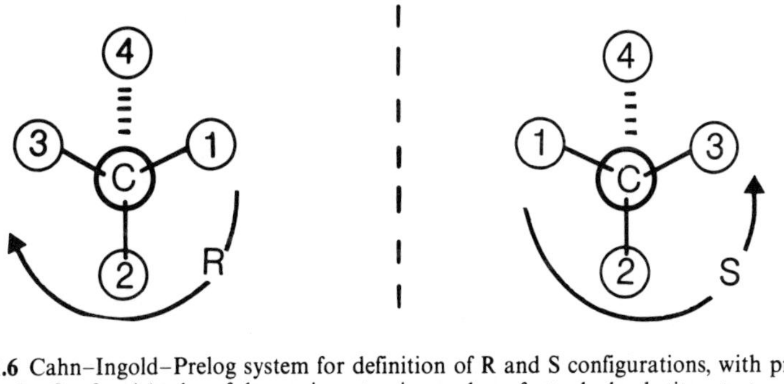

Figure 1.6 Cahn–Ingold–Prelog system for definition of R and S configurations, with priority sequence $1 > 2 > 3 > 4$ (order of decreasing atomic number of attached substituent atoms).

Figure 1.6. The priority sequence is assigned according to the atomic number of each atom attached to the asymmetric carbon, with higher atomic number (or higher mass number in the case of isotopes) acquiring higher priority. When two identical atoms are attached, their substituents are then inspected in turn, always moving an equal distance along each branch from the asymmetric centre, until a substituent of higher atomic number (or, failing that, a greater number of substituents of equal highest atomic number) is found. Using this system, the (+) or (d) or (D) isomer of glyceraldehyde is assigned the absolute configuration (R).

Compounds with more than one asymmetric centre in the molecule can have a number of stereoisomers: for *n* asymmetric centres, a maximum of 2^n isomers is possible. Within the set of stereoisomers, some will consist of pairs in which all asymmetric centres are inverted (enantiomers): other stereoisomers will be related by differing in configuration in at least one, but not all, of the asymmetric centres present, and are called diastereoisomers. For example, the tetrose $OHCCH(OH)CH(OH)CH_2OH$ has four stereoisomers, consisting of

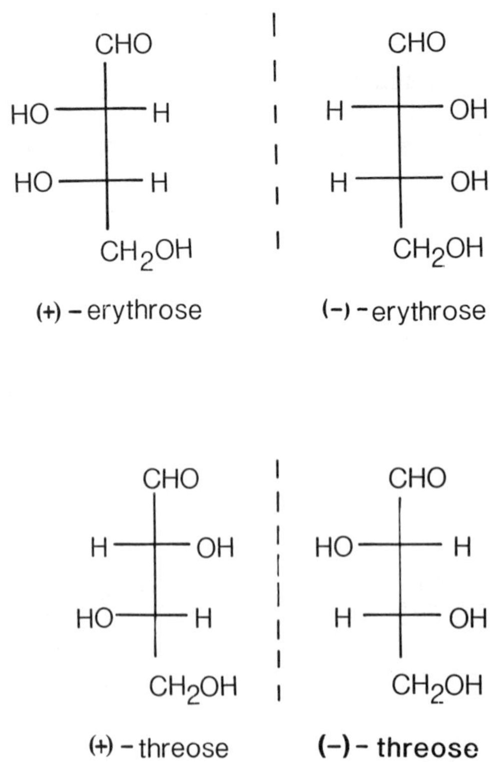

Figure 1.7 Stereochemistry of the tetroses. Each enantiomer of erythrose is a diastereoisomer of each enantiomer of threose.

two pairs of enantiomers. Each enantiomer is a diastereoisomer of the other pair of enantiomers (Figure 1.7). Molecules with more than one asymmetric centre which have identical sets of substituents about a central line can have an internal plane of symmetry. The isomer which has this internal, mirror plane is called a meso form, is said to be internally compensated and is achiral: the molecule and its mirror image are superimposable and this form does not exhibit optical activity. For example, the tartaric acids (Figure 1.8) consist of a meso form (achiral), diastereomeric with a pair of enantiomers (d and l forms).

Atoms other than carbon can also provide centres of asymmetry. Nitrogen can be asymmetric in quaternary ammonium compounds and tertiary amine oxides, which are configurationally stable. In tertiary amines with a free lone pair, the energy barrier to inversion is usually too small for enantiomers to be separable at ambient temperatures, but may be increased in less conformationally mobile cyclic amines. Similarly, cyclic sulphoxides can sometimes be resolved and pentavalent phosphorus compounds can be resolved in cases where pseudorotation (i.e. the process whereby there is a rearrangement of nuclear positions in a trigonal bipyramidal complex by way of an intermediate square pyramidal complex) is sterically inhibited.

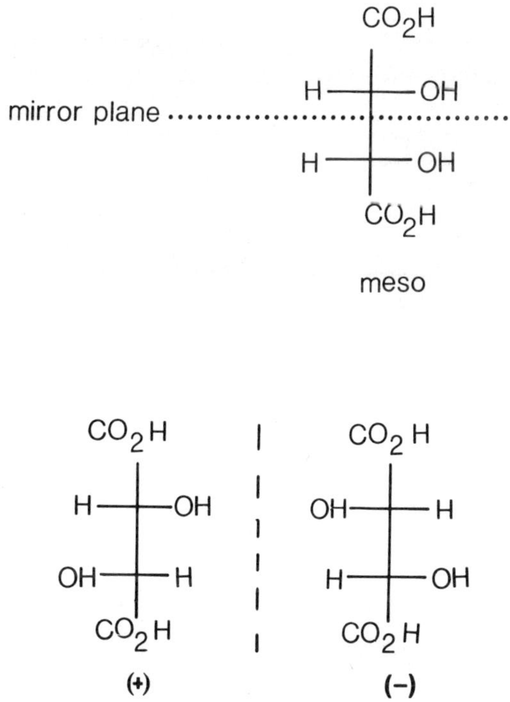

Figure 1.8 Stereochemistry of the tartaric acids.

1.4 Twisted structures

Molecules in which there is a permanent and rigid twist in the planes of atom connectivities display chirality.

1.4.1 *Cumulenes*

Cumulenes are compounds containing two or more successive double bonds. In allenes, the two alkylidene fragments connected to the central carbon atom occupy perpendicular planes. Provided that the two substituents within each alkylidene fragment are not identical to each other, the molecule will have two non-superimposable mirror image forms (enantiomers), even though there is not an asymmetric carbon atom present (Figure 1.9).

Buta-1, 2, 3-trienes have two 90° twists, resulting in coplanarity of the terminal substituents. Consequently, these cumulated trienes will show only geometric isomerism, as found in substituted ethylenes. In cumulated tetraenes, a net 90° twist between the termini will again result in the possibility of enantiomeric pairs.

Racemization in chiral cumulenes can be effected by thermal, photochemical or metal catalysis, as is the case for geometric isomerization in simple olefins.

1.4.2 *Spiro compounds*

Compounds in which a pair of rings is joined at a single, common (spiro) atom, will display chirality when the rings are substituted so as to distinguish

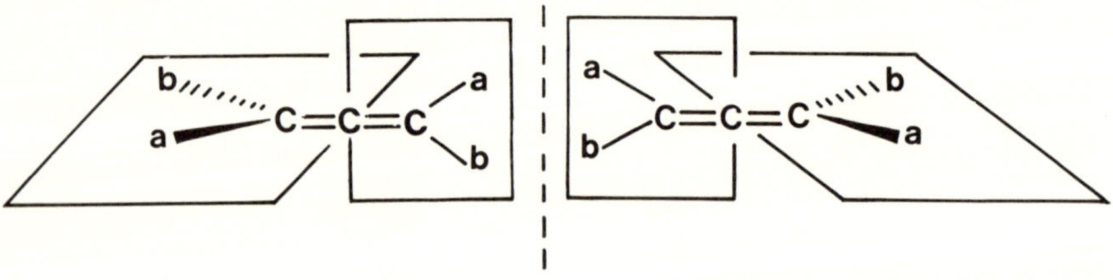

Figure 1.9 Enantiomeric allenes.

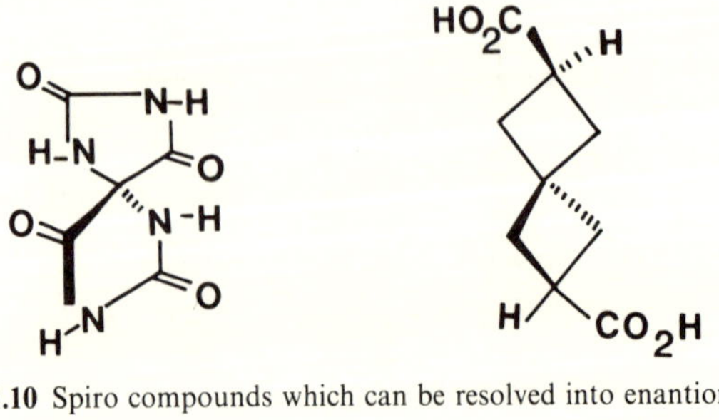

Figure 1.10 Spiro compounds which can be resolved into enantiomers.

between their two faces. The spiro atom may be carbon, or any other tetravalent or tetraco-ordinate species (Figure 1.10).

Inversion at spiro atoms is generally impossible without first breaking at least one bond to generate a radical or ionic intermediate.

1.5 Hindered rotation

1.5.1 *Atropisomers*

Atropisomerism occurs when complete rotation about a single bond in a molecule is prevented by the bulk of non-identical neighbouring substituents, so that a pair of enantiomers can be obtained.

In molecules in which two ring systems are joined by a single bond, if complete rotation about this bond is prevented by steric effects then the two ring systems will occupy different planes. Enantiomeric pairs will occur if the two ring systems are differently substituted so that planes of symmetry are absent. An example of enantiomeric, hindered biphenyls is shown in Figure 1.11*a*. The hindered binaphthyl derivative, gossypol (Figure 1.11*b*), is a chiral natural product produced in large amounts by the cotton plant. Many flavonoids are also capable of chirality as a result of restricted rotation about the bond between substituted benzene and chroman rings. Even in the monocyclic compound (Figure 1.11*c*), steric effects of the ortho substituents restrict the rotation of the branched side chain, creating atropisomers.

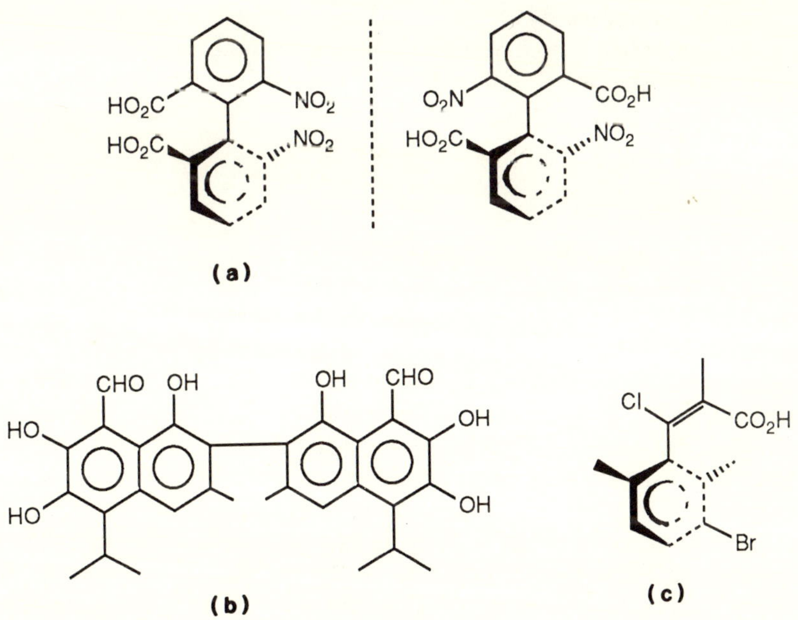

Figure 1.11 Molecules which are optically active due to hindered rotation about a single bond.

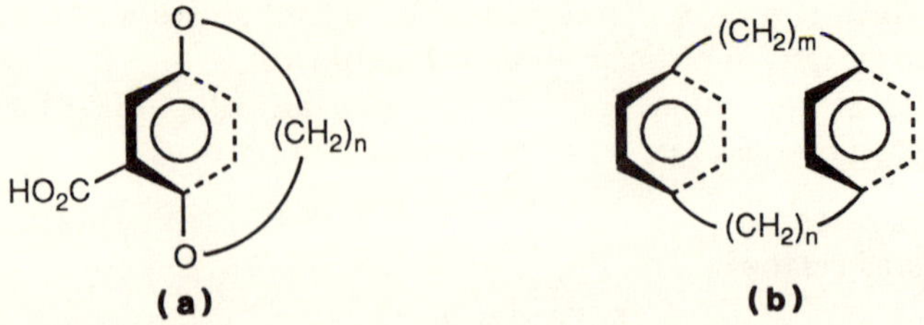

Figure 1.12 Structures of ansa compounds and paracyclophanes with restricted rotation.

Interconversion of atropisomers can often be accomplished by heating, supplying sufficent thermal energy to overcome the rotational energy barrier. This is observed for the enantiomeric biphenyls (Figure 1.11*a*), but the enantiomers of gossypol are not racemized even on melting at 200°.

1.5.2 *Ansa compounds*

In ansa molecules, a 'strap' or 'basket handle' is present, bridging across a ring system and the complete rotation of the ring about the points of junction to the handle is sterically prevented. Provided that the molecule is substituted so as to distinguish between the two edges of the ring, enantiomeric forms can occur: e.g. in the ansa compound (Figure 1.12*a*), the enantiomers are optically stable when $n = 8$, racemize on heating when $n = 9$ and are unresolvable when $n = 10$.

Similar effects are seen in cyclophanes, where two ring systems are doubly connected face-to-face by means of a pair of bridges. For example, ring rotations are prevented in the paracyclophanes (Figure 1.12*b*) when the bridge lengths are short ($n, m = 2, 3$). The introduction of at least one ring substituent will remove the plane of symmetry from these molecules and lead to chirality.

1.6 Molecular overcrowding

In molecules which possess a high degree of rigidity, usually due to the presence of one or more (aromatic) rings, two substituents may be placed so that they are trying to occupy the same space. In such overcrowded conditions, the molecule will deform in such a way as to minimize the high-energy steric interactions, often resulting in a twisted or bent structure lacking in conformational flexibility and possessing chirality. Examples include the dimethyl phenanthrene (Figure 1.13*a*) and hexahelicene (Figure 1.13*b*). The latter, in which the chromophore is itself dissymmetric with a helical twist, has an extremely large specific rotation ($[\alpha]_D^{25} = 3700°$).

The ease of racemization in these compounds is related to the extent of overlap and the degree of flexibility of the molecule. Thus, whereas the dimethyl phenanthrene (Figure 1.13*a*) will racemize in solution at room temperature, hexahelicene only partly racemizes on melting at 266°.

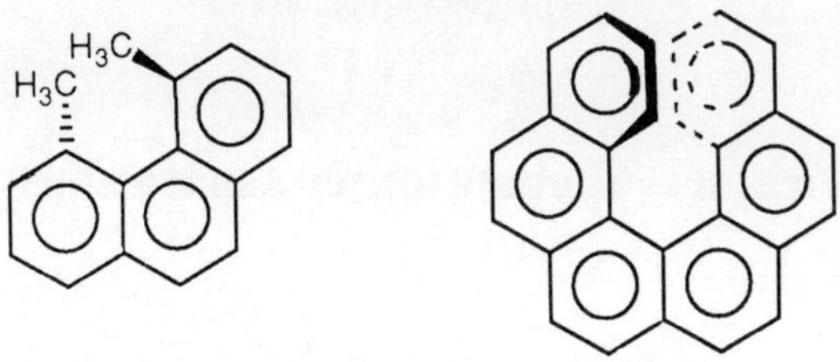

(a) (b)

Figure 1.13 Molecular overcrowding in polycyclic aromatics.

References

Ciardelli, F. and P. Salvadori (Eds) (1973) *Fundamental Aspects and Recent Developments in Optical Rotatory Dispersion and Circular Dichroism*. Heyden, London.

Crabbe, P. (1972) *Optical Rotatory Disperion in Chemistry and Biochemistry*. Academic Press, New York.

Eliel, E. (1962) *Stereochemistry of Carbon Compounds*. McGraw-Hill, New York.

Hallas, G. (1967) *Organic Stereochemistry*. McGraw-Hill, New York.

Herz, W. (1964) *The Shape of Carbon Compounds*. W.A. Benjamin, New York.

Kagan, H. (1979) *Organic Stereochemistry*. Edward Arnold, London.

Ramsey, O.B. (1981) *Stereochemistry: Nobel Prize Topics in Chemistry*. Heyden, London.

2 The importance of enantiomer separations

W.H. DE CAMP

2.1 Introduction

Since the discovery of the optical isomerism of tartaric acid by Louis Pasteur in 1848, the significance of stereoisomerism in relation to biological activity has been recognized. It was soon seen that the separation of a racemic mixture into its component stereoisomers presented a challenge of immensely greater magnitude than did the development of a stereospecific synthesis. As a result, efforts to resolve such mixtures were largely bypassed by scientists for more than a century. The rare instance of a successful separation was often treated either as happenstance or as the success of trial and error. In either event, a systematic approach to the problem was not seen as feasible, except in a few special cases. Clearly, commercial exploitation was unthinkable.

The result of this perception, particularly among scientists working on the development of new drugs, was that their research efforts were directed more by the technical feasibility of the experiment than by a concern with the biological effect of the drug. Questions of whether clinical efficacy and safety were greater in one member of an enantiomeric pair were asked only when a synthetic route which was both stereospecific and economical was available.

The successful development of chiral stationary phases for high-performance liquid chromatography (HPLC) in the late 1970s altered this situation. The separation of optically pure material from a racemic mixture in amounts adequate for clinical investigations on a routine basis became feasible. Enantiomeric purity could be determined for the bulk drug, for its formulations, and in biological fluids. The investigator could now ask the question, 'Why should I separate a racemic mixture?', knowing that it was not an exercise in rhetoric.

Regulatory agencies, such as the US Food and Drug Administration (FDA), generally focus primarily upon considerations of either or both clinical efficacy and consumer safety in determining whether to allow a drug to be marketed. These two criteria, therefore, represent the first focus of this chapter. The chiral environment found *in vivo* will be considered in relation to the biological activity of a drug. It will be demonstrated why this presents a problem in the approval of drugs for marketing. The case of thalidomide will be discussed in detail as an example of a problem that was, at least, complicated by the ignorance of stereochemical effects. Second, the evidence that a variety of other conclusions about the behaviour of a drug *in vivo* may be affected by the isomeric ballast (Ariëns, 1984, 1986; Ariëns and Wuis, 1987) present in a racemic mixture will be explored. In particular, it will be demonstrated that the use of racemates can lead both to erroneous models of pharmacokinetic behaviour and to the potential for opportunities to manipulate pharmacological activity. Finally, the implications of these factors for the design of pre-clinical and clinical tests will briefly be considered. How experiments may be designed which will answer, with as little ambiguity as possible, the question, 'Is a stereochemically pure drug more effective and/or less toxic than the racemate?' will then be considered.

The question of the importance of separations of enantiomeric agrochemicals, food additives, pesticides and so on will not be specifically addressed. However, many of the points raised here with respect to drugs are relevant to these other fields. The situation is still the same, i.e., a racemic mixture and its component stereoisomers may exhibit different biological activity.

2.2 Enantiomers versus racemates: questions of efficacy and safety

Molecules which are enantiomeric differ in properties only insofar as the property being investigated also has a characteristic of handedness. Thus, enantiomers do not differ in melting point because there is no chirality in the molecular vibrations which increase as temperature increases, thus causing the solid to melt. In contrast, the rotation of polarized light does have a specific handedness, and enantiomeric molecules behave differently with respect to this property. Similarly, reaction with a chiral substrate (e.g. an enzyme, a chiral HPLC stationary phase) is a process with handedness, leading to diastereomeric combinations which differ in properties. Such differences necessarily underlie all resolutions, even when the diastereomers are only transient in their existence, as is the case in chiral HPLC.

The physiological environment *in vivo* is inherently chiral, as a molecule of a drug must interact with enzymes, cell surfaces and so on, all of which are asymmetric at the molecular level. Therefore, enantiomeric molecules are expected to differ in their behaviour under such conditions. But, just as the differences between diastereomeric molecules in chemical or physical properties, such as solubility, cannot be predicted, the differences in physiological effect between enantiomers *in vivo* also cannot be predicted.

Whether in consideration of the desired or undesired effect of a drug, the opinion appears to be generally held among pharmaceutical scientists that the effects are *necessarily* separable between the enantiomers. While this may be true in some instances, for instance, ketamine, described by White *et al.* (1980), such an assumption is not generally justified. Lacking any agonist or antagonist interactions between the enantiomers, any effect which differs between the enantiomers should be intermediate in magnitude when the racemate is tested. Consequently, investigation of the potential for differences in efficacy, toxicity, or even pharmacological effect should be undertaken as part of pre-clinical studies.

The definition of a "drug" in the US Food, Drug and Cosmetic Act does not specifically consider the question of its stereochemical composition. Whether a drug substance is considered to be the racemate or a pure stereoisomer has, in general, been left up to the judgment of the firm at the time that an application to market a new drug is submitted. From the discussion to follow, however, it appears that the most conservative route to be followed by a manufacturer in the course of drug development is to consider both enantiomers, as well as the racemate, to be potential drugs, and to carry out at least limited pre-clinical and/or clinical studies on all three forms. In contrast, the equally conservative attitude which should be adopted initially by a regulatory agency is that only one of the three stereoisomeric combinations is most desirable (i.e. greatest efficacy or least undesired side-effects). Thus, the marketing of a racemate in preference to a pure stereoisomer should be justified by the data submitted.

The tragedy of thalidomide in the early 1960s led to amendments to the US Food, Drug, and Cosmetic Act which established the FDA's authority to evaluate the effectiveness of drugs and to require applications for investigational studies. Given the fact that the routine resolution of racemates was not feasible at that time, it is not surprising that, in addressing these issues, the Kefauver–Harris Amendments of 1962 did not contain a specific requirement that the clinical effectiveness of a racemic drug be evaluated relative to the pure stereoisomers of which the mixture is composed. While thalidomide was never marketed in the US, it entered commerce in Europe with no apparent consideration of the relation between its biological activity and its stereochemistry (*Sunday Times* (London) Insight team, 1979). Although alleged to be non-toxic, it was eventually found to be both neurotoxic (causing peripheral neuritis) and teratogenic (causing a variety of fetal abnormalities, especially a birth defect known as phocomelia, in which the hands or feet start immediately from the shoulder or hip). The tragedy of its teratogenicity stimulated research into this problem.

The developmental research for thalidomide took place without the technological tool of chiral HPLC. Even though a stereospecific synthesis for thalidomide was ultimately developed (Casini and Ferappi, 1964; Shealy *et al.*, 1965, 1968), the racemate was obtained by the synthetic methods used at the time of its initial use in Germany (Chemie Grunenthal GmbH, 1957). Resolution of clinically useful supplies of the pure enantiomers does not appear to have been described in the open literature. The results of several investigations of the safety and efficacy of the stereoisomers of thalidomide (all carried out after the fact of the tragedy) have been published. The results of these studies suggest that the enantiomers of thalidomide differ significantly in their biological activity. This has, in turn, led others to conclude that the teratogenic effect of thalidomide is found in only one enantiomer, and to speculate that the tragedy could have been avoided if the other had been marketed (De Ranter, 1984; Drayer, 1986). A critical review of the experimental data clearly shows that this conclusion, while possibly true, is not fully supported by the published studies. It is, however, clear that experimental designs were not adequate to unambiguously determine the enantiomeric differences.

A stereospecific synthesis for thalidomide is available, starting from glutamic acid or its derivatives. No step in the reaction sequence involves the chiral centre, so the absolute configuration of the product is known (Casini and Ferappi, 1964; Shealy *et al.*, 1965, 1968). Dextrorotatory thalidomide has the D-configuration (Shealy *et al.*, 1968). This corresponds to the R-configuration, using the nomenclature of Cahn *et al.* (1966). The stereochemical purity of a synthetic batch must, obviously, be identical to that of the starting material. A crystal structure has been reported for racemic thalidomide (Trotter and Allen, 1971), as well as a 4-bromo derivative (Furberg and Petersen, 1965). Both crystal structures were done on the racemic compound. The conformation of the molecule was essentially identical in both cases. The racemate has been resolved by HPLC, using a chiral stationary phase (Blaschke *et al.*, 1980).

The earliest reported investigation of stereoisomerism and biological activity in thalidomide showed no difference in teratogenic action at an oral dose of 150 mg/kg from days 7 to 12 inclusive in New Zealand white rabbits (Fabro *et al.*, 1967). However, this same report claimed that the LD_{50} in SAS ICI albino mice was greater by a factor of approximately 20 for the racemate relative to either of the pure stereoisomers. A later study (Blaschke *et al.*, 1979) showed significant differences between the enantiomers, with the teratogenic activity concentrated in the S(−) isomer. This study used SWS mice and Natal rats, rather than the New Zealand white rabbits which were known to be the most sensitive to such effects, as well as intraperitoneal instead of oral administration. An investigation of the effects of thalidomide on the graft versus host reaction in chick embryos showed that S(−) and racemic thalidomide had a significant immunosuppressant action, while R(+)-thalidomide had none (Field *et al.*, 1966). The evidence which seems most indicative of the teratogenic action being restricted to S(−)-thalidomide comes from a series of studies on its hydrolysis products (Ockenfels and Köhler, 1970; Köhler *et al.*, 1971; Meise and Köhler, 1971, 1972; Meise *et al.*, 1973). This conclusion should be accepted with caution, since the mechanism of action remains uncertain.

A review article (Simonyi, 1984) observed that the ratio of malformed to normal fetuses in one earlier study (Fabro *et al.*, 1967) was less than half that found for the racemate. While no specific comment was made on this observation in either source, the implication is clear that the pure stereoisomers are mutually synergistic with respect to both toxicity and teratogenicity.

Unfortunately, none of the published studies has succeeded in answering the questions without ambiguity. The definitive experiment does not yet appear to have been done. The published investigations have focused on stereochemical aspects of the hypnotic and teratogenic effects of thalidomide, to the neglect of its neurotoxicity, which is also an undesirable side-effect. It may be an over-simplification to assume that desirable and undesirable actions must be separable between enantiomers. Yet, in order to perform the necessary experiments, it is necessary to test such an hypothesis using the pure enantiomers.

2.3 Enantiomers versus racemates: questions of pharmacokinetics

Even if one enantiomer is not merely isomeric ballast, with no biological effect, it is clear that, within the body, the drug exists in a chiral environment where its release, absorption, transport, action, degradation, and elimination may involve interactions with enzymes, cell surfaces and so on. This fact means that two enantiomeric molecules should be acted on differently by the body. The factors which differ between enantiomers are not limited to pharmacological effects. We will consider selected examples which demonstrate that pharmacokinetic models for racemic drugs are inherently invalid if they assume a single substrate for such actions.

A primary example of such effects is the family of non-steroidal anti-inflammatory drugs (NSAIDs). With a single exception (naproxen and its sodium salt), all of the NSAIDs approved for marketing in the USA are marketed as racemic mixtures. The enantiomers of 2-aryl propionic acids frequently show stereoselectivity in their disposition kinetics (Hutt and Caldwell, 1983, 1984, 1985, and references cited therein). Furthermore, metabolic inversion of the R enantiomer to the S form has been demonstrated for many members of this family of drugs. Tiaprofenic acid is an exception (Singh *et al.*, 1986).

The different pharmacokinetic behaviour of enantiomers appears to have been a contributor to the adverse reactions which led to the withdrawal of benoxaprofen from the market in 1982. Inversion of R(–)-benoxaprofen to its S(+) enantiomer in humans following oral administration of either the racemate or the pure R-enantiomer has been demonstrated (Bopp *et al.*, 1979). Subsequent *in vitro* studies suggested that the inversion occurs as the drug passes through the intestinal wall (Simmonds *et al.*, 1980). The contribution of this inversion to the decreased rate of metabolism and excretion in elderly patients, which led to the hepatotoxicity which prompted the drug's withdrawal from the market, does not appear to have been established (Marshall, 1985; Prescott, 1986). As in the case of thalidomide, it is interesting to speculate about the possibility that this unfortunate incident would not have occurred if the pure stereoisomer had been developed for the market.

Disopyramide is another drug which shows stereoselective pharmacokinetics. Its binding to plasma protein has been shown to be both stereoselective and concentration-dependent (Giacomini *et al.*, 1986). This combination of kinetic factors leads to data which are unexplainable by a model which assumes that the drug is a single component. Such an assumption, unfortunately, appears to be common practice (Ariëns and Wuis, 1987).

In at least one case (for a drug not yet marketed in the USA) the enantiomeric ratio of a drug has been varied to improve its therapeutic effects (Tobert *et al.*, 1981). Indacrinone is a relatively long-acting, high-ceiling diuretic. Although both enantiomers have uricosuric activity, the (−) enantiomer is more potent as a natriuretic agent. A balance between the natriuretic and uricosuric effects was found for a 4:1 ratio of (+):(−) enantiomers.

A thorough understanding of the pharmacokinetics of a racemic drug, therefore, implies knowledge of the behaviour of the pure stereoisomers. Indeed, given adequate information about the kinetic behaviour of both enantiomers of a drug as well as the racemate, it is tempting to speculate as to the extent to which its pharmacokinetics might be able to be varied by the use of partially, rather than fully resolved drug.

The use of combinations of drugs to improve therapeutic efficacy is not unusual. FDA's regulations require that, in such a combination,

'...each component makes a contribution to the claimed effects and the dosage of each component...is such that the combination is safe and effective...' [21 CFR 300.50]

The potential for application of such a regulation to racemates seems obvious, although not explicitly stated. Extension of this regulation to racemates has not yet been formally proposed.

US regulatory requirements include a requirement that the bioavailability of the drug be demonstrated. When pharmacokinetic models differ between enantiomers, it seems obvious that establishing the bioavailability of the drug from a racemate is a much more complex task, which cannot be accomplished without separation of the enantiomers and investigations of their pharmacokinetics as individual molecular entities.

2.4 Experimental considerations for investigation of pure enantiomers

The experimental design of a clinical study to compare the efficacy of one drug as against another is well established. Such studies are required to be adequate and well-controlled, with such characteristics as

'The study uses a design that permits a valid comparison with a control to provide a quantitative assessment of drug effect....' [21 CFR 314.126(b) (2)]

The purpose of the control is clearly stated. The control drug should be administered in a clinically effective dose and regimen. However, in order that a comparison be valid, it must be similar in both dose level and stereochemical composition. Comparison of the biological effect of a pure stereoisomer with that of a racemate at the same dose level does not meet this condition.

Consider the case of two enantiomers being compared to the racemate. Let us designate a drug formulated from the racemic mixture as RS, and, therefore, that from the pure stereoisomers as R or S. Administering all three at the same dose in mg/kg may be a simple experimental design, but the results are likely to be uninformative. To be sure, if R and S are equal in both effectiveness and toxicity, and synergistic effects of one enantiomer on the other are lacking, an identical response is expected, regardless of whether R, S, or RS is administered. But an observation of unequal response does not lead to an unambiguous conclusion for the pharmacokinetic reasons discussed above. An observation of greater effect on the part of R or S than for RS leads correctly to a conclusion of stereoselectivity in biological effect only if a linear dose–effect relationship has been demonstrated and the opposite enantiomer is wholly lacking in activity.

The racemic drug RS should be considered to be a 50:50 fixed composition mixture of two potentially active substances. Therefore, R and S should be compared both with RS at the same dose level and with RS at twice the dose level (RS × 2). At the RS × 2 dose level, each enantiomer is present at the same level as in the enantiomerically pure drug. Differences in response can then be attributed to the presence of the opposite enantiomer. Thus, if one enantiomer is indeed stereoisomeric ballast, either with respect to efficacy or adverse effects, this will be revealed by an observation of different results between the R versus RS × 2 and S versus RS × 2 experiments. Given such evidence, both with respect to desirable and undesirable effects, development of the racemic drug for marketing would, in general, not be precluded on the basis of either safety or efficacy. If an economically feasible stereospecific synthesis were available, then development of the effective enantiomer would appear to be the preferable option from the point of view of both safety and efficacy.

If, for example, both R and S possess the desirable biological activity to an equal degree, then the RS × 2 experiment would show a greater effect (though not necessarily twice as great) than any of R, S, or RS, because the dose level would be doubled. If similar results were obtained in studies of toxic and other adverse effects, then the stereochemically pure drug could be concluded to lack an advantage over the racemate, from considerations of both efficacy and safety. In this situation (which would likely be unusual), it is conceivable that an argument might be made that an approval of the marketing of the racemate logically implied approval of both enantiomers as optically pure drugs.

It seems unlikely that any collection of evidence would allow the reverse situation, i.e., that approval of an optically pure stereoisomer would imply approval of the racemate. Yet this situation potentially exists. Consider a drug marketed as an optically pure compound, whether synthetic or natural in origin. Unless compendial requirements include at least one identification test which is sensitive to a difference between the optically pure material and the racemate, it is conceivable that a drug product manufactured from the racemate could reach the market on the basis of the compendial monograph alone. Usually optical rotation would suffice for such a test, although, given a drug with a sufficiently small specific rotation, other identity tests should be used. A specification for a single chromatographic peak, using a chiral column under conditions shown to be capable of separating the enantiomers would be one approach to this situation. However, measurement of physical properties known to be dependent upon the solid state structure (such as melting point or X-ray powder diffraction pattern) might be preferable, as the solid state structure generally differs between the racemate and the pure stereoisomer.

2.5 Conclusion

Good science requires that our conclusions be based upon experimental evidence which is derived from well-planned experiments. Such a level of planning cannot neglect the potential for differences in properties for enantiomers of a chiral molecule in a chiral environment. Thus, not only is it necessary to recognize the implications of stereochemistry for drug action, but it is also necessary that they be investigated, and therefore either the enantiomers must be separated, or they must be synthesized.

Good sense requires that the hazards associated with the use of any substance, or its components, be identified. It is expected that the toxicity of impurities, degradation products, and residues from manufacturing processes will be investigated as the development of a drug is pursued. The same standards should, therefore, be applied to the enantiomeric molecules in a racemate.

Whenever a drug can be obtained in a variety of chemically equivalent forms (such as enantiomers), it is both good science and good sense to explore the potential for *in vivo* differences between these forms.

References

Ariëns, E.J. (1984) Stereochemistry, a basis for sophisticated nonsense in pharmacokinetics and clinical pharmacology. *Eur. J. Clin. Pharmacol.* **26**, 663–668.

Ariëns, E.J. (1986) Stereochemistry: a source of problems in medicinal chemistry. *Med. Res. Rev.* **6**, 451–466.

Ariëns, E.J. and E.W. Wuis (1987) Bias in pharmacokinetics and clinical pharmacology. *Clin. Pharmacol. Ther.* **42**, 361–363.

Blaschke, G., H.P. Kraft, K. Fickentscher and F. Köhler (1979) Chromatographic racemic separation of thalidomide and teratogenic activity of its enantiomers. *Arzneim.-Forsch.* **29**, 1640–1642.

Blaschke, G., H.P. Kraft and H. Markgraf (1980) Optical resolution of thalidomide and other glutarimide derivatives. *Chem. Ber.* **113**, 2318–2322.

Bopp, R.J., J.F. Nash, A.S. Ridolfo and E.R. Shepard (1979) Stereoselective inversion of (R)-(−)-benoxaprofen to the (S)-(+)-enantiomer in humans. *Drug. Metab. Dispos.* **7**, 356–359.

Cahn, R.S., C.K. Ingold and V. Prelog (1966) Specification of molecular chirality. *Angew. Chem. Int. Ed.* **5**, 385–415.

Casini, G. and M. Ferappi (1964) Preparation of one optical antipode of 2-phthalimidoglutarimide. *Farmaco, Ed. Sci.* **19**, 563–565.

Chemie Grunenthal GmbH (1957) *Aminopiperidine-2, 6-dione derivatives*. Brit. Pat. 768, 821, Feb. 20, 1957.

De Ranter, C.J. (1984) Crystals, X-ray crystallography, and drugs. In *X-ray Crystallography and Drug Action*, ed. A.S. Horn and C.J. De Ranter, Clarendon Press, Oxford, 3–4.

Drayer, D.E. (1986) Pharmacodynamic and pharmacokinetic differences between drug enantiomers in humans: an overview. *Clin. Pharmacol. Ther.* **40**, 125–133.

Fabro, S., R.L. Smith and R.T. Williams (1967) Toxicity and teratogenicity of optical isomers of thalidomide. *Nature* **215**, 269.

Field, E.O., J.E. Gibbs, D.F. Tucker and K. Hellman (1966) Effect of thalidomide on the graft versus host reaction. *Nature* **211**, 1308–1310.

Furberg, S. and C.S. Petersen (1965) The crystal structure of N-(alpha-glutarimido)-4-bromophthalimide. *Acta. Chem. Scand.* **19**, 253–254.

Giacomini, K.M., W.L. Nelson, R.A. Pershe, L. Valdivieso, K. Turner-Tamiyasu and T.F. Blaschke (1986) *In vivo* interaction of the enantiomers of disopyramide in human subjects. *J. Pharmacokinet. Biopharm.* **14**, 335–356.

Hutt, A.J. and J. Caldwell (1983) The metabolic chiral inversion of 2-arylpropionic acids—a novel route with pharmacological consequences. *J. Pharm. Pharmacol.* **35**, 693–704.

Hutt, A.J. and J. Caldwell (1984) The importance of stereochemistry in the clinical pharmacokinetics of the 2-arylpropionic acid non-steroidal anti-inflammatory drugs. *Clin. Pharm.* **9**, 371–373.

Hutt, A.J. and J. Caldwell (1985) The metabolic chiral inversion of 2-arylpropionic acids—a novel route with pharmacological consequences a correction. *J. Pharm. Pharmacol.* **37**, 288.

Köhler, F., W. Meise and H. Ockenfels (1971) Teratogenicity of thalidomide metabolites. *Experientia* **27**, 1149–1150.

Marshall, E. (1985) Guilty plea puts Oraflex case to rest. *Science* **229**, 1071.

Meise, W. and F. Köhler (1971) Embryotoxic activity of N-phthaloyl-DL-isoglytamine. *Z. Naturforsch. B* **26**, 1081–1082.

Meise, W. and F. Köhler (1972) Synthesis and teratological testing of some thalidomide metabolites. *Pharmazie* **27**, 418–419.

Meise, W., H. Ockenfels and F. Köhler (1973) Teratologic determination of hydrolyzed products of thalidomide. *Experientia* **29**, 423–424.

Ockenfels, H. and F. Köhler (1970) Das L-isomere als teratogenes Prinzip der N-Phthalyl-DL-glutaminsäure. *Experientia* **26**, 1236–1237.

Prescott, L.F. (1986) Effects of non-narcotic analgesics on the liver. *Drugs* **32** (Supp. 4), 129–147.

Shealy, Y.F., C.E. Opliger and J.A. Montgomery (1965) D- and L-thalidomide. *Chem. Ind.* 1030–1031.

Shealy, Y.F., C.E. Opliger and J.A. Montgomery (1968) Synthesis of D- and L-thalidomide and related studies. *J. Pharm. Sci.* **57**, 757–764.

Simmonds, R.G., T.J. Woodage, S.M. Duff and J.M. Green (1980) Stereospecific inversion of (R)-(−)-benoxaprofen in rat and man. *Eur. J. Drug. Metab. Pharmacokinet.* **5**, 169–172.

Simonyi, M. (1984) On chiral drug action. *Med. Res. Rev.* **4**, 359–413.

Singh, N.N., F. Jamali, F.M. Pasutto, A.S. Russell, R.T. Coutts and K.S. Drader (1986) Pharmacokinetics of the enantiomers of tiaprofenic acid in humans. *J. Pharm. Sci.* **75**, 439–442.

Sunday Times (London) Insight team (1979) *Suffer the Children: The Story of Thalidomide*. The Viking Press, New York.

Tobert, J.A., V.J. Cirillo, G. Hitzenberger, I. James, J. Pryor, T. Cook, A. Butinx, I.B. Holmes and P.M. Lutterbeck (1981) Enhancement of the uricosuric properties of indacrinone by manipulation of the enantiomeric ratio. *Clin. Pharmacol. Ther.* **29**, 344–350.

Trotter, J. and F.H. Allen (1971) Crystal and molecular structure of thalidomide, N-(alpha-glutarimidyl)phthalimide. *J. Chem. Soc. B* 1073–1079.

White, P.F., J. Ham, W.L. Way and A.J. Trevor (1980) Pharmacology of ketamine isomers in surgical patients. *Anesthesiology* **52**, 231–239.

3 Chiral liquid chromatography: past and present

A. PRYDE

3.1 Historical perspectives

Around 1812, the French physicist Biot discovered the existence of optical activity in crystals of α-quartz (O'Loane, 1980). Credit for the first resolution of enantiomers from a racemate goes to Louis Pasteur who, in 1848, separated manually the non-identical crystals of the enantiomers of sodium ammonium tartrate (Pasteur, 1848). In 1874, Le Bel and Van't Hoff independently proposed the 'asymmetric carbon atom' to account for the chirality of organic compounds. Le Bel (1874) favoured a square pyramidal arrangement of substituents around an apical carbon atom whereas Van't Hoff (1874) correctly postulated a tetrahedral arrangement around a central carbon atom. Not surprisingly, early attempts to obtain resolution of racemates utilized readily-available, naturally-occurring chiral materials such as wool, silk, quartz and polysaccharides such as starch or cellulose. At the turn of the century, Wilstätter (1904) postulated that it should be possible to resolve racemic dyes in solution by enantioselective adsorption onto wool.

Partial resolution of an optically active aniline dye using wool was achieved by Ingersoll and Adams (1922) but a similar claim by Porter and Ihrig (1923) could not be confirmed by independent studies (Brode and Adams, 1926; Henderson and Rule, 1939). Recently, Konrad and Musso (1984) found no resolution of Porter and Ihrig's dyes on wool although partial resolution was obtained on powdered silk or potato starch used as chromatographic adsorbents. An investigation into the use of naturally-occurring chiral materials as chromatographic adsorbents led to the conclusion that potato starch (and cellulose triacetate, to be discussed later) was useful but that wool, silk and cellulose were unsuitable (Konrad and Musso, 1986). Column chromatography with lactose as the adsorbent was successfully employed by Henderson and Rule (1939) for the resolution of a racemic camphor derivative and by Prelog and Wieland (1944) for the resolution of Tröger's base. Early uses of quartz in chiral resolution include the partial resolution of racemic cobalt complexes by Tsuchida and co-workers (1936) who allowed saturated solutions of the complexes to cool over d- or l-quartz. Shortly thereafter Karagunis and Coumoulos (1938) chromatographed a racemic cobalt complex on optically active quartz.

The classical method for the resolution of enantiomers involves reaction of the racemate with an optically active reagent to form a pair of diastereomers which have different physical properties. After separation of the diastereomers, for example by fractional crystallization or by chromatography (using achiral stationary and mobile phases), the individual enantiomers have to be retrieved from the diastereomers. Apart from being tedious, the method requires the availability of a suitable functional group in the racemate as well as sufficient quantities of optically pure reagent. In addition, there is always the problem of racemization during diastereomer formation or retrieval of the individual enantiomers. Direct chromatographic methods for chiral resolution (i.e. those which do not require prior formation of diastereomers) are based on the formation of transient diastereomeric adsorbates by making available in either the stationary phase or the mobile phase an optically active component (the chiral selector).

The first direct resolution of enantiomers by gas chromatography (GC) was achieved by Gil-Av *et al.* (1966) using an optically active stationary phase coated onto a 100-m capillary column. However, compared to HPLC, GC has the disadvantage that the higher operating temperatures required may lead to racemization of both the chiral stationary phase (with concomitant loss of chiral selectivity) and of the enantiomers to be separated. In the chromatographic methods, the covalent bond of a true diastereomer is replaced by the transient interaction between the chiral selector and the enantiomers of the racemate. If one enantiomer interacts more strongly with the chiral selector then it will be more strongly retained on the column and hence chromatographic selectivity ($\alpha > 1$) is achieved.

Whereas the direct resolution of enantiomers was relatively rare even two decades ago, there now exists an abundance of different LC solutions to this problem. This situation is a result of two factors, namely the breakthrough in HPLC technology in the late 1960s along with the insights gained into the chiral selection mechanism by Pirkle and others. The advent of modern HPLC in the late 1960s was made possible by the development of appropriate hardware (such as pumps, low dead-volume injectors and detectors) for fast, efficient LC. Of particular relevance to the chiral stationary phases (CSPs) currently being exploited were the advances made in the HPLC packing materials. Polymeric materials such as cellulose or polystyrene resins, being compressible, were not suitable for the higher operating pressures of HPLC and furthermore, their mass transfer characteristics were poor. These problems were overcome with the advent of the mechanically-stable, superficially porous supports in which a thin layer of silica was grafted on to the surface of glass beads, thereby reducing diffusion distances. These materials were further improved with the advent of commercially-available small diameter, porous silica particles (3–5 μm). The increasing popularity of HPLC in the 1970s has been well documented (Snyder and Kirkland, 1974; Pryde and Gilbert, 1979). The advantages of the HPLC packing materials have been incorporated into CSPs where modern phases are available with silica-bonded cellulose, proteins or synthetic chiral selectors.

3.2 Three-point rule

Of more importance than these advances in HPLC technology were the insights into the chiral selection mechanism obtained by Pirkle (1979). Pirkle stated that for chiral resolution to take place, there must be at least three simultaneous interactions between a CSP and one solute enantiomer, whereby one or more of the interactions must be stereochemically dependent. Typical intermolecular interactions involved in the chiral recognition process are π–π interactions, hydrogen bonds, dipole–dipole, hydrophobic and steric interactions. The first mention of this three-point interaction model was made by Easson and Sedman (1933) when discussing the interaction of racemic drugs with receptors as well as enzyme–substrate interactions. Ogston (1948) invoked the three-point interaction model to explain the stereospecificity in the reaction of an enzyme with a racemate.

The first application of the three-point rule in chromatography was made by Dalgliesh (1952) who resolved amino acid enantiomers by paper chromatography and postulated a three-point interaction between the chiral cellulose stationary phase and the solute enantiomers. Senoh and coworkers (Kotake *et al.*, 1951) had also resolved amino acid enantiomers by paper chromatography and attributed the resolution to the chirality of the cellulose. An early use of a CSP rationally designed to utilize three-point interactions was reported by Baczuk and co-workers (1971) who bonded L-arginine to Sephadex and used the material to resolve β-3, 4-dihydroxyphenylalanine (DOPA) enantiomers. The three-point rule appears to be valid for those cases where the chiral selection mechanism is known in detail. These cases involve interactions between low-molecular-weight solutes and well-defined chiral stationary phases or chiral mobile phase additives. As discussed below, these include the Pirkle CSPs, ion pair and zwitterion pair chromatography and Cram's crown ether CSPs.

In chiral ligand exchange chromatography, Davankov has argued against the three-point rule in favour of a two-point interaction (Davankov and Kurganov, 1983) although Pirkle and Pochapsky (1987) have suggested that this might be an extension of the three-point rule rather than an exception. For more complex CSPs such as cellulose and cyclodextrin the situation is more complicated. As there is difficulty in identifying three discrete interactions, the three-point rule is of limited value in rationalizing the observed enantioselectivities in such cases.

3.3 Development of commercially available or easily accessible direct chiral resolution methods

In the following subsections the development of each of the major direct chiral resolution methods which are now commercially available or easily accessible will be traced. These methods are dealt with in the approximate order in which they became readily available, so that an impression can be gained of how the practitioner's scope for carrying out chiral liquid chromatography has steadily widened since the late 1970s.

3.3.1 *Chiral ligand exchange chromatography*

Chiral ligand exchange chromatography (LEC) was the first direct chiral resolution method that became easily accessible to the practising analyst. LEC has proved a powerful technique for the chiral resolution of underivatized amino acids. In the original system described by Davankov and Rogozhin (1971), L-proline was chemically bonded to polystyrene beads. Using an aqueous mobile phase containing Cu^{2+} ions, the racemic solute DL-proline formed diastereomeric metal co-ordination complexes with the bonded amino acid and the copper ions. As the ternary complexes formed with the individual solute enantiomers differed in energy, chromatographic resolution was achieved. Davankov has reviewed the possibilities for optimizing chiral resolution with CSPs in LEC by varying the resin, metal ion, bonded chiral group and the nature, pH and ionic strength of the buffer (Davankov, 1980; Davankov and Kurganov, 1983).

The poor mass transfer characteristics and long analysis times of the original polystyrene-based materials were improved in subsequent generations of chiral ligand exchange materials such as the silica-bonded materials described by Lindner (1982). The usefulness of the technique for both analytical and preparative scale separations has led to the appearance of several commercial CSPs for LEC these utilize silica-bonded amino acids and are recommended for use with aqueous Cu^{2+} buffered mobile phases (Chiralpak WE, WH, WM; Daicel: Nucleosil Chiral-1; Macherey–Nagel). The scope of chiral LEC was extended by Karger and coworkers who were the first to use an achiral stationary phase (ODS-silica) along with an aqueous mobile phase containing the chiral selector (LePage *et al.*, 1979; Lindner *et al.*, 1979). The latter was an optically active zinc or cadmium complex of a hydrophobic triamine, and dansyl amino acids were well separated under alkaline conditions.

3.3.2 *Synthetic multiple interaction CSPs*

Although chiral LEC was a major step forward, the technique has the disadvantage of being more or less restricted to the resolution of amino acids. However, the next direct chiral resolution method to find common use had a much broader spectrum, i.e. was applicable to a wider variety of compounds. The first synthetic multiple interaction CSPs to become commercially successful were those which were initially synthesized by Pirkle. Pirkle has been at the forefront of the development of such phases ever since and has played a leading role in introducing a more mechanistic approach to chiral separations. Before embarking on the synthesis of stationary phases for chiral LC, Pirkle had employed chiral fluoroalcohols such as 2, 2, 2-trifluoro-1-(9-anthryl) ethanol as chemical shift reagents in NMR. The chiral fluoroalcohols were shown to undergo a three-point interaction with enantiomers giving diastereomeric, chelate-like solvates in solution which allowed the determination of enantiomeric purity and absolute configuration (Pirkle and Sikkenga, 1977).

In his first CSP, Pirkle (1979) bonded the chiral anthryl alcohol to silica and used it to resolve a wide range of racemates (e.g. sulphoxides, lactones and derivatives of amines, amino acids and alcohols). An important interaction in the chiral selection process was the π–π interaction between the anthryl π-base substituent and a π-acid group on the solute enantiomer (e.g. a nitroaromatic group) while hydrogen bonding between the stationary phase and the enantiomeric solute was also important. Pirkle's detailed understanding of the interaction between the CSP and an enantiomeric solute led to the possibility of predicting the elution order of individual enantiomers. In generalized form the process of chiral selection is illustrated in Figure 3.1. In Figure 3.1*a*, the chiral selector in the stationary phase forms three interactions with one enantiomer whereas only two interactions are possible with its mirror image (Figure 3.1*b*). The latter enantiomer is therefore less strongly bound by the CSP and will elute first. Pirkle utilized the reciprocity of the chiral selection process to prepare a second generation of CSPs. His original anthryl alcohol CSP gave good resolution of dinitrobenzoyl phenylglycine enantiomers and

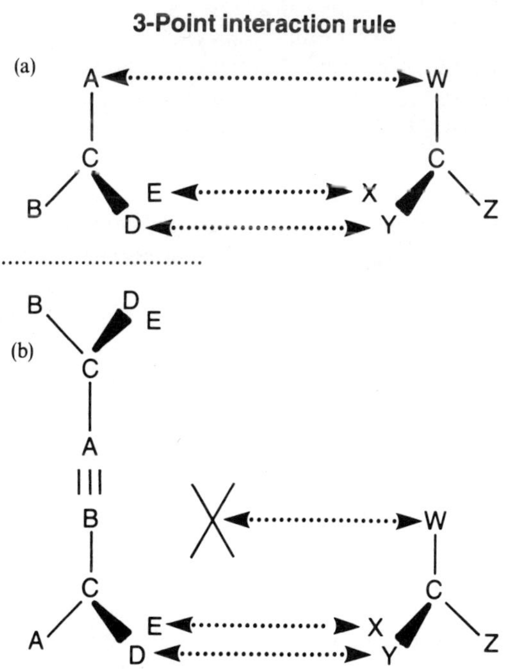

Figure 3.1 Generalized representation of the three-point rule.

hence dinitrobenzoyl phenylglycine was bonded to the surface of silica (Pirkle *et al.*, 1980). The resulting CSPs possessed such wide versatility for chiral resolution that two materials with N-(3, 5-dinitrobenzoyl)phenylglycine ionically or covalently bonded to silica are commercially available (J.T. Baker; Regis). The materials are used for normal-phase chromatography, for example with hexane–isopropanol mobile phases. An intrinsic advantage of having the chiral selector in the stationary phase is that only small quantities of the chiral selector are required, in contrast to diastereomer formation or methods in which the chiral selector is a component of the mobile phase. Furthermore, as the Pirkle CSPs are well suited for preparative-scale separations (Pirkle and Hamper, 1987), the presence of a bonded chiral selector is advantageous because in cases where the chiral selector is a component of the mobile phase, it must be subsequently separated from the resolved enantiomer fractions.

3.3.3 *Protein CSPs*

Pirkle CSPs were referred to as broad spectrum CSPs. However, there are fairly strict requirements on what type of racemate can be resolved into its enantiomers by any one type of synthetic multiple interaction CSP. For instance, the N-(3, 5-dinitrobenzoyl)phenylglycine CSP is only effective for racemic compounds which, at the least, contain an electron-rich aromatic moiety. Therefore, a large proportion of the published applications are for compounds containing a naphthyl-group. It was not until the advent of commercially-available protein-immobilized CSPs that it was possible to resolve racemates from widely different compound classes on a single column.

Two proteins have been used as chiral selectors in LC and both have given rise to commercially available packing materials. Albumin is the principal plasma binding protein for weakly acidic drugs and α_1-acid glycoprotein (α_1-AGP), the major plasma binding protein for basic drugs. Stewart and Doherty (1973) reported the chromatographic resolution of DL-tryptophan using bovine serum albumin (BSA) bonded to agarose and thereby confirmed an earlier report using equilibrium dialysis which suggested that L-tryptophan had an affinity for albumin 100 times greater than that of D-tryptophan (McMenamy and Oncley, 1958). Following on from the work of Allenmark and co-workers (1982), a CSP with BSA covalently bonded to 7 μm silica became commercially available (Resolvosil; Macherey-Nagel). The BSA material can be used with aqueous buffer mobile phases for a wide range of analytical-scale chiral separations (Allenmark, 1986) (including sulphoxides, benzodiazepines and coumarins such as warfarin) as well as to monitor the stereoselective microbial conversion of racemic substrates (Allenmark *et al.*, 1986).

As the protein-bound materials have low sample capacity they have not been used for preparative-scale separations. The main interactions between BSA and the solute enantiomers are hydrophobic and electrostatic interactions although hydrogen bonding or charge transfer interactions could also contribute to chiral selection. In working with protein stationary phases, optimization of chiral resolution can be achieved by varying the buffer pH or ionic strength and shorter retention times are obtained by adding an organic modifier such as propanol to the mobile phase. Albumin has also been used as a chiral mobile phase additive in reversed phase chromatography with ODS-silica for the chiral resolution of carboxylic acids (Pettersson *et al.*, 1986). In 1983, Hermansson reported a silica-bonded α_1-AGP chiral material for LC and obtained useful resolutions of a series of basic drugs. As with the BSA material, hydrophobic and electrostatic interactions were involved in the retention mechanism, along with hydrogen bonding (Hermansson, 1983). The α_1-AGP/silica material is now commercially available (EnantioPac; LKB).

3.3.4 *Cyclodextrin CSPs*

The next significant advance in chiral LC was the commercial introduction of cyclodextrin CSPs. Although not superior to protein CSPs in their range of applicability, they were chemically and physically much more robust.

The cyclodextrins are a group of cyclic oligosaccharides which have found wide application in chiral LC. α, β and γ-cyclodextrins, containing 6, 7 or 8 D-glucose units respectively, each possess a chiral cavity proportional to the number of glucose units present (e.g. 8 Å for β-cyclodextrin) and the materials are capable of forming inclusion complexes with a variety of racemates. The inclusion complex is formed between the hydrophobic groups of the guest molecule and the relatively hydrophobic chiral cavity of the cyclodextrin and involves various hydrophobic and hydrogen bonding forces. Initially the materials were used in water-soluble form to generate diastereomeric inclusion complexes which could be separated by fractional crystallization to give enantiomer enrichment. Modern commercially-available materials for HPLC include cyclodextrin and acetylated cyclodextrin bonded to 5 μm spherical silica (Cyclobond range; Astec). The materials are used in the reversed phase mode with aqueous alcohol mobile phases (Armstrong and DeMond, 1984) and have been successfully employed in chiral separations of prostaglandins, sulphonamides and vitamins. Partial chiral resolution of mandelic acid was also achieved using β-cyclodextrin as a chiral mobile-phase additive in conjunction with an achiral reversed-phase column (Debowski *et al.*, 1982).

3.3.5 *Cellulose CSPs*

CSPs employing cellulose or its derivatives as the chiral selector have only recently been introduced commercially. Nonetheless they have their origins in work carried out over a decade ago.

Leading on from early attempts at using naturally-occurring chiral polysaccharides for chiral resolution, Hesse and Hagel (1973) made a breakthrough with the preparation of microcrystalline cellulose triacetate, a material with wide applicability for chiral resolution. Hesse and Hagel noted that chiral recognition by cellulose triacetate was lost if the material was recrystallized, presumably by destroying the chiral cavities in the helical structure of the polysaccharide. Inclusion complexes are considered to be formed between the solute enantiomers and the helical chains of cellulose triacetate. As the mechanism of chiral selection on cellulose triacetate columns is not well understood, resolution of racemates on cellulose materials is still empirical. However, despite this, the materials have been used in practice for a wide range of analytical and semi-preparative scale separations. Cellulose triacetate powder and various cellulose derivatives coated on silica (cellulose triacetate, tribenzoate and triarylcarbamate) are now commercially available (Chiralcel materials; Daicel). Typical mobile phases include aqueous alcohol mixtures through to hexane/isopropanol. Okamoto has shown that the material obtained by bonding cellulose triarylcarbamate to silica had similar chromatographic efficiency to the coated material but with the advantage that more polar solvents could be used without dissolving the cellulose carbamate (Okamoto *et al.*, 1987).

3.3.6 *Synthetic polymer CSPs*

CSPs in which the chiral selector is a synthetic polymer have only recently become commercially available although the basic idea for their use is not recent.

The origins of these materials in fact lie in the theory proposed by Linus Pauling (1940) to account for the natural formation of antibodies specific for a given antigen. Pauling proposed that the antibody protein adopts a tertiary structure complementary to that of the antigen template. The idea was later applied to the preparation of silica (Dickie, 1949; Majors and Rogers, 1969). Silica prepared in the presence of methyl orange or related azo dyes was shown to adsorb specifically the given template molecule and in 1949 Dickie mentioned that he was investigating 'the possibility that optical isomers can be separated with such adsorbents'. Later, Wulff and Sarhan (1972) prepared a series of enzyme-analogue polymers for chiral LC. Wulff's method involved synthesizing highly cross-linked co-polymers which incorporated chiral carbohydrate template molecules. Subsequent removal of the template by hydrolysis yielded polymers with chiral cavities which were used to resolve racemates of the template molecules.

The polymers have also been bonded to silica (Wulff *et al.*, 1986) but the chromatographic efficiency of such materials is fairly poor. Okamoto and coworkers polymerized triphenylmethyl methacrylate in the presence of a chiral anionic initiator and obtained a polymer whose chirality was due solely to its helicity (Okamoto *et al.*, 1979; Okamoto and Hatada, 1986). Modern materials for chiral LC in which the polymer is bonded to silica are commercially available (Chiralpak OT and OP; Daicel). In the OP material the phenyl groups are replaced by pyrazine. Recently the chiral template approach was applied to another type of CSP. Aubel and Rogers (1987) reported initial results regarding the possibility of reorienting the tertiary structure of silica-bonded BSA in the presence of a single enantiomer.

More recently still, synthetic polymer CSPs have been prepared commercially by a different approach. The use of optically active monomers as building blocks for polymers suitable for chiral LC has been described (Blaschke, 1974). Development of these materials has led to the commercial exploitation of a silica-bonded chiral polymer prepared from N-acryloyl-(S)-phenylalanine ethyl ester (Chiraspher, E. Merck).

3.3.7 *Chiral ion pair chromatography*

Chiral ion pair chromatography has been readily available for longer than all direct chiral resolution methods apart from chiral ligand exchange chromatography and LC on Pirkle CSPs. However, chiral ion pairing agents have never had heavy commercial promotion.

An ion pair partition system was described for the chiral resolution of enantiomeric amines (Pettersson and Schill, 1981). The non-polar mobile phase contained the chiral counter ion, (+)-10-camphor sulphonic acid and a polar stationary phase such as silica was used. Dalgleish's three-point rule applied to the ion pairs formed. The interactions included the expected electrostatic one between solute and counter ion, along with hydrogen bonds and hydrophobic ones. For this reason, water had to be excluded rigorously from the mobile phase as it would have interacted with the hydrogen bonding groups of the ion pair components. Developing their original studies on zwitterion pair chromatography, Knox and Jurand (1982) extended the technique to chiral separations. The chiral selector, L-leucyl-L-leucyl-L-leucine was added to the aqueous methanol mobile phase used in conjunction with a reversed phase (ODS-silica) packing material. The chiral selector was adsorbed on to the hydrophobic stationary phase where it could form quadrupolar ion pairs with zwitterionic racemates present in the mobile phase. Knox showed that two of the three interactions required for chiral selection by the Dalgleish/Pirkle model were provided by the two electrostatic interactions involved in the formation of the quadrupolar ion pairs (rather than the single electrostatic interaction for a conventional dipolar ion pair).

3.3.8 *Other direct chiral resolution methods*

To describe the development of each type of commercially available or easily accessible chiral resolution method does not fully describe the development of chiral LC. There have been several important pieces of work which, although they have for various reasons not led directly to a commercial CSP, have had a significant influence on the overall development of chiral LC. In this category are various reports of the application of charge transfer chromatography for chiral separations (Klemm and Reed, 1960; Mikes *et al.*, 1976; Lochmüller and Ryall, 1978). In 1960, Klemm and Reed adsorbed α-(2, 4, 5, 7-tetranitro-9-

fluorenylidineaminooxy) propionic acid (TAPA) to silica and used the material to resolve chiral aromatics, and Mikes *et al.* later used coated and bonded TAPA to resolve helicenes by HPLC.

Also important were the chiral crown ethers synthesized by Cram and coworkers (1975). These cyclic polyethers contained two 1,1′-dinaphthyl groups which imparted a rigid structure to the chiral crown ether 'host'. CSPs were synthesized with the crown ether attached either to polystyrene resin (Sogah and Cram, 1976) or to silica (Sousa *et al.*, 1978). Enantiomeric alkylammonium salts (the 'guests') formed diastereomeric complexes of differing energy with the chiral cavity of the crown ether and could therefore be resolved chromatographically. High selectivity factors ($\alpha > 20$) were achieved for the resolution of amino acids (Sogah and Cram, 1976). The chiral selection mechanism in the host–guest complexation process is well understood and the elution order of enantiomers can usually be predicted from molecular models. Although well suited to the resolution of chiral ammonium salts, the chiral crown ethers are unfortunately difficult to synthesize and hence these CSPs have not been commercialized.

3.4 Present status of chiral liquid chromatography

It is evident from the foregoing that, from modest beginnings, chiral LC has witnessed an explosive growth in popularity within the last decade (Armstrong, 1987; Blaschke, 1980; Däppen *et al.*, 1986; Davankov *et al.*, 1983*b*; Krull, 1978; Pirkle and Pochapsky, 1987). The reason for this, apart from the breakthrough

Table 3.1 Landmarks in the development of chiral LC

1848	Pasteur : manual separation of the enantiomeric crystals of sodium ammonium tartrate
1936	Tsuchida *et al.*: use of d- or l-quartz in chiral resolution
1939	Henderson and Rule: resolution of a racemic camphor derivative by chromatography on lactose
1959	Dalgliesh: postulation of the three-point rule in the paper chromatography of amino acids
1966	Gil-Av *et al.*: direct resolution of enantiomers by GC
1971	Davankov and Rogozhin: introduction of chiral ligand exchange chromatography
1972	Wulff and Sarhan: preparation of enzyme analogue polymers for chiral LC
1973	Hesse and Hagel: preparation of cellulose triacetate for chiral resolution
1973	Stewart and Doherty: use of agarose-bonded bovine serum albumin for chiral resolution
1974	Blaschke: synthesis of chiral polymers from optically active monomers, for chiral LC
1975	Cram and co-workers: development of host–guest chromatography using chiral crown ethers
1977	Pirkle and Sikkenga: use of chiral chemical shift reagents in NMR
1979	Pirkle and House: synthesis of first silica-bonded CSP and application in chiral LC
1979	Okamoto *et al.*: synthesis of helical polymers for chiral LC
1982	Allenmark *et al.*: use of agarose-bonded bovine serum albumin in chiral LC
1983	Hermansson: use of silica-bonded α_1-acid glycoprotein for chiral resolution
1984	Armstrong and DeMond: preparation of silica-bonded cyclodextrins

in HPLC technology, lies in the range of LC modes currently available to tackle a chiral separation. The various types of interaction between the solute enantiomer and the stationary phase (such as π–π, electrostatic, hydrogen bonding) can be exploited using commercially available CSPs or chiral mobile phase additives. In many cases the chiral selection mechanism is well understood and can even be used to predict elution orders of enantiomers but even in those systems where the enantioseparations are more empirical, the materials still prove useful in practice. The key landmarks in the development of modern chiral LC are collected chronologically in Table 3.1.

References

Allenmark, S. (1986) Optical resolution by liquid chromatography on immobilized bovine serum albumin. *J. Liq. Chromatogr.* **9**, 425–442.

Allenmark, S., B. Bomgren and H. Boren (1982) Direct resolution of enantiomers by liquid affinity chromatography on albumin-agarose under isocratic conditions. *J. Chromatogr.* **237**, 473–477.

Allenmark, S., B. Bomgren and H. Boren (1986) Enantioselective microbial degradation of some racemates studied by chiral liquid chromatography. *Enzyme Microb. Technol.* **8**, 404–408.

Armstrong, D.W. (1987) Optical isomer separation by liquid chromatography. *Anal. Chem.* **59**, 84A–91A.

Armstrong, D.W. and W.J. DeMond (1984) Cyclodextrin bonded phases for the liquid chromatographic separation of optical, geometrical, and structural isomers. *J. Chromatogr. Sci.* **22**, 411–415.

Aubel, M.T. and L.B. Rogers (1987) Effects of pretreatment on the enantioselectivity of silica-bound proteins used as high-performance liquid chromatographic stationary phases. *J. Chromatogr.* **408**, 99–113.

Baczuk, R.J., G.K. Landram, R.J. Dubois and H.C. Dehm (1971) Liquid chromatographic resolution of racemic β-3,4-dihydroxyphenylalanine. *J. Chromatogr.* **60**, 351–361.

Blaschke, G. (1974) Chromatographie racemischer Mandelsäure an Polyacrylsäure-estern und -amiden optisch aktiver Ephedrinderivate. *Chem. Ber.* **107**, 237–252.

Blaschke, G. (1980) Chromatographische Racemattrennung. *Angew. Chem.* **92**, 14–25.

Brode, W.R. and R. Adams (1926) Optically active dyes III. Physical properties, dyeing reactions and mechanism of dyeing. *J. Amer. Chem. Soc.* **48**, 2193–2201.

Cram, D.J., R.C. Helgeson, L.R. Sousa, J.M. Timko, M. Newcomb, P. Moreau, F. De Jong, G.W. Gokel, D.H. Hoffman, L.A. Domeier, S.C. Peacock, K. Madan and L. Kaplan (1975) Chiral recognition in complexation of guests by designed host molecules. *Pure Appl. Chem.* **43**, 327–349.

Dalgliesh, C.E. (1952) The optical resolution of aromatic amino-acids on paper chromatograms. *J. Chem. Soc.*, 3490–3942.

Däppen, R., H. Arm and V.R. Meyer (1986) Applications and limitations of commercially available chiral stationary phases for high-performance liquid chromatography. *J. Chromatogr.* **373**, 1–20.

Davankov, V.A. (1980) Resolution of racemates by ligand exchange chromatography. In *Advances in Chromatography*, ed. J.C. Giddings, E. Grushka and P.R. Brown, Marcel Dekker, New York, **18**, chapter 4.

Davankov, V.A. and A.A. Kurganov (1983) The role of achiral sorbent matrix in chiral recognition of amino acid enantiomers in ligand-exchange chromatography. *Chromatographia* **17**, 686–690.

Davankov, V.A. and S.V. Rogozhin (1971) Ligand chromatography as a novel method for the investigation of mixed complexes: stereoselective effects in α-amino acid copper(II) complexes. *J. Chromatogr.* **60**, 280–283.

Davankov, V.A., A.A. Kurganov and A.S. Bochkov (1983) Resolution of racemates by high-performance liquid chromatography. In *Advances in Chromatography*, ed. J.C. Giddings, E. Grushka and P.R. Brown, Marcel Dekker, New York, **22**, chapter 3.

Debowski, J., D. Sybilska and J. Jurczak (1982) β-Cyclodextrin as a chiral component of the mobile phase for separation of mandelic acid into enantiomers in reversed-phase systems of high-performance liquid chromatography. *J. Chromatogr.* **237**, 303–306.

Dickie, F.H. (1949) The preparation of specific adsorbents. *Proc. Nat. Acad. Sci.* **35**, 227–229.

Easson, L.H. and E. Stedman (1933) Studies on the relationship between chemical constitution and physiological action. V. Molecular dissymmetry and physiological activity. *Biochem. J.* **27**, 1257–1266.

Gil-Av, E., B. Feibush and R. Charles-Sigler (1966) Separation of enantiomers by gas liquid chromatography with an optically active stationary phase. *Tet. Letts.* 1009–1015.

Henderson, G.M. and H.G. Rule (1939) A new method of resolving a racemic compound. *J. Chem. Soc.* 1568–1573.

Hermansson, J. (1983) Direct liquid chromatographic resolution of racemic drugs using α_1-acid glycoprotein as the chiral stationary phase. *J. Chromatogr.* **269**, 71–80.

Hesse, G. and R. Hagel (1973) Eine vollständige Racemattrennung durch Elutions-Chromatographie an Cellulose-tri-acetat. *Chromatographia* **6**, 277–280.

Ingersoll, A.W. and R. Adams (1922) Optically active dyes. *J. Amer. Chem. Soc.* **44**, 2930–2937.

Karagunis, G. and G. Coumoulos (1938) A new method of resolving a racemic compound. *Nature* **142**, 162–163.

Klemm, L.H. and D. Reed (1960) Optical resolution by molecular complexation chromatography. *J. Chromatogr.* **3**, 364–368.

Knox, J.H. and J. Jurand (1982) Separation of optical isomers by zwitterion-pair chromatography. *J. Chromatogr.* **234**, 222–224.

Konrad, G. and H. Musso (1984) Ueber die enantioselektive Adsorption chiraler Azofarbstoffe an Seide, Wolle, Stärke und Cellulose. *Chem. Ber.* **117**, 423–426.

Konrad, G. and H. Musso (1986) Ueber die Enantiomerentrennung durch Chromatographie an natürlichen Polymeren. *Liebigs Ann. Chem.* 1956–1967.

Kotake, M., T. Sakan, N. Nakamura and S. Senoh (1951) Resolution into optical isomers of some amino acids by paper chromatography. *J. Amer. Chem. Soc.* **73**, 2973–2974.

Krull, I.S. (1978) The liquid-chromatographic resolution of enantiomers. In *Advances in Chromatography*, ed. J.C. Giddings, E. Grushka and P.R. Brown, Marcel Dekker, New York, **16**, chapter 6.

Le Bel, J.A. (1874) Sur les relations qui existent entre les formules atomiques des corps organiques et le pouvoir rotatoire de leurs dissolutions. *Bull. Soc. Chim.* **22**, 337–347.

LePage, J.N., W. Lindner, G. Davies, D.E. Seitz and B.L. Karger (1979) Resolution of the optical isomers of dansyl amino acids by reversed phase liquid chromatography with optically active metal chelate additives. *Anal. Chem.* **51**, 433–435.

Lindner, W.F. (1982) Resolution of optical isomers by gas and liquid chromatography. In *Chemical Derivatisation in Analytical Chemistry*, ed. F. Lawrence and R.W. Frei, Plenum Press, New York, chapter 4.

Lindner, W., J.N. LePage, G. Davies, D.E. Seitz and B.L. Karger (1979) Reversed-phase separation of optical isomers of dns-amino acids and peptides using chiral metal chelate additives. *J. Chromatogr.* **185**, 323–344.

Lochmüller, C.H. and R.R. Ryall (1978) Direct resolution of enantiomers by high-performance liquid chromatography on a bonded chiral stationary phase. *J. Chromatogr.* **150**, 511–514.

McMenamy, R.H. and J.L. Oncley (1958) The specific binding of L-tryptophan to serum albumin. *J. Biol. Chem.* **233**, 1436–1447.

Majors, R.E. and L.B. Rogers (1969) Effects of structural changes of the pretreating agent on adsorption by modified silica adsorbents. *Anal. Chem.* **41**, 1058–1065.

Mikes, F., G. Boshart and E. Gil-Av (1976) Resolution of optical isomers by high-performance liquid chromatography, using coated and bonded chiral charge-transfer complexing agents as stationary phases. *J. Chromatogr.* **122**, 205–221.

Ogston, A.G. (1948) Interpretation of experiments on metabolic processes, using isotopic tracer elements. *Nature*, 963.

Okamoto, Y. and K. Hatada (1986) Resolution of enantiomers by HPLC on optically active poly(triphenylmethyl methacrylate). *J. Liq. Chromatogr.* **9**, 369–384.

Okamoto, Y., K. Suzuki, K. Ohta, K. Hatada and H. Yuki (1979) Optically active poly(triphenylmethyl methacrylate) with one-handed helical conformation. *J. Amer. Chem. Soc.* **101**, 4763–4765.

Okamoto, Y., R. Aburatani, S. Miura and K. Hatada (1987) Chiral stationary phases for HPLC: Cellulose tris (3, 5-dimethylphenylcarbamate) and tris (3, 5-dichlorophenylcarbamate) chemically bonded to silica gel. *J. Liq. Chromatogr.* **10**, 1613–1628.

O'Loane, J.K. (1980) Optical activity in small molecules, nonenantiomorphous crystals, and nematic liquid crystals. *Chem. Rev.* **80**, 41–61.

Pauling, L. (1940) A theory of the structure and process of formation of antibodies. *J. Amer. Chem. Soc.* **62**, 2643–2657.

Pasteur, L. (1848) Recherches sur les relations qui peuvent exister entre la forme cristalline, la composition chimique et le sens de la polarisation rotatoire. *Ann. Chim. Phys.* **24**, 442–459.

Pettersson, C. and G. Schill (1981) Separation of enantiomeric amines by ion-pair chromatography. *J. Chromatogr.* **204**, 179–183.

Pettersson, C., T. Arvidsson, A.-L. Karlsson and I. Marle (1986) Chromatographic resolution of enantiomers using albumin as complexing agent in the mobile phase. *J. Pharm. Biomed. Anal.* **4**, 221–235.

Pirkle, W.H. and B.C. Hamper (1987) The direct preparative resolution of enantiomers by liquid chromatography on chiral stationary phases. In *Preparative Liquid Chromatography*, ed. B.A. Bidlingmeier, Elsevier, Amsterdam, chapter 7.

Pirkle, W.H. and D.W. House (1979) Chiral high-pressure liquid chromatographic stationary phases. 1. Separation of the enantiomers of sulfoxides, amines, amino acids, alcohols, hydroxy acids, lactones, and mercaptans. *J. Org. Chem.* **44**, 1957–1960.

Pirkle, W.H. and T.C. Pochapsky (1987) Chiral stationary phases for the direct LC separation of enantiomers. In *Advances in Chromatography*, ed. J.C. Giddings, E. Grushka and P.R. Brown, Marcel Dekker, New York, **27**, chapter 3.

Pirkle, W.H. and D.L. Sikkenga (1977) The use of chiral solvation agents for nuclear magnetic resonance determination of enantiomeric purity and absolute configuration of lactones. Consequences of three-point interactions. *J. Org. Chem.* **42**, 1370–1374.

Pirkle, W.H., D.W. House and J.M. Finn (1980) Broad spectrum resolution of optical isomers using chiral high-performance liquid chromatographic bonded phases. *J. Chromatogr.* **192**, 143–158.

Porter, C.W. and H.K. Ihrig (1923) Asymmetric dyes. *J. Amer. Chem. Soc.* **45**, 1990–1993.

Prelog, V. and P. Wieland (1944) Ueber die Spaltung der Tröger'schen Base in optische Antipoden, ein Beitrag zur Stereochemie des dreiwertigen Stickstoffes. *Helv. Chim. Acta* **27**, 1127–1134.

√ Pryde, A. and M.T. Gilbert (1979) *Applications of High Performance Liquid Chromatography*. Chapman and Hall, London.

√ Snyder, L.R. and J.J. Kirkland (1974) *Introduction to Modern Liquid Chromatography*, Wiley-Interscience, New York.

Sogah, G.D.Y. and D.J. Cram (1976) Total chromatographic optical resolutions of α-amino acid and ester salts through chiral recognition by a host covalently bound to polystyrene resin. *J. Amer. Chem. Soc.* **98**, 3038–3041.

Sousa, L.R., G.D.Y. Sogah, D.H. Hoffmann and D.J. Cram (1978) Host–guest complexation. 12. Total optical resolution of amine and amino ester salts by chromatography. *J. Amer. Chem. Soc.* **100**, 4569–4576.

Stewart, K.K. and R.F. Doherty (1973) Resolution of DL-tryptophan by affinity chromatography on bovine serum albumin–agarose columns. *Proc. Nat. Acad. Sci.* **70**, 2850–2852.

Tsuchida, R., M. Kobayashi and A. Nakamura (1936) The configuration of chlorobisdimethylglyoximoammine-cobalt. *Bull. Chem. Soc. Jap.* **11**, 38–40.

Van't Hoff, J.H. (1874) Sur les formules de structure dans l'espace. *Arch. Nederl. Sci. Exactes Natur.* **9**, 445–454.

Wilstätter, R. (1904) Ueber einen Versuch zur Theorie des Färbens. *Ber. Deutsch. Chem. Ges.* **37**, 3758–3760.

Wulff, G. and A. Sarhan (1972) The use of polymers with enzyme-analogous structures for the resolution of racemates. *Angew. Chem. Intnl. Ed.* **11**, 341.

Wulff, G., H.-G. Poll and M. Minarik (1986) Enzyme-analogue built polymers. XIX. Racemic resolution on polymers containing chiral cavities. *J. Liq. Chromatogr.* **9**, 385–405.

Part 2 Chiral derivatization

4 Chiral derivatization

M. AHNOFF and S. EINARSSON

4.1 Introduction

Derivatization before chromatographic separation may serve several purposes, such as to increase detectability and to improve chromatographic properties. Derivatization of chiral substances is carried out with both chiral and achiral reagents. The latter may serve to block functional groups which may cause excessive band broadening on chiral columns (e.g. amines), or to introduce groups which interact favourably with the chiral stationary phase (e.g. dinitrobenzoyl-derivatives, Pirkle and Finn, 1983). In analytical applications, derivatization with achiral reagents may be needed for detection purposes.

Derivatization of chiral substances with chiral reagents yields diastereomers. The diastereomers are physico-chemically different also in an achiral environment and are therefore, in principle, separable on conventional achiral columns. For the reagent (R)-A and an enantiomeric mixture of the substrate B we obtain:

$$(R)\text{-}A + (R,S)\text{-}B \rightarrow (R)\text{-}A(R)\text{-}B + (R)\text{-}A(S)\text{-}B$$

The degree of separation of (R)-A(R)-B and (R)-A(S)-B depends on the chiral structures A and B and the linking between the chiral centres, as well as on the chromatographic system employed. The elution order of the derivatized enantiomers will be reversed if the other reagent enantiomer, (S)-A, is used.

$$(S)\text{-}A + (R,S)\text{-}B \rightarrow (S)\text{-}A(S)\text{-}B + (S)\text{-}A(R)\text{-}B$$

The elution order of the B enantiomers thus can be directed through the selection of the reagent enantiomer. Access to both enantiomeric forms of the reagent is therefore valuable. For the investigation of the degree of chromato-

Abbreviations

AITC, 2,3,4-tri-O-acetyl-α-D-arabinopyranosyl isothiocyanate; Boc-, t-butoxycarbonyl; CDI, 1, 1-carbonyldimidazole, DANE, 1-(4-dimethylamino-1-naphthyl)ethyl amine; DCC, N, N′-dicyclohexylcarbodiimide; FLEC, 1-(9-fluorenyl)ethyl chloroformate; Fmoc, 9-fluorenylmethoxycarbonyl; GITC, 2, 3, 4, 6-tetra-O-acetyl-β-D-glucopyranosyl isothiocyanate; HETE, hydroxyeicosatetraenoic acid; HOBT, 1-hydroxybenzotriazole; MMNA, α-methoxy-α-methyl-(1-naphthyl)acetic acid; MTPA α-methoxy-α-(trifluoromethyl(phenylacetic acid; NCA, N-carboxy anhydride; NEIC, 1-(1-naphthyl)ethyl isocyanate; OPA, ortho-phthalaldehyde; PEIC, phenylethyl isocyanate (α-methylbenzyl isocyanate); SINEC, succinimido 1-(α-naphthyl)ethyl carbamate; SIPEC, succinimido 1-phenylethyl carbamate; TATG, 2, 3, 4, 6-tetra-O-acetyl-1-thio-β-glucopyranoside; WSC, 1-(3-dimethylaminopropyl)-3-ethylcarbodiimide; Z, benzyloxycarbonyl.

graphic resolution of a pair of enantiomers, for example when different reagents are evaluated, it is sufficient to use a racemic mixture of the reagent. An optically pure reagent is, however, required to determine the elution order and the enantiomeric composition of a chiral substance. When an optical impurity is present in the reagent, a certain fraction of the molecules of either substrate enantiomer will form a diastereomer which is falsely measured as coming from the other enantiomer. The same results if measurable racemization takes place during the reaction. If the reagent is not 100% optically pure and its enantiomeric composition is known, then the results can be corrected, but with a certain loss in precision. The optical purity of the reagent is therefore a limiting factor in the determination of the enantiomeric ratio of a component. The purity can be determined by derivatizing an optically pure substance, or by chiral chromatography. The reagent may be purified by chromatographic methods, as well as by classical non-chromatographic techniques (Jacques *et al.*, 1981; Newman, 1981–84).

Diastereomer formation is used for different purposes such as preparative isolation of enantiomers, for assessment of optical purity, for determination of absolute configuration and for enantioselective assays in biological samples. These widely different applications emphasize different aspects of the methodology. While sensitivity and selectivity are of primary importance for determinations in biological samples, this is not critical in preparative chromatography. Smooth cleavage for recovery of enantiomers from separated diastereomers is essential in preparative work, but less important in analytical applications where ease of diastereomer formation is more important. Requirements on the chromatographic resolution are also different.

Factors that affect the chromatographic resolution R are the relative retention of the two diastereomers (the separation factor α), the mean retention k' and the actual column efficiency N. The resolution as a function of N, α, and k' has been expressed as follows by Snyder and Kirkland (1979).

$$R = (1/4)^*(\alpha - 1)^*(k'/(1 + k'))^*(N)^{1/2}$$

In preparative chromatography high α-values are desirable as they allow a heavier sample load on the column, without serious peak overlapping due to reduction in N. In analytical applications, where columns with relatively high separation efficiency are used to separate the diastereomers from other sample components, a separation factor of 1.1–1.2 is sufficient to obtain adequate resolution. When one enantiomer is to be determined in the presence of large excess of the antipode, higher α-values may be needed. In this case it is desirable that the smaller peak elutes in front of the larger because of the peak tailing commonly observed (Figure 4.3). The desired elution order can be obtained by choosing the proper enantiomer of the reagent.

Crystallization of covalent diastereomers for preparative resolution of enantiomers has been practised since the beginning of the century. With the emergence of gas chromatography, formation and separation of diastereomers

Table 4.1 Reagents: chiral structure

Reagent	References
1 1-Phenylethyl amine 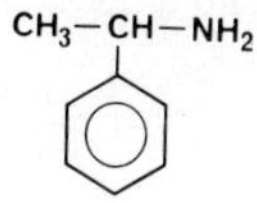	Maître *et al.* (1984), Michelsen *et al.* (1985), Stoltenborg *et al.* (1981), McKay *et al.* (1979), Helmchen *et al.* (1977), Nakazawa *et al.* (1979)
2 1-Phenylethyl isocyanate 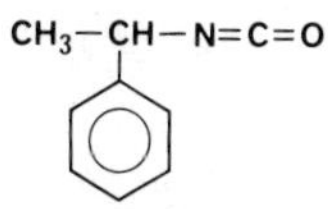	Miller *et al.* (1984), Christ and Walle (1985), Blaschke *et al.* (1985), Gulaid *et al.* (1985), Wilson and Walle (1984), Schönenberger *et al.* (1986), Williams (1984), Thompson *et al.* (1982), Oatis *et al.* (1983), Hsyu and Giacomini (1986), Pflugmann *et al.* (1987)
3 Succinimido-1-phenylethyl carbamate 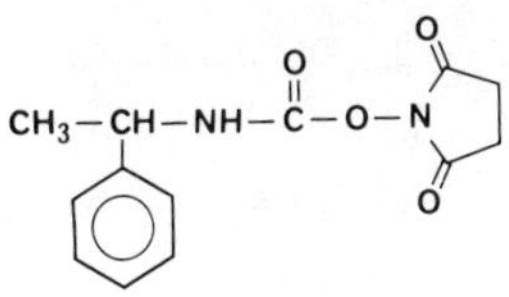	Iwaki *et al.* (1987)
4 1-Phenylethyl isothiocyanate 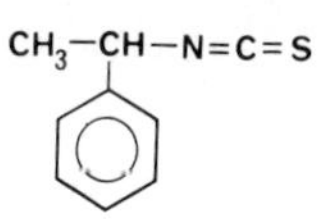	Gal and Sedman (1984)
5 1-(4-Nitrophenyl)ethyl amine 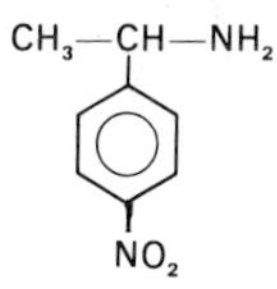	Scott *et al.* (1976), Valentine *et al.* (1976)
6 Levobase, dextrobase 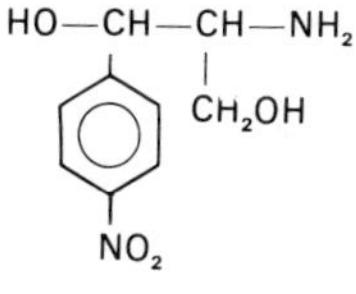	Slégel *et al.* (1987), Ladányi *et al.* (1987)

Continued

Table 4.1 (Contd.)

	Reagent	References
7	α-Methoxy-α-(trifluoromethyl)- phenyl acetic acid 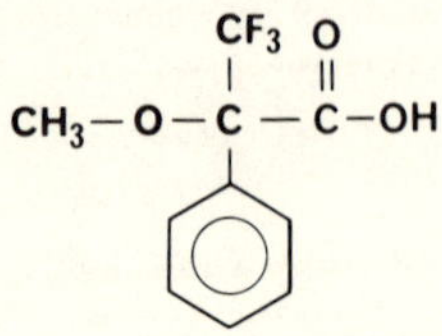	Thakker *et al.* (1977, 1984), Bergot *et al.* (1978), McKay *et al.* (1979), Miller *et al.* (1984), Bladeren *et al.* (1985)
8	O-Methyl mandelic acid 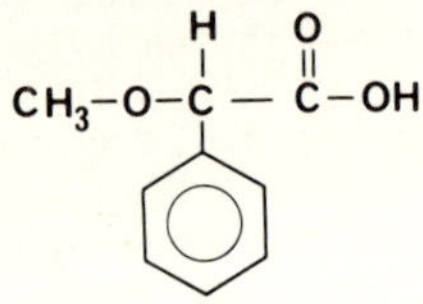	Helmchen and Strubert (1974), Hoffmann *et al.* (1984)
9	Ferrocenylethyl amine	Shimada *et al.* (1987*a*), Helmchen *et al.* (1977)
10	Ferrocenylpropyl amine 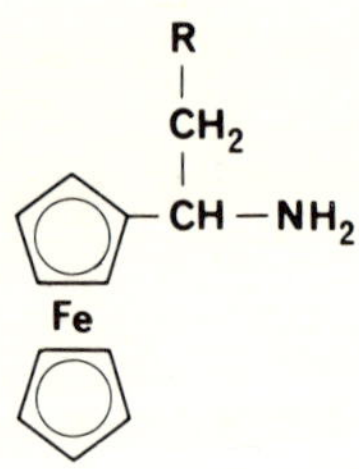	Shimada *et al.* (1987*a*)
11	Benoxaprofen chloride 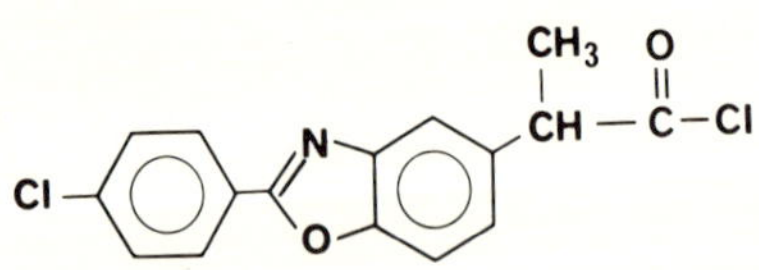	Spahn *et al.* (1984), Weber *et al.* (1984), Pflugmann *et al.* (1987)
12	1-Naphthylethylamine 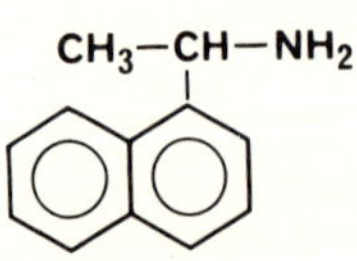	Helmchen *et al.* (1977), Bergot *et al.* (1978), Hutt *et al.* (1986), Avgerinos and Hutt (1987)
13	1-Naphthylethyl isocyanate 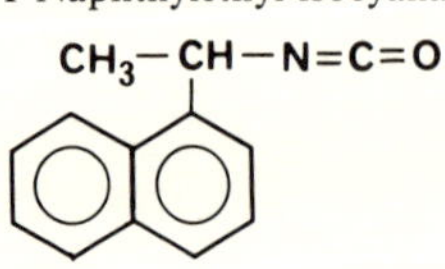	Pirkle and Hoekstra (1974), Pirkle and Simmons (1983), Gübitz and Mihellyes (1984), Michelsen *et al.* (1985), Darman and Thernot (1986), Dunlop and Neidle (1987), Gietl *et al.* (1988)

Continued

Table 4.1 (Contd.)

Reagent	References
14 Succinimido-1-(α-naphthyl)ethyl carbamate 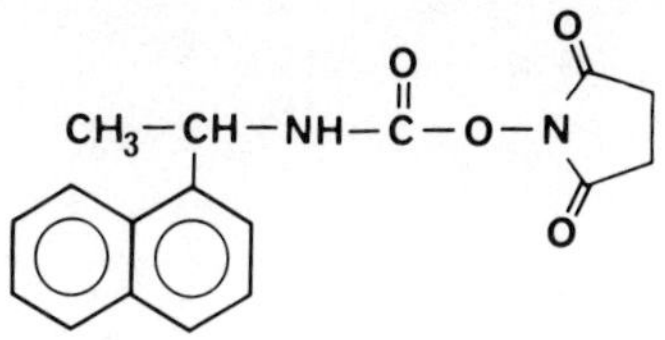	Iwaki *et al.* (1987)
15 2-(β-Naphthyl)propionyl chloride 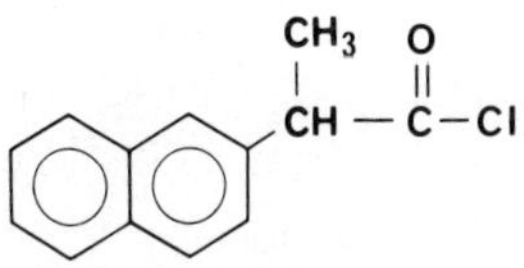	Shimizu *et al.* (1982)
16 DANE 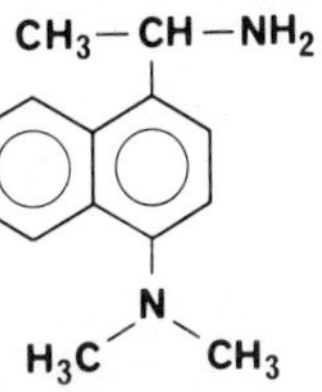	Goto *et al.* (1980, 1982*b*)
17 α-Methoxy-α-methyl-α-(1-naphthyl)acetic acid 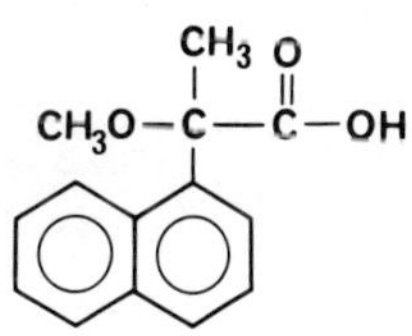	Goto *et al.* (1977, 1978)
18 α-Methoxy-α-methyl-α-(1-naphthyl)acetoxy succinimide 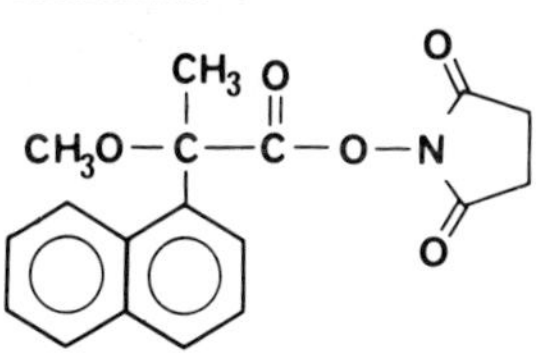	Goto *et al.* (1979)
19 1-(9-Fluorenyl)ethyl chloroformate 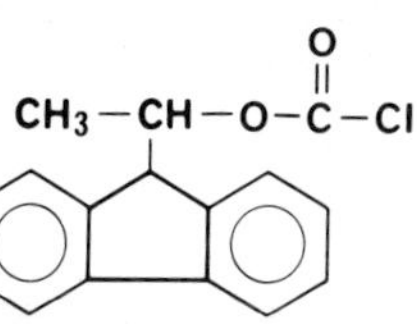	Einarsson *et al.* (1987*b*)

Continued

Table 4.1 (Contd.)

Reagent	References
20 1-(1-Anthryl)ethylamine	Goto *et al.* (1986)
21 1-(2-Anthryl)ethylamine 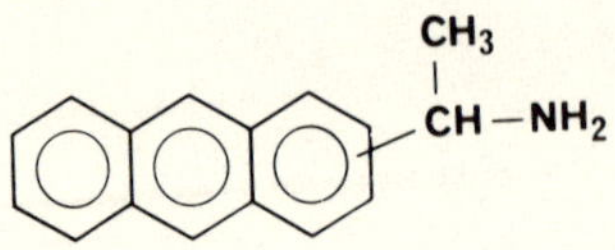	Goto *et al.* (1986)
22 Methylbinaphthalenecarbonyl nitrile 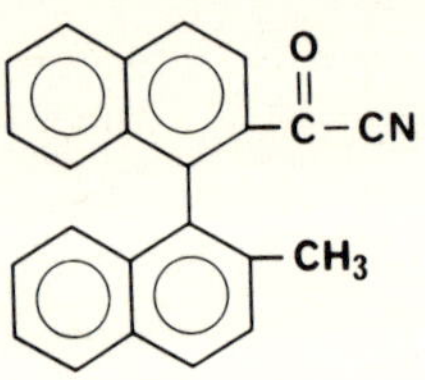	Goto *et al.* (1982*a*)
23 Neomenthyl isothiocyanate 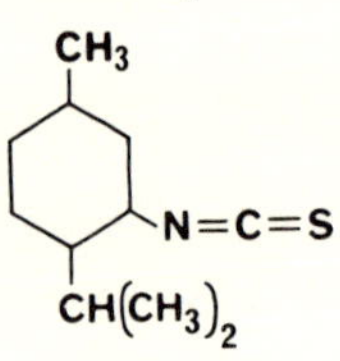	Nambara *et al.* (1978)
24 Menthyl chloroformate 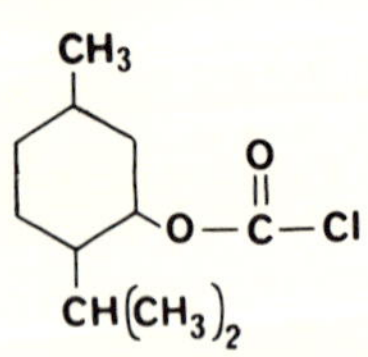	Pirkle and Hauske (1977*b*), Arnould *et al.* (1980), Brash *et al.* (1985), Seeman *et al.* (1985)
25 Menthoxyacetyl chloride 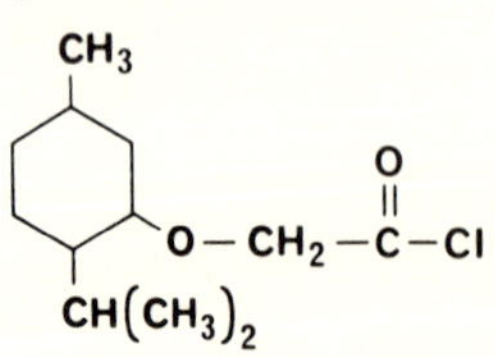	Yang *et al.* (1977), Yang and Fu (1984), Yoneda *et al.* (1985)
26 Camphorsulphonyl chloride 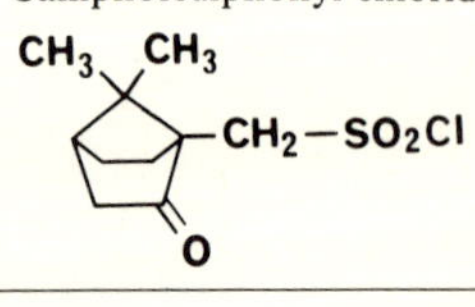	Furukawa *et al.* (1975, 1977), Souter (1976), Aberhart *et al.* (1985)

Continued

Table 4.1 (Contd.)

Reagent	References
27 Norbornyl isothiocyanate 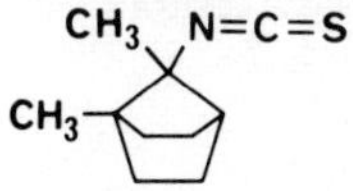	Nambara *et al.* (1978)
28 Camphanic acid chloride 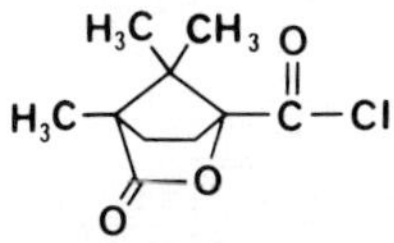	Ruud-Christensen and Salvesen (1984), Johnström *et al.* (1984)
29 Dehydroabietyl isocyanate 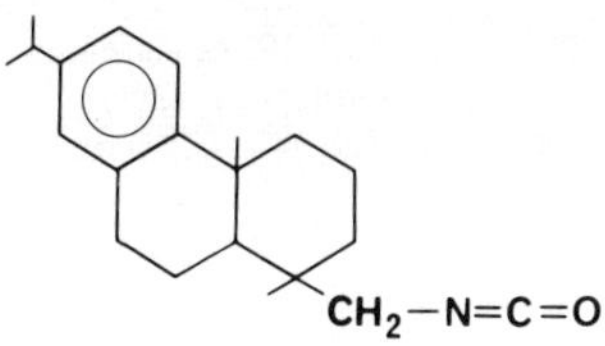	Corey and Hashimoto (1981), Falk *et al.* (1984), Wollard (1986)
30 3β-Acetoxyetienic acid 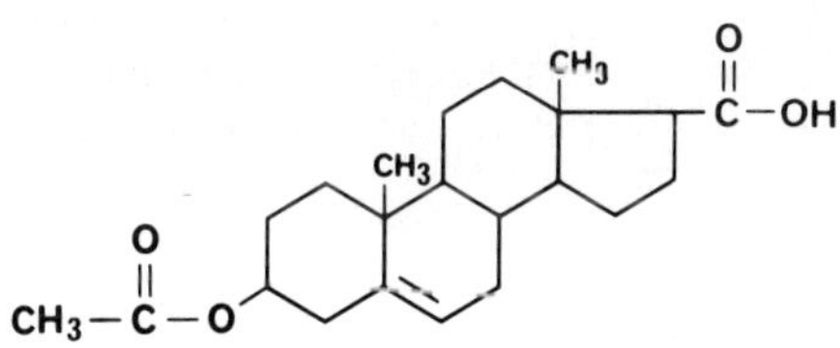	Bergot *et al.* (1978)
31 Tartaric anhydrides 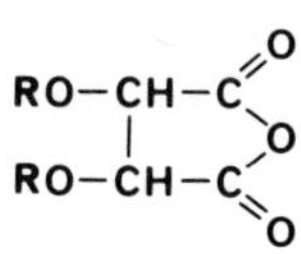	Lindner *et al.* (1984), Demian and Gripshover (1987)
32 AITC 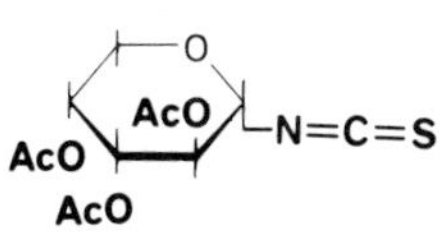	Kinoshita *et al.* (1981), Nimura *et al.* (1981), Sedman and Gal (1983), Miller *et al.* (1984)
33 GITC 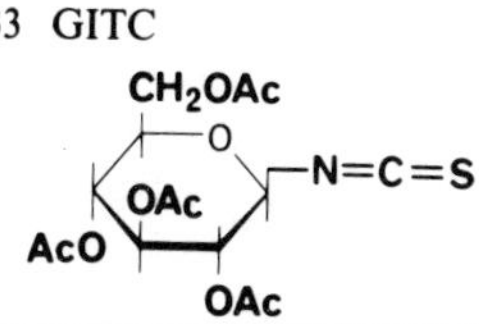	Nimura *et al.* (1980, 1981, 1984), Kinoshita *et al.* (1981), Sedman and Gal (1983), Gal *et al.* (1984*a*, 1986*a*, *b*), Miller *et al.* (1984), Christ and Walle (1985), Kirchhoefer *et al.* (1985), Walle *et al.* (1985)

Continued

Table 4.1 (Contd.)

Reagent	References
34 Amino acid-NCA 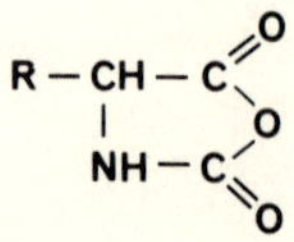	Hirschmann *et al.* (1967), Manning and Moore (1968), Takaya *et al.* (1981)
35 Boc-amino acid anhydrides 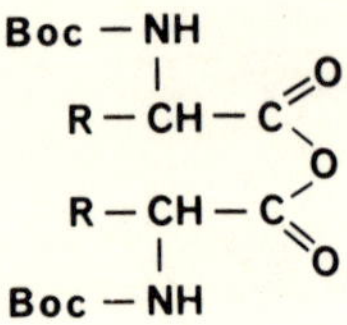	Hermansson (1982), Hermansson and Bahr (1982)
36 Boc-Leu-Su 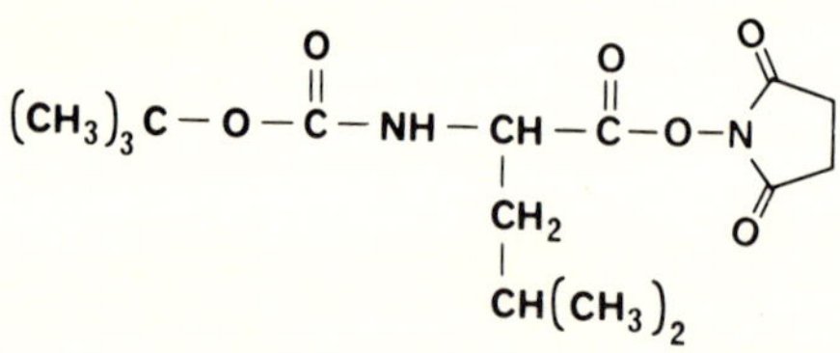	Mitchell *et al.* (1978), Cahill *et al.* (1980), Lankmayr *et al.* (1980), Nachtmann (1980), Chen and Contario (1984)
37 L-Leucinamide 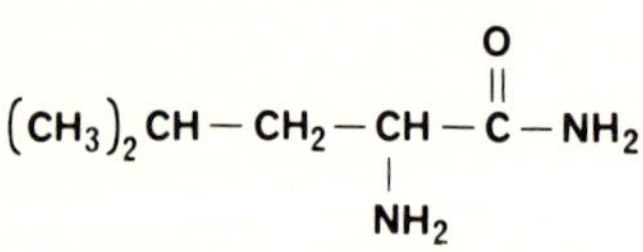	Björkman (1985, 1987). Spahn (1987), Foster and Jamali (1987)
38 Z-L-Proline 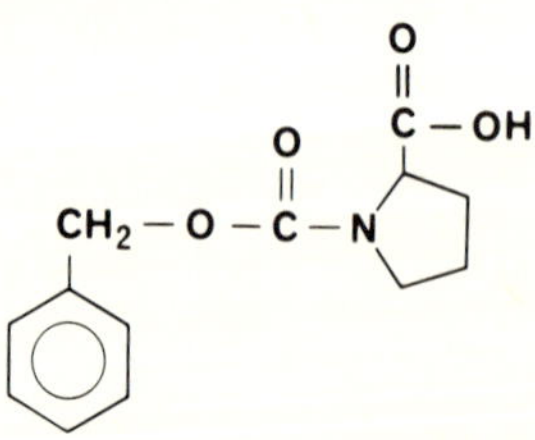	Banfield and Rowland (1983, 1984)
39 N-TFA-prolyl chloride 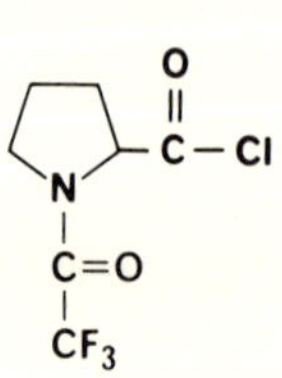	Hermansson and Bahr (1980)

Continued

Table 4.1 (Contd.)

	Reagent	References
40	Naphthsulphonylprolyl chloride 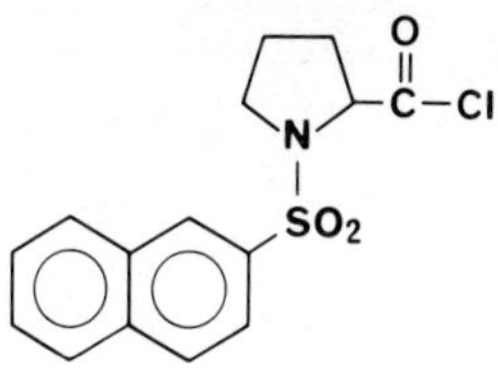	Shimizu *et al.* (1986)
41	*p*-Nitrophenylsulphonylprolyl chloride 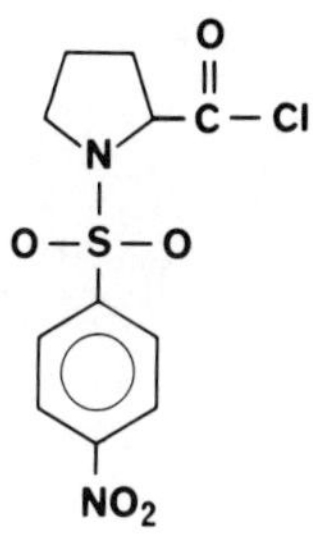	Clark and Barksdale (1984), Barksdale and Clark (1985)
42	θ-(p-Nitrobenzyl)-tyrosine methyl ester 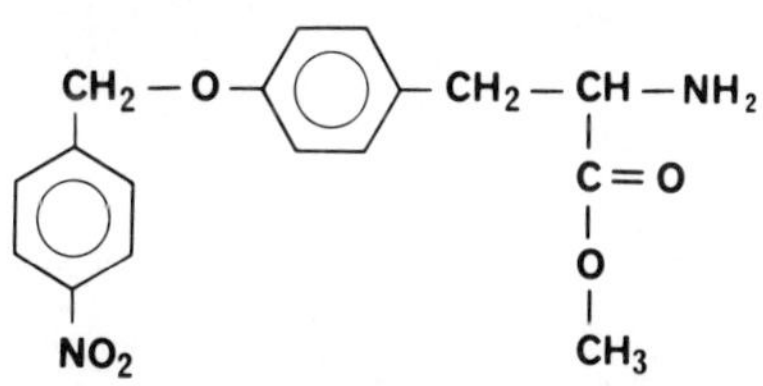	Görög *et al.* (1986)
43	N-Fmoc-L-isothiocyanophenylalanine methyl ester 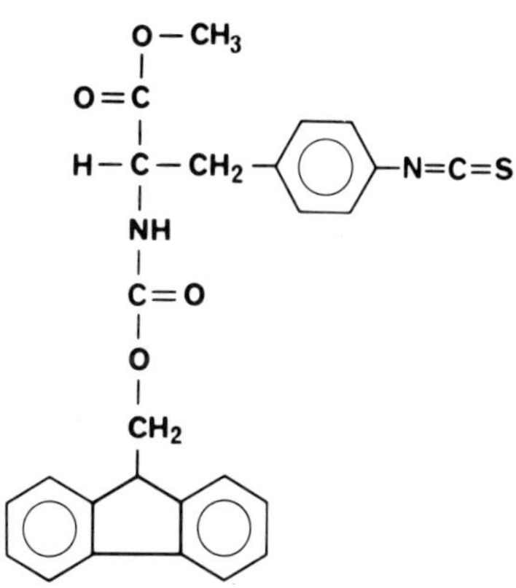	Davis and Mohammed (1984)

Continued

Table 4.1 (Contd.)

Reagent	References
44 OPA/Boc-L-cysteine 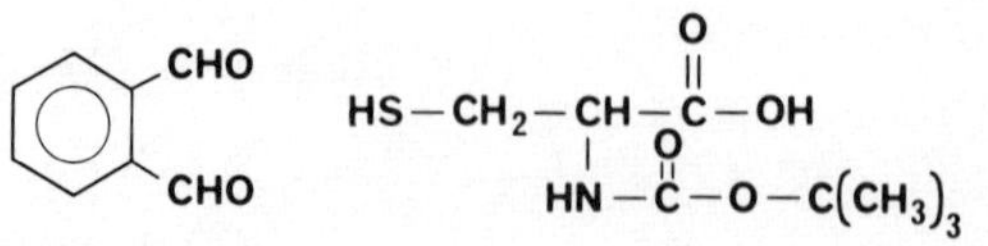	Buck and Krummen (1984, 1987)
45 OPA/N-acetyl-L-cysteine 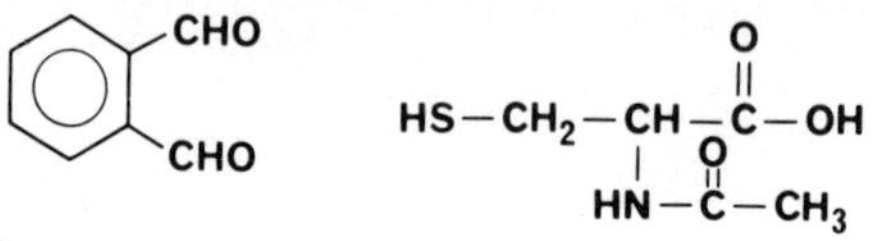	Aswad (1984), Nimura and Kinoshita (1986), Lam (1986), Buck and Krummen (1987), Wuis *et al.* (1987)
46 OPA/N-acetyl-D-penicillamine 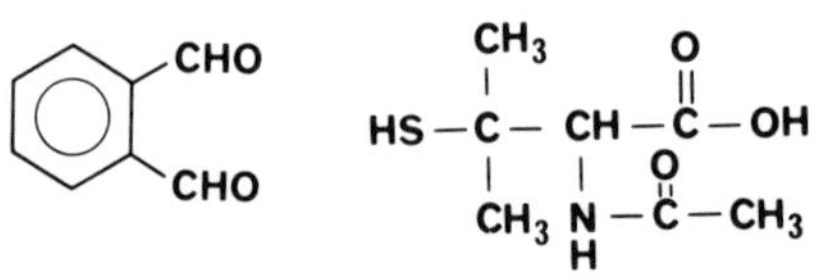	Buck and Krummen (1987)
47 OPA/TATG 	Einarsson *et al.* (1987*a*)
48 2-Octanol 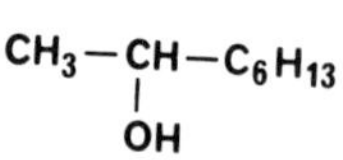	Johnsson *et al.* (1979), Lee *et al.* (1984)
49 2-Bromooctane 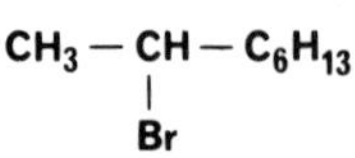	Wainer (1980)
50 Tetrahydro-5-oxo-2-furan-carboxylic acid 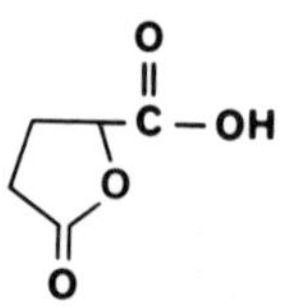	Doolittle and Heath (1984)

Continued

Table 4.1 (Contd.)

	Reagent	References
51	1-Fluoro-2,4-dinitrophenyl-5-L-alanine amide 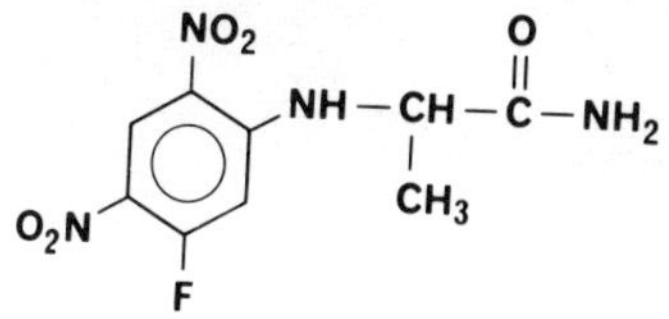	Marfey (1984), Aberhart (1985), Ruterbories and Nurok (1987)
52	4,5-Diphenyl-2-oxazolidone-3-carbamyl chloride 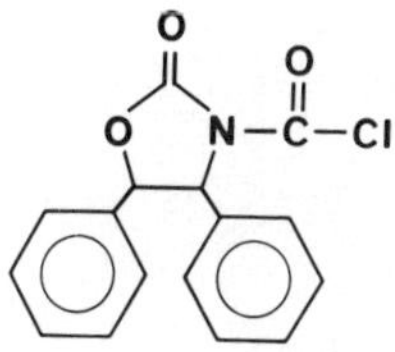	Pirkle and Simmons (1983)

became a practical analytical method for enantiomers. The high separation efficiency obtained with capillary columns has enabled determination of enantiomers in complex samples (Halpern, 1977; König, 1987). The recent development of liquid chromatography has, however, resulted in a rapidly increasing number of applications using this separation technique. Most published papers on diastereomer formation for separation with LC have thus been reported in the past 10 years, and a number of reviews have appeared (Pirkle and Finn, 1983; Souter, 1985; Lindner and Pettersson, 1985; Lindner, 1987; Gal, 1987, 1988).

The material in this chapter has been structured to allow different ways of reading. In section 4.3, chiral reagents are grouped according to their reactive group. Reagents are listed in Table 4.1 according to their chiral structure. Table 4.2 (pp. 64–68) contains key information on applications, grouped according to the chemical structure of the compound to be derivatized.

4.2 Diastereomer formation

The chiral substances most commonly resolved as diastereomers are amines, alcohols and carboxylic acids. The main types of derivatives formed are depicted in Figure 4.1. The amines are the easiest to derivatize, a fact which is reflected in the variety of different reagents available. The amino group is therefore usually the site for derivatization in bifunctional compounds such as amino acids and amino alcohols. Amines, alcohols and carboxylic acids also act as chiral precursors for chiral derivatizing reagents. Isocyanates, for

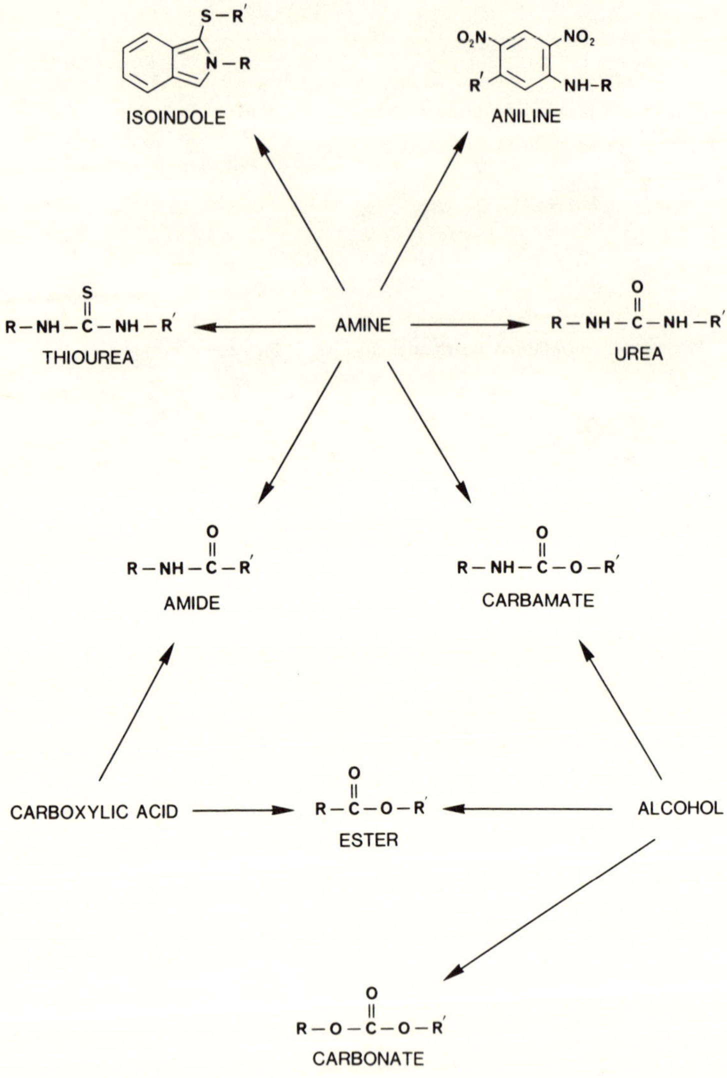

Figure 4.1 Main types of derivatives formed from amines, carboxylic acids and alcohols in reactions with chiral reagents.

example, which are useful for making carbamate diastereomers of alcohols, are synthesized from amines and phosgene. The reaction of alcohols with phosgene yields chloroformates which are used to derivatize amines to form diasteromeric carbamates.

In gas chromatography the formation of esters has been preferred to that of amides for resolution of carboxylic acids, because of the less favourable chromatographic properties of amides. The opposite situation has arisen in liquid chromatography. Diastereomeric amides are relatively easily resolved due to hindered rotation around the amide bond and consequently the vast majority of carboxylic acids are separated as amides. Different ways to mediate the formation of amides are illustrated in Figure 4.2.

For the determination of unknown concentrations of enantiomers, the derivatization should preferably be quantitative, i.e. 90–100%, in order to ensure good accuracy. A high reaction yield at the same time ensures that the

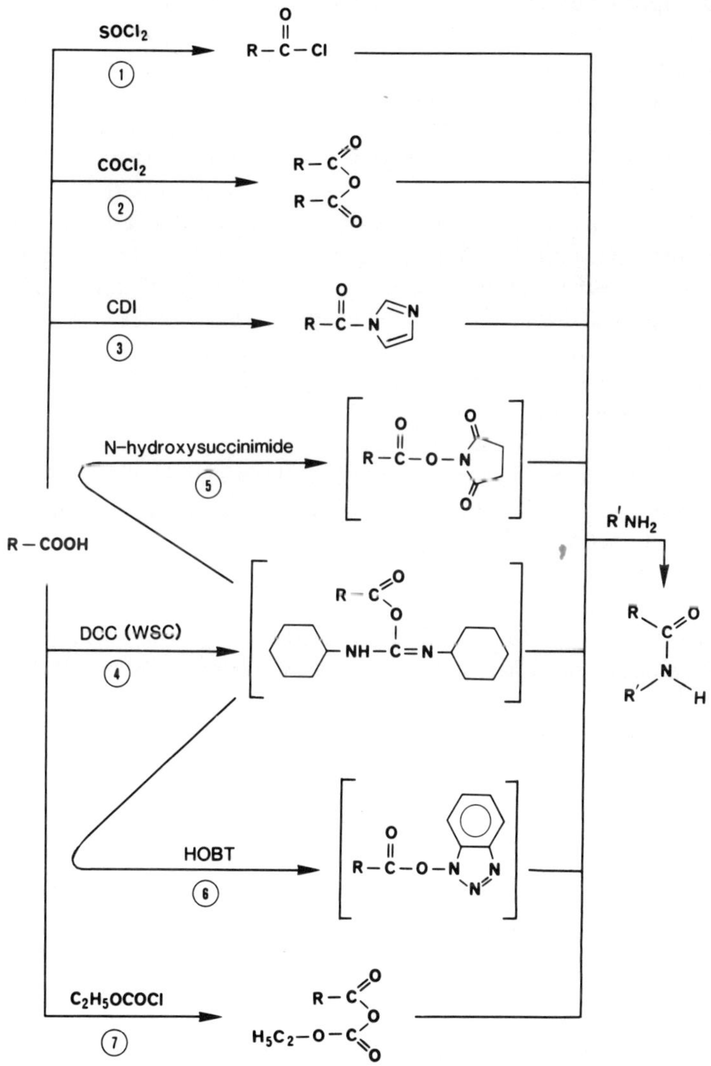

Figure 4.2 Formation of amides. Chemical structures given in brackets represent reactive intermediates.

enantiomeric ratio is not changed by so-called kinetic resolution which, when present, has larger influence on relative yields when a reaction is incomplete. The reagent should certainly be in excess of the substrate. For small amounts of the substrate, the ratio between amounts of reagent and substrate is not important. Instead, the absolute concentration of the reagent in the reaction mixture is of primary importance as it will determine the rate of diastereomer formation. In case the reagent is not continuously consumed by e.g. solvolytic reactions and its concentration together with other critical factors such as pH remains approximately constant, the formation of diastereomers for simple reactions often follow pseudo first-order kinetics, i.e. the substrate concentration declines exponentially with time, with a half-life which is independent of the initial substrate concentration. A properly selected reaction time will then ensure quantitative yields, as long as the substrate is not itself involved in undesired side reactions. Lower reagent concentrations will normally require correspondingly longer reaction times. For a reaction where the reagent is continuously consumed by solvolysis, it is essential that the initial rate of diastereomer formation is high enough, so that quantitative conversion is obtained while the reagent concentration is still high. If not, a prolonged reaction time will add little to the reaction yield. The recovery of slow-reacting substrates can be increased by using higher reagent concentration. Therefore when using reagents, such as isocyanates, chloroformates, dinitroaryl halides and acylating reagents, in aqueous media, it is a great help to know the actual rate of hydrolysis of the reagent.

An elevated temperature may be used to shorten reaction times. Moderate changes normally have marginal influence on relative reaction rates of different competing reactions.

If extensive clean-up is carried out after derivatization and before the chromatographic determination, it is advisable to check the recovery of each enantiomer. The formed diastereomers may have slightly different properties, e.g. in liquid–liquid distribution, as shown by Hermansson and Bahr (1982).

4.3 Reagents

4.3.1 *Acylating reagents*

Chiral acylating reagents are derived from carboxylic acids which in most cases have a chiral carbon in α-position to the carboxyl group. Many chiral carboxylic acids are available commercially as enantiomers and can be converted to reactive forms by the user. If a higher purity is needed, this might be obtained by fractional crystallization of the diastereomeric salt of the acid with an optically pure amine, such as (+)-1-phenylethyl amine (Goto *et al.*, 1978), or cinchonidine (Shimizu, 1982). Formation of diastereomeric amides with 1-phenylethylamine, followed by preparative chromatography and hydrolysis of the amide bond, has also been used (Weber *et al.*, 1984).

Carboxylic acids may be used as free acids in acylation reactions in aqueous solutions or in organic solvents, using a carbodiimide as condensing agent (Figure 4.2). This is a standard procedure for preparative derivatization, and for analytical derivatization of carboxylic acids with amine reagents (see section 7.3.2), but less frequently used for analysis of amines and alcohols using carboxylic reagents. Banfield and Rowland (1983) investigated the derivatization reaction between warfarin, a 4-hydroxycoumarin derivative, and benzyloxycarbonyl-L-proline, using dicyclohexylcarbodiimide (DCC) as condensing agent and imidazole as catalyst. They applied the reaction to analysis of plasma samples. DCC is not effective in acylation of sterically hindered secondary amines, such as N-alkylated amino acids. Bis(2-oxo-oxazolidinyl)- phosphinic chloride (BOP-Cl) has been used for this purpose (Tung *et al.*, 1986).

Acylating reagents are used in the form of acyl chlorides, anhydrides, imidazolides and N-hydroxysuccinimide esters. Acyl chlorides are highly reactive and are used to derivatize sterically hindered secondary alcohols and amines. As they are often of limited long-term stability they are commonly prepared before use by treating the carboxylic acid with thionyl chloride. Acyl chlorides are more susceptible to racemization than less reactive acyl reagents. Nichols reported already in 1973 that derivatization of chiral secondary amines with N-TFA-prolyl chloride (Table 4.1, 39) gave erroneous results due to racemization. Unlike N-TFA-prolyl chloride, closely similar imidazolides, used in gas chromatography, have been claimed to be free from racemization (Gal, 1987). Imidazolide reagents may be prepared from the carboxylic acid by treatment with carbonyldiimidazole (Figure 4.2, reaction 3).

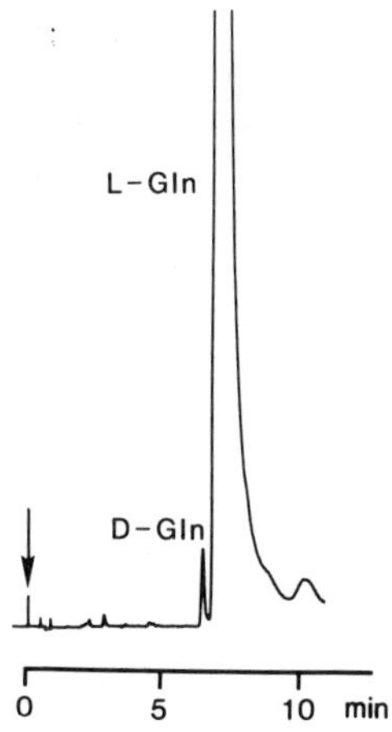

Figure 4.3 Determination of D-glutamine (0.06%) in L-glutamine by reversed-phase separation of their corresponding carbamates formed by derivatization with (+)-1-(9-fluorenyl)ethyl chloromate (FLEC). (From Einarsson *et al.*, 1987*b*).

Racemization at the chiral carbon in α-position can also be avoided by choosing reagents lacking an α-hydrogen atom, such as α-methoxy-α-(trifluoromethyl)-α-phenyl acetylchloride (Table 4.1, 7). Anhydrides may be prepared by treatment with phosgene. Hermansson and Bahr (1982) used anhydrides of N-protected amino acids (Table 4.1, 35) to derivatize propranolol at the secondary amino group. The N-protecting group was subsequently removed to obtain better chromatographic resolution of the diastereomers. Typical solvents for reactions with acyl chlorides, imidazolides and anhydrides are dichloromethane and toluene, but alcoholic and aqueous solvents may also be used for derivatization of (primary) amines. Excess reagent normally has to be removed before chromatography. Acyl chlorides have been used to resolve dihydrodiols of polyaromatic hydrocarbons. An example is given in Figure 4.4.

A special class of acylating reagents are the N-carboxy anhydrides of amino acids (Table 4.1, 34), used for enantioselective determination of amino acids according to the Manning and Moore procedure (Manning and Moore, 1968; Takaya *et al.*, 1981). Dipeptides are formed with very little racemization

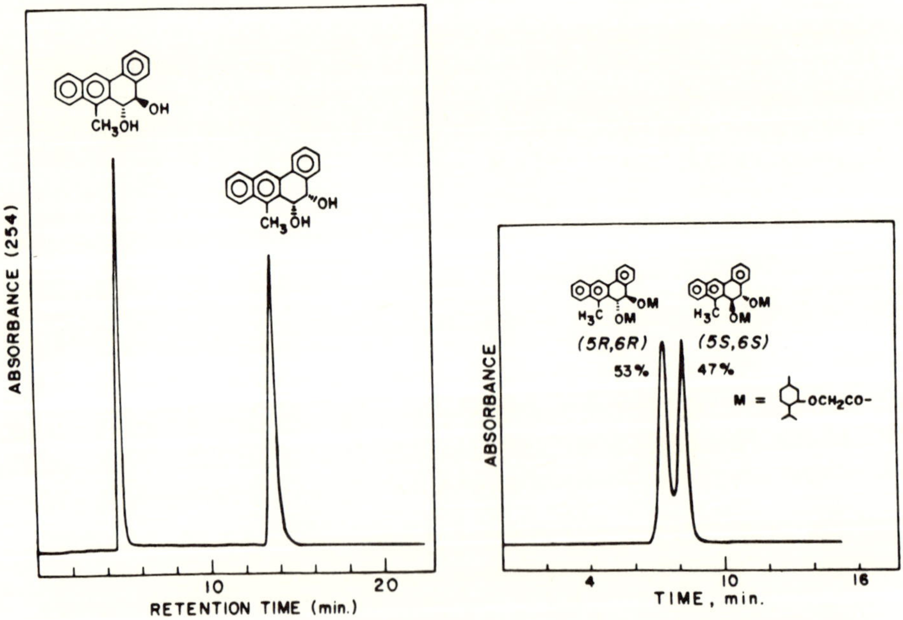

Figure 4.4 Separation of 5,6-dihydrodiol metabolites of 7-methylbenz[a]anthracene (7-MBA). Left: Reversed-phase separation of underivatized *trans*- and *cis*-isomers. The first peak represents the *trans*-(5R, 6R)- and *trans*-(5S, 6S)-isomers. The second peak is from the *cis*-(5R, 6S)- and *cis*-(5S, 6R)-isomers.
Right: Normal phase separation of the di(−)-menthoxyacetates of 7-MBA *trans*-5, 6-dihydrodiol. (From Yang and Fu, 1984).

(Hirschmann *et al.*, 1967). The reagent, which is very moisture-sensitive, is added as crystals directly to the aqueous substrate solution. In the acylation reaction, the free amino group is formed from the N-carboxy anhydride, and care has to be taken to avoid further reaction to higher peptides.

Acylation with esters of N-hydroxysuccinimide (Table 4.1, 36) proceeds under mild conditions, and is used for racemization-free coupling of amino acids in aqueous solution (Mitchell, 1978). The reagents can be prepared using a carbodiimide as coupling agent (Figure 4.2, reactions 4 and 5) and are sufficiently stable to allow purification with e.g. chromatography and long-term storage. Several succinimide esters of N-protected amino acids are available commercially. The moderate reactivity of succinimide esters make them useful mainly for derivatization of the sterically unhindered primary amines and (primary and secondary) amino acids.

Sternson (1981) has presented a discussion of general aspects of acylation reactions for precolumn derivatization, including correlations between chemical structure and reactivity of reagents, which can be recommended for further reading.

4.3.2 *Amine reagents*

Chiral amine reagents have mainly been used to derivatize carboxylic acids, such as N-protected amino acids, terpenoid acids and α-arylpropionic acids. The latter group includes several non-steroidal anti-inflammatory drugs, and many of the analytical applications of amine reagents are for such substances. As these substances are stereospecific in action and also undergo metabolic chiral inversion, it is essential to have enantioselective methods for their determination in body fluids. Chiral amine reagents have also been used to derivatize alcohols, such as alkylated glycerols, after conversion of the alcohols to chloroformates by phosgene treatment (Michelsen *et al.*, 1985).

Condensation reactions between carboxylic acids and amines are outlined in Figure 4.2. For carboxylic substrates, activation with thionyl chloride to the corresponding chlorides has been used. Later applications tend to use milder methods. 1, 1′-carbonyl-diimidazole (CDI) was used in a chlorinated solvent to convert 2-arylpropionic acids to their imidazolides, which were then allowed to react with the chiral reagent phenylethyl amine (Maître *et al.*, 1984; Stoltenborg *et al.*, 1981; Pedrazzini *et al.*, 1987; Sioufi *et al.*, 1987). Typical reaction times were 10 min for the first reaction and 30–60 min for the second reaction. Görög and co-workers (1986) used N, N′-dicyclohexylcarbodiimide in methanol as coupling agent, and studied the reaction recovery and the formation of reaction side products. They accepted non-quantitative reaction yields as long as the enantiomeric ratio was not affected, and emphasized the simplicity of the procedure; the reaction mixture was injected without further treatment.

Higher reaction yields and smaller artifact formation were obtained with a combination of a carbodiimide (WSC = 1-(3-dimethylaminopropyl)-3-ethylcarbodiimide) and 1-hydroxybenzotriazole (HOBT) used by Hutt and co-workers (Hutt *et al.*, 1986; Avgerinos and Hutt, 1987) for the derivatization of 2-arylpropionic acids with 1-(1-naphthyl)ethyl amine (Table 4.1, 12). Reactions were carried out in dichloromethane for 1.5–2 h and the solvent was removed before chromatography.

The reagents 1-phenylethylamine (α-methylbenzylamine) and 1-(1-naphthyl)ethylamine were introduced early and are still widely used owing to favourable separation characteristics of the corresponding amides. Closely related are (*p*-nitrophenyl)ethyl amine (Table 4.1, 5) and 1-(dimethylamino-1-naphthyl)ethylamine (DANE, Table 4.1, 16), the latter introduced as a fluorescence labelling reagent (Goto *et al.*, 1980). A desire to improve further detection properties has recently led to the introduction of 1-(anthryl)ethylamines (Table 4.1, 20, 21). 1-Ferrocenylethylamine (Table 4.1, 9), which was evaluated as resolving agent by Helmchen and co-workers in 1977, but was found to be less efficient than e.g. 1-(2-naphthyl)ethylamine, has recently been reintroduced by Shimada *et al.* (1987a) owing to its interesting properties in conjunction with electrochemical detection.

Two chiral amine reagents which are available with high purity and at low cost were introduced by Slégel, Ladányi and co-workers: levobase, (1R, 2R)-(−)-1-(4-nitrophenyl)-2-amino-1, 3-propanediol, and dextrobase, the (1S, 2S)-enantiomer (Table 4.1, 6). Different carboxylic acids were coupled using dicyclohexylcarbodiimide and separated on straight-phase TLC (Slégel *et al.*, 1987) and reversed-phase LC (Ladányi *et al.*, 1987). The presence of hydroxyl groups that may participate in intramolecular hydrogen bonds make these reagents interesting alternatives to e.g. phenylethyl amine.

Amino acid methyl esters may be used as chiral resolving agents for N-protected amino acids. The O-(4-nitrobenzyl)ether of tyrosine methyl ester (Table 4.1, 42) was introduced as a UV-absorption labelling reagent by Görög *et al.* (1986).

A promising technique for the analytical conversion of carboxylic acids into amides was introduced by Björkman (1985). He used ethylchloroformate to convert enantiomers of indoprofen, a 2-arylpropionic acid, extracted from human plasma, to the corresponding carboxylic–carbonic mixed anhydride (Figure 4.2, reaction 7). The reaction was carried out in acetonitrile containing triethylamine and was very fast (30 s at room temperature). The mixed anhydride was then allowed to react with L-leucinamide (Table 4.1, 37) for 2 min. The formed carbamates were partitioned into ethyl acetate. Racemization could not be observed. Attractive features of this derivatization technique are: the reaction is fast, little disturbance is seen from reaction by-products (ethanol, CO_2, HCl and N-ethoxycarbonyl-leucinamide), reaction yields seem to be high, and the chiral reagent is stable and of high purity. The method has also been applied to the determination of ketoprofen enantiomers in plasma (Björkman, 1987; Forster and Jamali, 1987). Spahn (1987) reported

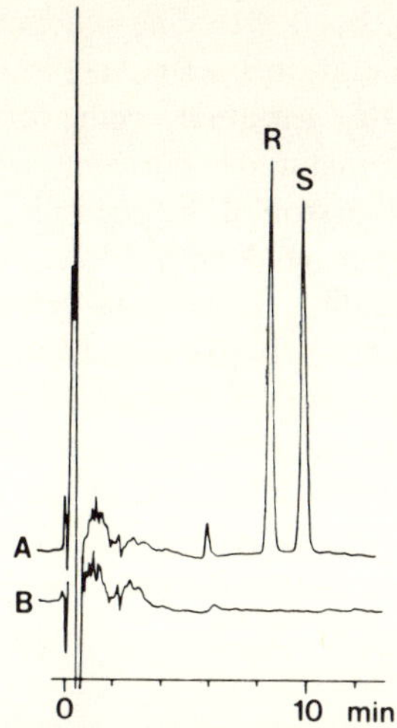

Figure 4.5 Resolution of the L-leucinamide derivatives of (R, S)-2-(1-(3-fluoro)biphenyl)propionic acid-(fluorbiproten) using reversed-phase chromatography and UV-detection at 254 nm. (*A*) 37.5 ng per peak, (*B*) blank; produced by mixing the reagents without addition of a 2-arylpropionic acid. (From Spahn, 1987). The derivatization method of Björkman (1985, 1987) was used.

the separation of diastereomers of eight 2-arylpropionic acids on a reversed-phase LC-system using this technique, while benoxaprofen was adequately resolved on a straight-phase system (see also Figure 4.5). Coupling with ethyl chloroformate should certainly be possible using other chiral amines than L-leucinamide.

4.3.3 *Isocyanates and activated carbamates*

Isocyanates are generally prepared from the corresponding amine by the action of phosgene on the hydrochloride salt of the amine in dry toluene. Three different chiral isocyanates have been used for resolution of racemic amines and alcohols: phenylethyl isocyanate (PEIC, Table 4.1, 2), naphthylethyl isocyanate (NEIC, Table 4.1, 13) and dehydroabiethyl isocyanate (Table 4.1, 29). Phenylethyl- and naphthylethyl isocyanates are commercially available and so is the amine precursor of dehydroabietyl isocyanate. Isocyanates form urea derivatives with amines, and carbamates (urethanes) by reaction with alcohols. The reactions are usually carried out in dry methylene chloride, chloroform or toluene. The reaction of PEIC (0.1 M in methylene chloride) with amines is complete within 30 minutes at room temperature (Gulaid *et al.*, 1985). Alcohols react more slowly and are derivatized at elevated temperatures. Thirty minutes' reaction time has been reported for derivatization of various alcohols with PEIC (1 M, in methylene chloride) at 70° (Gal *et al.*, 1981). The reactions with alcohols are in some cases catalysed by addition of N, N-dimethylethanolamine (Michelsen *et al.*, 1985) or 4-dimethylaminopyridine (Corey and Hashimoto, 1981). The difference in reactivity makes it possible to derivatize amine groups selectively in aminoalcohols such as β-blockers, with no trace of the carbamate being formed (Gulaid *et al.*, 1985). Phenolic groups, however, react under the same conditions as the amines but more slowly. Partial derivatization of the phenolic group was observed when 4-hydroxypropranolol was reacted with PEIC (Wilson and Walle, 1984). Single derivatives corresponding to the urea-derivatives could, however, be obtained by hydrolysing the carbamate bond in 0.1 M hydrochloric acid. Although isocyanates react rapidly with water, Dunlop and Neidle (1987) were able to use NEIC for derivatization of amino acids in aqueous samples in high yields. After mixing the reagent (dissolved in dry acetone) and the buffered aqueous sample (pH 9.0), the solution was vigorously stirred for 30s. By-products were removed by centrifugation (precipitate of the symmetrical naphthylethyl urea) and partition into cyclohexane/ether (naphthylethyl-amine).

The optical purity and possible racemization of PEIC during derivatization of alcohols has been evaluated by derivatization of (−)-menthol, and subsequent separation with gas chromatography (Gal *et al.*, 1981). Enantiomeric impurity and racemization was concluded to amount to less than 1%. A corresponding study with NEIC, using aspartic acid which is available in high optical purity (Dunlop and Neidle, 1987), suggested that the reagent had an optical purity of 99.7%. Racemization therefore seems to be negligible when isocyanates are reacted with amines and alcohols even under prolonged periods at elevated temperatures.

Phenylethyl isocyanate has been used for determination of β-blockers in plasma samples (Gulaid *et al.*, 1985; Hsyu and Giacomini, 1986) and in urine (Wilson and Walle, 1984; Hsyu and Giacomini, 1986). The samples are made alkaline and the substances are isolated by partition into diethyl ether or ethyl acetate. The derivatization is then carried out after evaporation of the organic phase. The separations are usually made with reversed phase chromatography. Better resolution of 4-hydroxypropranolol was observed with normal phase chromatography ($R = 1.9$) compared with reversed phase chromatography ($R = 0.62$), while the resolution of propranolol remained the same in both systems (Wilson and Walle, 1984). The derivatives of PEIC are generally detected with UV. In cases where the sample molecules exhibit native fluorescence this has been utilized for higher sensitivity and selectivity.

Naphthylethyl isocyanate, which was introduced by Pirkle and Hoekstra (1974), has been used for resolution of a variety of substances such as amino acids (Dunlop and Neidle, 1987), β-blockers (Gübitz and Mihellyes, 1984, Darmon and Thenot, 1986), alcohols (Michelsen *et al.*, 1985; Pirkle and Hoekstra, 1974) and oxazolidones (Pirkle and Simmons, 1983). Separation factors in the order of 2–3 were achieved for various oxazolidone derivatives on silica. Good resolution of NEIC-derivatized neutral and basic amino acids ($\alpha = 1.1-1.4$) was generally obtained by reversed phase chromatography, whereas separation of acidic amino acids was more difficult (Dunlop and Neidle, 1987). NEIC-derivatives are in most cases fluorescent (ex/em = 285/330 nm), which makes the reagent suitable for measurements at low concentrations (Darmon and Thenot, 1986).

Dehydroabietyl isocyanate (Table 4.1, 29) has mainly been used for determination of hydroxyeicosatetraenoic acid (HETE) (Corey and Hashimoto, 1981; Falck *et al.*, 1984; Wollard, 1986). The reagent was found to be superior to both PEIC and NEIC for the resolution of 5-HETE. The reagent was used for preparative resolution of 5-HETE diastereomers on silica, whereafter the 5-HETE enantiomers were recovered upon treatment of the isolated carbamates fractions with trichlorosilane (Corey and Hashimoto, 1981).

Racemic amino acids and α-amino alcohols have been resolved as urea derivatives afforded by reaction with succinimido carbamates (Iwaki *et al.*, 1987). The two reagents were produced from 1-phenylethylamine (Table 4.1, 3) and 1-(α-naphthyl)ethylamine (Table 4.1, 14) by reaction with disuccinimido carbonate. The succinimido carbamates react rapidly in alkaline aqueous solutions at room temperature with primary and secondary amines. During the derivatization the reagent undergoes hydrolysis to produce the parent amine and N-hydroxysuccinimide. Provided that the concentration of reagent is kept sufficiently high the hydrolysis does not adversely affect the derivatization yields as the rate of urea formation is much faster than the reaction with water. Resolutions of amino acid diastereomers performed with reversed-phase chromatography yielded slightly better separation factors for the naphthylethyl-reagent ($\alpha = 1.18$) compared with the phenylethyl-reagent ($\alpha = 1.15$).

4.3.4 *Isothiocyanates*

The principal uses of isothiocyanates have been for peptide sequencing (Edman, 1950) and amino acid analysis (Hendrikson and Meredith, 1984). Chiral isothiocyanates have been synthesized from the corresponding amines by reaction with thiophosgene (Davies and Mohammed, 1984) or carbon disulphide followed by reaction with ethyl chloroformate (Nambara *et al.*, 1978). The two most widely used reagents, 2, 3, 4, 6-tetra-O-acetyl-β-D-glucopyranosyl isothiocyanate (GITC, Table 4.1, 33) and 2, 3, 4-tri-O-acetyl-α-D-arabinopyranosyl isothiocyanate (AITC, Table 4.1, 32), were made from the corresponding acetobromosugars by reactions with silver thiocyanate (Nimura *et al.*, 1980; Kinoshita *et al.*, 1981). Isothiocyanates are more stable than the corresponding isocynates.

Isothiocyanates react with primary and secondary amines to form thiourea derivatives. The derivatizations are commonly performed in acetonitrile, methylene chloride or dimethylformamide, often in the presence of triethylamine, and reported reaction times are 10–60 min at room temperature. Epinephrine and norepinephrine were observed to react faster with GITC and AITC in dimethylformamide than in acetonitrile (Nimura *et al.*, 1981).

GITC and AITC were originally used for resolution of amino acid esters (Nimura *et al.*, 1980). Separation factors in the range 1.17–1.79 were obtained for free amino acids using reversed-phase chromatography (Kinoshita *et al.*, 1981), and with gradient elution the D- and L-form of 16 amino acids were resolved in 2 h (Nimura *et al.*, 1984). The detection limit for the derivatives was 5 ng, using UV detection at 250 nm (Kinoshita *et al.*, 1981). Various amines of pharmaceutical interest have been successfully resolved with reversed-phase chromatography as GITC derivatives (Nimura *et al.*, 1981; Gal, 1984, 1986; Sedman and Gal, 1983; Gal and Brown, 1986; Kirchhoffer *et al.*, 1985; Allgire *et al.*, 1985). The AITC and GITC reagents were generally found to perform better than phenylethyl isocyanate and (−)-α-methoxy-α-(trifluoromethyl)-α-phenylacetyl chloride for a number of amphetamines (Miller *et al.*, 1984). The enantiomers of hydroxypropranolol sulphate isolated from human plasma have been determined using GITC (Walle *et al.*, 1985). The derivatives were stable for at least 24 h, and the minimum detectable quantity was 20 ng. As alternatives to GITC, which temporarily disappeared from the market, Gal and co-workers introduced phenylethyl isothiocyanate (Table 4.1, 4) and, more recently, (1-naphthyl)ethyl isothiocyanate (Gal, 1988).

The reagent N-Fmoc-L-*p*-isothiocyanophenylalanine methyl ester (Table 4.1, 43) was synthesized for assessment of chiral purity in peptide synthesis (Davies and Mohammed, 1984). Racemization was found to occur at the thiazolinone stage after cleavage from the peptide. This was, however, not the case when thiourea derivatives of free amino acids were converted to the thiohydantoin derivative. The reagent should be suitable for fluorescence detection owing to the fluorine moiety.

4.3.5 *Chloroformates*

Chloroformates are easily prepared by treating the corresponding alcohol with phosgene. Both alkyl and aryl chloroformates are made this way, often with a tertiary base as an acid acceptor (Matzner *et al.*, 1964). Chloroformates react readily with primary and secondary amines in moderately alkaline aqueous solutions with formation of carbamates (urethanes). Reactions with thiols and phenols under similar conditions yield the corresponding mercapto-carbonates and carbonates. Simultaneous hydrolysis produces the parent alcohol, carbon dioxide and hydrogen chloride. Alcohols and tertiary amines may be reacted with chloroformates in organic solvents such as toluene or methylene chloride in the presence of a suitable hydrogen acceptor (Takadate *et al.*, 1983; Ahnfelt *et al.*, 1982).

Application of (−)-menthyl chloroformate (Table 4.1, 24) to gas chromatographic resolution of esterified amino acids and hydroxyl compounds appears to be the first use of a chloroformate as derivatization reagent for chromatography (Westley and Halpren, 1968). In liquid chromatography (−)-menthyl chloroformate has been used for resolution of hydroxy acids (Brash *et al.*, 1985) and for preparative resolution of the secondary amines fenfluramine

(Arnould *et al.*, 1980) and nornicotine (Seeman *et al.*, 1985). The results indicate racemization-free derivatization and deblocking of the reagent (Arnould *et al.*, 1980). Optical purity of $> 99\%$ was reported for the nornicotine enantiomers after separation and acid catalysed hydrolysis of the carbamates. Base catalysed hydrolysis was reported to result in racemization (Seeman *et al.*, 1985).

Free amino acids and amines have been resolved as fluorescent disastereomers after reaction with 1-(9-fluorenyl)ethylchloroformate (FLEC, Table 4.1, 19). An enantiomeric purity of $\geq 99.94\%$, measured by determination of the D-form in an optically pure L-amino acid, confirms the result from methyl chloroformate of racemization-free formation of carbamates (Figure 4.3). The diastereomers of the D- and L-form of 17 amino acids were resolved by reversed phase chromatography in 70 minutes. The imino acids proline and hydroxyproline can be selectively derivatized if primary amino acids are first allowed to react with OPA. The different spectral properties of the OPA-derivatives as compared to the FLEC-derivatives allowed selective detection of the imino acids, using fluorescence emission at 315 nm (Einarsson *et al.*, 1987*b*).

4.3.6 *Reagents based on o-phthalaldehyde and chiral thiols*

o-phthalaldehyde (OPA) forms isoindoles when reacted with primary amines in the presence of a thiol (Simons and Johnson, 1976). The isoindoles are in general both highly fluorescent (Roth, 1971) and amenable to electrochemical detection (Joseph and Davies, 1983). The reaction which is selective for primary amines is normally complete within a few minutes at room temperature. Thiols may also be determined by derivatization with OPA in the presence of an excess of a suitable primary amine. The reactants are non-fluorescent which eliminates problems often connected with interferences from excess of the reagent. The derivatives have limited stability, which makes exact timing of the analytical procedure necessary. The derivatization procedure is widely used for determination of amino acids (Deyl *et al.*, 1986).

By selecting a chiral thiol for the derivatization with OPA, diastereomers are obtained from chiral amines. Successful separations were reported using thiol-containing N-protected amino acids (Table 4.1, 44–46), or an O-acetylated thioglucose (Table 4.1, 47). The reagents proved to have an optical purity of at least 99.9%, by determination of the D-enantiomer in pure L-amino acids (Buck and Krummen, 1987, Einarsson *et al.*, 1987*a*), Racemization during the reaction therefore seems negligible.

OPA-N-acetyl-L-cysteine was introduced for resolution of aspartic acid (Aswad, 1984) and later used for separation of protein amino acids (Nimura and Kinoshita, 1986; Lam, 1986; Buck and Krummen, 1987). Reversed-phase chromatography effectively resolved the common protein amino acids with the exception of lysine (and proline which does not react). The reagent OPA-

2, 3, 4, 6-tetra-O-acetyl-1-thio-β-glucopyranoside (TATG) was used for resolution of amino acids (Einarsson, 1987*a*). A mean separation factor α of 1.27 was obtained with reversed-phase chromatography. Evaluation of the electrochemical behaviour of the derivatives demonstrated a half wave potential of 0.65–0.75 V.

Various amino alcohols, amino acids and amines have been resolved using OPA with Boc-L-cysteine, N-acetyl-D-penicillamine and N-acetyl-L-cysteine (Buck and Krummen, 1984, 1987). The optimal thiol was different depending on the structure of the substrate. With OPA-Boc-L-cysteine, lysine was resolved, and a mean resolution factor R of 2.45 was obtained for the protein amino acids (Buck and Krummen, 1984). A skeletal muscle relaxant (4-amino-3-*p*-chlorophenylbutyric) acid was resolved as diastereomers of OPA-N-acetyl-L-cysteine. An exceptionally long reaction time (25 min at 80°) was required for maximum response (Wuis *et al.*, 1987).

OPA combined with a chiral thiol is potentially useful for resolution of a number of different chiral primary amines and thiols. However, the fluorescent properties of the derivatives is dependent on the structure of substrate. This may be overcome by using electrochemical detection, which is less sensitive for differences in the structure of the derivatives (Allison *et al.*, 1984).

4.3.7 *Other reagents*

Arylhalides, such as Sanger's reagent (1-fluoro-2, 4-dinitrobenzene), are well known reagents for amino acids, peptides and other amines. A chiral reagent of this type 1-fluoro-2,4-dinitrophenyl-5-L-alanine amide, was introduced by Marfey in 1984 (Table 4.1, 51). High α-values were obtained for separation of L- and D-amino acids, although the chiral centres are separated by five atoms, evidently because of the rigid conformation of these atoms. Sensitive UV/VIS absorption detection was achieved (molar absorptivity was near 3.0×10^4 at 338 nm and 1.1×10^4 at 414 nm). A reaction time of 1 h was used. A few applications of this interesting reagent have appeared (Aberhart *et al.*, 1985, Ruterbories and Nurok, 1987, see also Figure 4.6). As pointed out by Marfey, other chiral reagents can be made by coupling other amino acids instead of L-alaninamide to the reagent precursor, 1, 5-difluoro-2, 4-dinitrobenzene.

A number of diastereomeric acylureas derived from aryl-substituted 2-oxazolidones and 1-(1-naphthyl)ethyl isocyanate were synthesized in a systematic study of their chomatographic separability on silica (Pirkle and Simmons, 1983). The most effective separation was achieved with the *cis*-diphenyl substituted 2-oxazolidone. This compound was converted into its carbamyl chloride by reaction with phosgene and used for resolution of a number of racemic amines. Acylureas were produced on a preparative scale by refluxing the amines with cis-4, 5-diphenyl-2-oxazolidone-3-carbonyl chloride (Table 4.1, 52) and an equivalent amount of triethylamine in benzene. Separation factors of the different diastereomers ranged from 1.17 to 4.14.

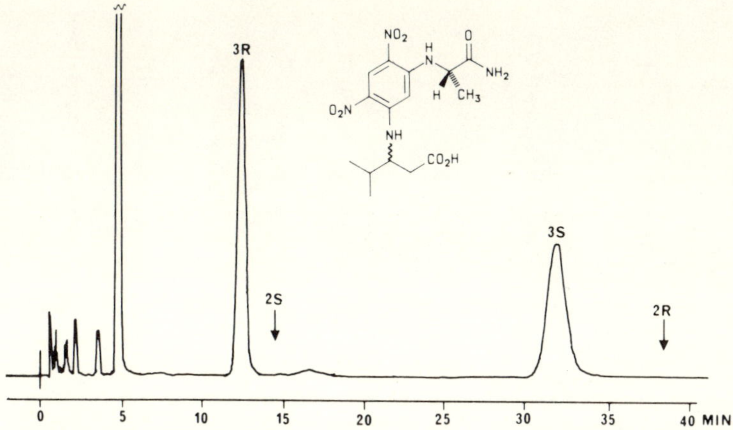

Figure 4.6 Reversed-phase separation of (3R)- and (3S)-leucine after derivatization with l-fluoro-2,4-dinitrophenyl-5-L-alanine-amide (Marfey's reagent). The elution times of (2S)- and (2R)-leucine are also incicated. UV-absorbance was measured at 340 nm. (From Aberhart *et al.*, 1985).

4.4 Separation mechanisms

The choice of reagent for diastereomer formation has a main influence on the resolution achieved, the selectivity being dependent on the structure of the chiral moiety and the type of chemical bond formed. Different approaches taken for derivatization of the most common functional groups are illustrated in Figure 4.1. Differential migration of diastereomers in a chromatographic system is based on differential interactions between the sample molecules and the stationary phase as well as the solvent. Systematic studies of the chromatographic behaviour of diastereomers may be of help in the prediction of optimal chiral reagents and allow the assignment of absolute configuration of chiral compounds. Even if the nature of interactions involved in the retention differs in the various modes of chromatography, some common features such as proximity of the chiral centres and conformational immobility around the centres are seen to enhance the resolution of diastereomers. The effect on the resolution of increasing the distance between the chiral centres has been demonstrated by Scott and co-workers (1976), in the separation of a number of acyclic isoprenoid acid enantiomers derivatized was *p*-nitrophenylethylamine. The separation factor dropped from 2.2 with the secondary methyl chiral centre in the α-position to 1.0 when it was at carbon 5.

A significant proportion of diastereomer separations reported in recent years have been performed with reversed-phase chromatography (Table 4.2). The parameters, underlying the separation of diastereomers, have, however, not been systematically studied. On silica and alumina, detailed studies have been made on the relative retention of diastereomers. Separation of more than 100 diastereomer pairs on silica have been discussed in papers by Snyder,

Table 4.2 Selected applications

Compound	Reagent No.	Name	Application	Chromatography	Detection	References
AMINES	26	Camphorsulphonyl chloride		N	UV	Souter (1976)
	52	4, 5-Diphenyl-2-oxazolidone-3-carbamyl chloride		N		Pirkle and Simmons (1983)
	33	GITC				
	32	AITC				
	2	Phenylethyl isocyanate				
	7	MIPA		R	UV	Miller *et al.* (1984)
	24	Menthyl chloroformate		N		Pirkle and Hauske (1977)
Amphetamine	40	Naphthylsulphonyl-L-prolyl chloride		N, R	UV	Shimizu *et al.* (1986)
Amphetamines	11	Benoxaprofen chloride	plasma	N	Fl	Weber *et al.* (1984)
2, 5-Dimethoxy-4-methylamphetamine	18	MMNA succinimide ester	plasma	N	UV	Goto *et al.* (1979)
						Arnould *et al.* (1980)
Fenfluramine	24	Menthyl chloroformate	prep	R		Blaschke *et al.* (1985)
Flecainide	2	Phenylethyl isocyanate		N		Seeman *et al.* (1985)
Nornicotine	24	Menthyl chloroformate	prep	N		Helmchen and Strubert (1974)
Phenylethylamine	8	O-Methylmandelyl chloride		N		
Tertiary amines	2	Phenylethyl isocyanate		N	UV	Maibaum (1988)
Amino alcohols	33	GITC		R	UV	Gal (1986), Gal and Brown (1986)
	45	OPA/N-acetyl-L-cysteine				
	46	OPA/N-acetyl-D-penicillamine				
	44	OPA/Boc-L-cysteine	hydrolysate	R	Fl	Buck and Krummen (1987)
	4	Phenylethyl isothiocyanate		R	UV	Gal and Sedman (1984)
Acebutolol/diacetolol	2	Phenylethyl isocyanate		R	Fl	Gulaid *et al.* (1985)
Betaxolol	13	Naphthylethyl isocyanate	blood	R	Fl	Darmon and Thenot (1986)
β-Blockers	33	GITC				
	32	AITC		R	UV	Sedman and Gal (1983)
β-Blockers	13	Naphthylethyl isocyanate		R	UV/Fl	Gübitz and Mihellyes (1984)
Ephedrines	33	GITC		R	UV	Gal (1984)
Epinephrine	33	GITC	pharmaceutical formulations	R	UV	Kirchhoefer *et al.* (1985)
						Allgire *et al.* (1985)

Epinephrines	33	GITC				
	32	AITC		R	UV	Nimura *et al.* (1981)
Hydroxypropranolol sulphate	33	GITC	plasma	R	UV	Walle *et al.* (1985)
						Christ and Walle (1985)
Hydroxypropranolol	2	Phenylethyl isocyanate	urine	N, R	UV	Wilson and Walle (1984)
Metoprolol	2	Phenylethyl isocyanate	plasma	R	Fl	Pflugmann *et al.* (1987)
Metoprolol	19	FLEC		R	Fl	Einarsson *et al.* (1987)
Metoprolol, propranolol	11	Benoxaprofen chloride	urine	N	Fl	Pflugmann *et al.* (1987)
Oxprenolol	2	Phenylethyl isocyanate	biological fluids	N	UV	Dieterle and Faigle (1983)
Pindolol	2	Phenylethyl isocyanate	biological fluids	R	Fl	Hsyu and Giacomini (1986)
Propranolol	39	N-TFA-prolyl chloride	plasma	R	Fl	Hermansson and Bahr (1980)
β-Blockers	35	Boc-L-Ala anhydride				Hermansson and Bahr (1982)
	35	Boc-L-Leu anhydride	plasma	R	Fl	Hermansson (1982)
Propranolol	2	Phenylethyl isocyanate		R	Fl	Thompson *et al.* (1982)
						Oatis *et al.* (1983)
Free amino acids	26	Camphorsulphonyl chloride		N	UV	Furukawa *et al.* (1975, 1977)
	34	Leu-NCA, Glu-NCA		IE	UV	Manning and Moore (1968)
	34	Leu-NCA, Phe-NCA		R	UV	Takaya *et al.* (1981)
	36	Boc-L-Leu-Su		R	UV	Mitchell *et al.* (1978)
						Cahill *et al.* (1980)
	19	FLEC		R	Fl	Einarsson *et al.* (1987)
	51	1-Fluoro-2, 4-dinitrophenyl-5-L-alanine amide		R	UV	Marfey (1984), Aberhart *et al.* (1985)
	33	GITC				
	32	AITC		R	UV	Kinoshita *et al.* (1981)
	33	GITC		R	UV	Nimura *et al.* (1984)
	13	Naphthylethyl isocyanate		R	UV	Dunlop and Neidle (1987)
	14	SINEC				
	3	SIPEC		R	UV/Fl	Iwaki *et al.* (1987)
	44	OPA/Boc-L-cysteine		R	Fl	Buck and Krummen (1984, 1987)
	45	OPA/N-acetyl-L-cysteine		R	Fl	Aswad (1984), Lam (1986), Nimura and Kinoshita (1986), Buck and Krummen (1987)

Continued

Table 4.2 (*Contd.*)

Compound	Reagent No.	Reagent Name	Application	Chromatography	Detection	References
	46	OPA/N-acetyl-D-penicillamine		R	Fl	Buck and Krummen (1987)
	47	OPA/TATG		R	Fl/E	Einarsson *et al.* (1987)
Aspartic acid	45	OPA/N-acetyl-L-cysteine		R	Fl	van den Oetelaar *et al.* (1987)
Baclofen	45	OPA/N-acetyl-L-cysteine		R	Fl	Wuis *et al.* (1987)
Iodo thyronines (T_3, T_4)	36	Boc-L-Leu-Su	e.p.	R	UV	Lankmayr *et al.* (1980)
Penicillamine	36	Boc-L-Leu-Su	e.p.	R	UV	Nachtmann (1980)
Amino acid esters	17	MMNA		N, R	UV	Goto *et al.* (1977, 1978)
	27	Dimethylnorbornyl isothiocyanate		N	UV	Nambara *et al.* (1978)
	23	Neomenthyl isothiocyanate		N	UV	Nambara *et al.* (1978)
	33	GITC		R	UV	Nimura *et al.* (1980)
ALCOHOLS	13	Naphthylethylisocyanate	prep	N		Pirkle and Hoekstra (1974)
	50	Tetrahydro-5-oxo-2-furan-carboxylic acid	prep	N		Doolittle and Health (1984)
Alkyl glycerols	12	Naphthylethyl amine				
	13	Naphthylethyl isocyanate		N	UV	Michelsen *et al.* (1985)
B[a]P dihydrodiols	7	MTPA chloride	prep	N		Thakker *et al.* (1977, 1984), Yagi *et al.* (1977)
Dihydrodiols	7	MTPA chloride		R		Bladeren *et al.* (1985)
B[a]P dihydrodiols	25	Menthoxyacetyl chloride	prep	N		Lee and Harvey (1984)
Deacetyldiltiazem	15	Naphthylpropionyl chloride	e.p.	R	UV	Shimizu *et al.* (1982)
Deacetyldiltiazem	40	Naphthylsulphonyl-L-prolyl chloride	e.p.	N	UV	Shimizu *et al.* (1986)
Aminoalcohols	31	Tartaric anhydrides	e.p.	R	UV	Lindner *et al.* (1984)
1-Methyl-3-pyrrolidinol	31	Dibenzoyltartaric acid	e.p.	R	UV	Demian and Gripshover (1987)
Hydroxyeicosa-tetraenoic acid	29	Dehydroabiethyl-isothiocyanate		N	UV	Falck *et al.* (1984), Wollard (1986)
			prep	N		Corey and Hashimoto (1981)

HETE	24	Menthyl chloroformate		N	UV	Brash *et al.* (1985)
Hydroxyacids	22	Methylbinaphthalene carbonyl nitrile		N	Fl	Goto *et al.* (1982a)
Misonidazole	2	Phenylethylisocyanate	plasma	R	UV	Williams (1984)
B[a]A dihydrodiols	25	Menthoxyacetyl chloride	a.c.	N		Yang and Fu (1984), Yang *et al.* (1977)
Nipradilol	25	Menthoxyacetyl chloride	a.c.	N	UV	Yoneda *et al.* (1985)
Propoxyphylline	28	Camphanoyl chloride		R	UV	Ruud-Christensen and Salvesen (1984)
Warfarin	38	Z-L-proline	plasma	N	UV	Banfield and Rowland (1983, 1984)
CARBOXYLIC ACIDS	12	Naphthylethylamine	a.c.	N		Helmchen *et al.* (1977)
	6	Levobase/dextrobase		N	UV	Slégel *et al.* (1987)
	6	Levobase/dextrobase		R	UV	Ladányi *et al.* (1987)
2-Arylpropionic acids	37	L-Leucinamide	plasma	R	UV	Spahn (1987)
2-Arylpropionic acids	1	Phenylethylamine	liver prep.	N	UV	Maître *et al.* (1984)
2-Arylpropionic acids	12	Naphthylethylamine	urine	N	UV	Hutt *et al.* (1986, 1987)
Benoxaprofen	1	Phenylethylamine		N	UV	McKay *et al.* (1979)
Carprofen	1	Phenylethylamine	blood	N	UV/Fl	Stoltenborg *et al.* (1981)
Clinofibrate	1	Phenylethylamine		N	UV	Nakazawa *et al.* (1979)
Flunoxaprofen	1	Phenylethylamine	biological fluids	R	UV	Pedrazzini *et al.* (1987)
Ketoprofen/Indoprofen	37	L-Leucinamide	plasma	R	UV	Björkman (1985, 1987), Foster and Jamali (1987)
Pyrethroid acids	1	Phenylethylamine	prep	N	UV	Jiang and Soderlund (1982)
Ibuprofen	48	2-Octanol	plasma	N	UV	Lee *et al.* (1984)
Naproxen	48	2-Octanol		N	UV	Johnson *et al.* (1979)
Naproxen	16	DANE	serum	N	Fl	Goto *et al.* (1982)
Mandelic acid	49	2-Bromooctane		N	UV	Wainer (1980)
Pirprofen	1	Phenylethylamine			UV	Sioufi *et al.* (1987)
Terpenoid acids	12	Naphthylethyl amine		N	UV	Bergot *et al.* (1978)
Terpenoid acids	5	1-(4-Nitrophenyl)ethyl amine	e.p.	N	UV	Valentine *et al.* (1976)
Terpenoid acids	5	1-(4-Nitrophenyl)ethyl amine		N	UV	Scott *et al.* (1976)

Continued

Table 4.2 (*Contd.*)

Compound	Reagent No.	Reagent Name	Application	Chromatography	Detection	References
N-Protected amino acids	42	Nitrobenzyltyrosine methyl ester		N	UV	Görög *et al.* (1986)
	16	DANE		N	Fl	Goto *et al.* (1980)
N-acetyl amino acids	9	Ferrocenylethyl amine				
	10	Ferrocenylpropyl amine		R	E	Shimada *et al.* (1987)
N-acetyl amino acids	21	Anthrylethylamine		N	Fl	Goto *et al.* (1986)
THIOLS						
N-Acetyl-penicillamine		OPA/Arginine		R	Fl	Lam (1986)

Abbreviations: N, normal-phase chromatography; R, reversed-phase chromatography; IE, ion-exchange chromatography; UV, absorbance detection; Fl, fluorescence detection; E, electrochemical detection; e.p., enantiomeric purity; a.c., absolute configuration.

Palamareva and co-workers (Snyder, 1982; Snyder *et al.*, 1986; Palamareva and Snyder, 1984). The elution order was characterized by preferential retention of molecules with the least steric hindrance of the adsorbing polar groups. The data showed significant influence of the mobile phase composition on separation. The retention was explained in terms of solvent–solute localization, which refers to the tendency of a solute (solvent) to attach to specific sites on the adsorbent. The results indicate that a localizing solvent (e.g. acetonitrile) preferentially reduces the retention of the more retained solute. Maximum separation factors were therefore obtained with non-localizing mobile phases (e.g. dichloromethane).

The bonds in amides (Helmchen *et al.*, 1972, 1977, 1979) and carbamates (Pirkle and Hauske, 1977; Pirkle and Simmons, 1983) are planar, and the adsorption to the stationary phase is assumed to take place between the chiral moieties primarily by hydrogen bonding of the acyl oxygen and to a lesser extent by the hydrogen on the amide nitrogen, provided the chiral substituents are non-polar. Helmchen *et al.* found that the degree of shielding by the non-polar groups in secondary amides governed the order of elution. Better shielding and thereby less retention resulted when the substituents most efficient in steric hindrance were situated on opposite faces of the common plane provided by the amine bond. Polar groups such as hydroxyls located close to the carbonyl group were found to have a favourable effect on the resolution (Helmchen *et al.*, 1979). Carbamate-derivatives of secondary amines (methyl group on the nitrogen) were less well separated (Pirkle and Hauske, 1977; Pirkle and Simmons, 1983). The same effect was observed in separation of amphetamine against methamphetamine derivatized with 4-nitrophenylsulphonyl-(S)-prolyl chloride, with no resolution of methamphetamine (Barksdale and Clark, 1985). Interestingly, a reversed-phase separation of the derivatives showed the opposite behaviour, with successful resolution of only methamphetamine. Similar improved resolutions have been achieved with GITC derivatives of N-alkyl aminoalcohols when separated with reversed-phase chromatography (Gal and Brown, 1986).

A comparison has been made of different straight- and reversed-phase systems for separation of epimers of corticosteroids (Wikby *et al.*, 1978). The straight-phase supports were nitrile, silica and an amino-phase. The separation factors increased in the mentioned order, which parallels increased polarity of the supports. Comparison with the reversed-phase mode showed that an increased length of the alkyl group located at the centre of chirality (C-22) yielded improved separation in the reversed mode, while the resolution in the straight-phase mode remained unaffected.

4.5 Detection

Detection properties of enantiomers may change as a result of derivatization. Conjugated structures are often part of the chiral reagent or are created when new bonds are formed, and may contribute not only to the separation, but also to the detection of the formed diastereomers, by UV-absorption or by fluorescence. Chiral reagents such as naphthyl propionic acid and naphthylethylamine give derivatives which may be detected in UV at e.g. 254 or 280 nm or by fluorescence emission at 320–350 nm.

The thiourea structure formed by reaction between isothiocyanates and amines has a molar absorptivity of about 12 000 at 250 nm (Nimura *et al.*, 1980) and such diastereomers can be detected at fairly low concentration levels without the incorporation of a special chromophore.

Enantiomers with native UV-absorption can normally be expected to have equally good or better detectability after derivatization. Being more sensitive for changes in the close environment, fluorescence may sometimes change considerably in intensity as a result of derivatization. For example, underivatized metoprolol may be detected by fluorescence but the diastereomers formed by reaction with GITC exhibit poor fluorescence and can only be detected by UV (Balmér, 1987).

Some chiral reagents have been specially designed to allow sensitive detection. Anthracene has exceptionally high molar absorptivity ($2 \times 10^5\,\text{mol}^{-1}\,\text{cm}^{-1}$ at 254 nm) and has fairly high fluorescence quantum efficiency. Anthrylethylamines were introduced by Goto *et al.* (1986) for derivatization of N-protected amino acids. Reported detection limits were 100 f mol (S/N 10). Fluorene has a molar absorptivity of *c.* 2×10^4 at 260 nm and a high quantum yield (emission maximum at 315 nm). Fluorenylethyl chloroformate was introduced by Einarsson *et al.* (1987*b*). The reagent was used for derivatization of primary and secondary amino acids and amines.

Detection principles other than UV-absorption and fluorescence have been little exploited in conjunction with chiral derivatization. Ferrocene compounds have interesting electrochemical properties: they can be oxidized at a low potential and thereafter reduced at a second downstream electrode, in some cases with very high sensitivity and selectivity (Shimada *et al.*, 1987*b*). The chiral ferrocene reagents ferrocenylethylamine and ferrocenylpropylamine were introduced for electrochemical detection by Shimada *et al.*, 1987*a*).

When enantiomers, which have favourable native fluorescence or UV properties, are to be determined at low concentrations in e.g. biological samples, it may be advantageous to use so-called transparent reagents. Such reagents and their degradation products do not themselves fluoresce or absorb light at the actual wavelengths, and no fluorophores/chromophores are introduced into other reactive components in the sample. An example of the use of transparent reagents is the method of Björkman (1985) for determination of indoprofen in plasma after derivatization with ethyl chloroformate and leucinamide. For fluorescence detection, excitation and emission wavelengths were 275 and 433 nm, respectively, which gives very little interference from other co-extracted sample components.

The spectral properties of the two diastereomers are not necessarily identical. Therefore, a slight difference in measured peak areas for a derivatized racemate may be due to different detector response for the two diastereomers.

4.6 Applications

Table 4.2 lists applications of chiral derivatization for liquid chromatography, with a certain emphasis on recent papers and on analytical-scale separations.

4.6.1 *Resolution of 1-isopropylamino-2-propanols* (*β-blockers*)

Below are given examples on chiral derivatization, which have in common that they deal with β-blockers of the 1-isopropyl-amino-2-propanol type. Although these substances may differ in terms of lipophilicity and spectral properties, they are very similar regarding the chemistry involved in derivatization, and therefore can be treated as a group here.

Weber *et al.* (1984) resolved benoxaprofen racemate by preparative LC-separation of the (R)-(+)-α-methylbenzylamide diastereomers. The R-form was converted to (R)-benoxaprofen chloride (Table 4.1, 11) with an enantiomeric purity of 96.5%. Crystalline derivatives of β-blockers, used as reference substances, were prepared by Pflugmann *et al.* (1987) by reaction in methylene chloride with anhydrous sodium carbonate as a proton acceptor. Metoprolol, oxprenolol or propranolol were extracted from urine and derivatization with (R)-benoxaprofen chloride was performed at room temperature overnight. Separation was done on silica thin-layer plates and the chromatogram was recorded measuring fluorescence at 365 nm (excitation 313 nm). The pairs of diastereomers of the three β-blockers were adequately resolved. The detection limit was about 100 ng/ml in urine. The absolute recovery of propranolol after extraction and derivatization was 78%.

Hermansson (1982) prepared the symmetrical anhydrides of *tert*-butoxycarbonyl-L-leucine from commercial Boc-amino acids and phosgene. The reagents were stored at −20°. Propranolol was extracted from plasma and the samples were derivatized using 0.10 ml pure reagent and 0.25 ml of dichloromethane with triethylamine. The reaction with the alanine reagent was complete after 5 min at room temperature. The reaction with the leucine reagent was about six times slower. To remove the *tert*-butoxycarbonyl group from the derivatives, the samples were treated with neat trifluoroacetic acid for 5–15 min. The samples were extracted with diethyl ether and back-extracted into phosphoric acid. Separation was carried out on a reversed-phase LC-system. Peak tailing was reduced with dimethyloctylamine in the mobile phase. Separation factors of 1.3 and 1.7, respectively, were obtained for the alanine and leucine derivatives. Measurement of fluorescence emitted above 340 nm (native fluorescence of propranolol) allowed determination of drugs levels in plasma as low as 1 ng/ml.

A quite different approach for the analytical and preparative resolution of enantiomers of β-blockers was reported by Lindner *et al.* (1984). Disubstituted (R, R)- or (S, S)-tartaric acid anhydrides (Table 4.1, 31) were prepared and derivatization performed in aprotic media was effective on the secondary alcohol, giving substituted tartaric acid monoesters. The secondary amino group was hindered to react by addition of trichloroacetic acid which formed an ion-pair with the amine. The time needed for reaction at 50° was up to several hours. Very high separation factors were obtained on reversed-phase systems for diastereomeric pairs from the different disubstituted tartaric anhydrides. Separation was explained by an intramolecular interaction involving the free carboxylic group of the tartaric acid residue and the free amino group of the β-blocker. The reported enantiomeric purity of the reagent was > 99.9%. Racemization was less than 0.1% for reactions with propranolol. This, together with the high α-values achieved, makes derivatization with substituted tartaric acid anhydrides a powerful technique for the determination of enantiomeric purity of β-blockers. As the diastereomers are esters and amenable to hydrolysis, the technique is suited for preparative work.

Sedman and Gal (1983) investigated the resolution of various β-blockers as thiourea derivatives, using the isothiocyanate reagents GITC and AITC (Table 4.1, 32, 33). Separation factors, obtained on a reversed-phase system, were 1.21–1.28 for GITC-derivatives and 1.09–1.22 for AITC-derivatives. Christ and Walle (1985) applied the method to a β-blocker metabolite, 4′-hydroxypropranolol sulphate. The metabolite was isolated from plasma by ion-pair extraction and reversed-phase chromatography, freeze-dried and reacted with GITC in acetonitrile-water with triethylamine. After 5–10 min, the reaction mixture was separated by reversed-phase chromatography, with excellent resolution of the two enantiomers. UV detection at 313 nm permitted 20 ng of each enantiomer to be detected.

Darmon and Thenot (1987) recently reported a careful study on the use of naphthylethyl isocyanate for the enantioselective determination of betaxolol in whole blood. After extraction and solvent evaporation, derivatization was carried out in methanol at room temperature for 1.5 h. Excess reagent was evaporated and the derivatized sample was purified by extraction into chloroform. Reversed-phase chromatography gave baseline separation ($\alpha = 1.1$) of the enantiomers. Fluorescence detection at 330 nm (excitation wavelength 285 nm) allowed detection of 0.5 ng ml^{-1} blood.

β-Blockers may also be separated as carbamates after reaction with 1-(9-fluorenyl)ethyl chloroformate (FLEC, Table 4.1, 19) as shown by Einarsson *et al.* (1987*b*). The reaction was carried out in a buffer–acetone mixture for 30 min, whereafter excess reagent was allowed to react with hydroxyproline for another 2 min. Metoprolol was separated on a reversed-phase system, giving an α-value of 1.1.

The examples given above, picked out from a large number of papers (see Table 4.2) were taken to illustrate that a variety of methods is available for resolution of one type of closely related chiral substrates. For a certain application, the final choice can be directed by weighing different aspects such as type of sample, range of concentrations, ease of sample treatment, available instrumentation and own experience.

Acknowledgements

The authors thank Magnus Erickson, Hässle AB for useful comments on the manuscript. This work was supported by the Swedish Council for Technical Development.

References

Aberhart, D.J., J-A. Cotting and H-J. Lin (1985) Separation by high-performance liquid chromatography of (3R)- and (3S)-β-leucine as diastereomeric derivatives. *Anal. Biochem.* **151**, 88–91.

Ahnfelt, N.O., P. Hartvig and K.E. Karlsson (1982) Derivatization of amines with chloroformate esters for gas chromatographic analysis. *Chromatographia* **16**, 60–62.

Allgire, J.F., E.C. Juenge, C.P. Damo, G.M. Sullivan and R.D. Kirchhoefer (1985) High performance liquid chromatographic determination of D-/L-epinephrine enantiomer ratio in lidocaine–epinephrine local anesthetics. *J. Chromatogr.* **325**, 249–254.

Allison, L.A., G.S. Mayer and R.E. Shoup (1984) *o*-Phthalaldehyde derivatives of amines for high-speed liquid chromatography/electrochemistry. *Anal. Chem.* **56**, 1089–1096.

Arnould, M.C., B. Serkiz and J.P. Volland (1980) Separation of fenfluramine enantiomers by reverse-phase preparative liquid chromatography. *Analusis* **8**, 76–77.

Aswad, D.W. (1984) Determination of D- and L-aspartate in amino acid mixtures by high-performance liquid chromatography after derivatization with a chiral adduct of *o*-phthaldialdehyde. *Anal. Biochem.* **137**, 405–409.

Avgerinos, A. and A.J. Hutt (1987) Determination of the enantiomeric composition of ibuprofen in human plasma by high-performance liquid chromatography. *J. Chromatogr.* **415**, 75–83.

Balmér, K. (1988) Personal communication.

Banfield, C. and M. Rowland (1983) Stereospecific high-performance liquid chromatographic analysis of warfarin in plasma. *J. Pharm. Sci.* **72**, 921–924.

Banfield, C. and M. Rowland (1984) Stereospecific fluorescence high-performance liquid chromatographic analysis of warfarin and its metabolites in plasma and urine. *J. Pharm. Sci.* **73**, 1392–1396.

Barksdale, J.M. and C.R. Clark (1985) Liquid chromatographic determination of the enantiomeric composition of amphetamine and related drugs by diastereomeric derivatisation. *J. Chromatogr. Sci.* **23**, 176–180.

Bergot, B.J., R.J. Anderson, D.A. Schooley and C.A. Henrick (1978) Liquid chromatographic analysis of enantiomeric purity of several terpenoid acids as their 1-(1-naphthyl)ethyl-amide derivatives. *J. Chromatogr.* **155**, 97–105.

Björkman, S. (1985) Determination of the enantiomers of indoprofen in blood plasma by high-performance liquid chromatography after rapid derivatization by means of ethyl chloroformate. *J. Chromatogr.* **339**, 339–346.

Björkman, S. (1987) Determination of the enantiomers of ketoprofen in blood plasma by ion-pair extraction and high-performance liquid chromatography of leucinamide derivatives. *J. Chromatogr.* **414**, 465–471.

Bladeren, P.J. van. J.M. Sayer, D.E. Ryan, P.E. Thomas, W. Levin and D.M. Jerina (1985) Differential stereoselectivity of cytochromes P-450b and P 450c in the formation of naphthalene and anthracene 1, 2-oxides. *J. Biol. Chem.* **260**, 10226–10235.

Blaschke, G., U. Scheidemantel and B. Walther (1985) Herstellung, optische reinheit und konfiguration der flecainid-enantiomeren. *Chem. Ber.* **118**, 4616–4619.

Brash, A.R., A.T. Porter and R.L. Maas (1985) Investigation of the selectivity of hydrogen abstraction in the nonenzymatic formation of hydroxyeicosatetraenoic acids and leukotrienes by autoxidation. *J. Biol. Chem.* **260**, 4210–4216.

Buck, R.H. and K. Krummen (1984) Resolution of amino acid enantiomers by high-performance liquid chromatography using automated pre-column derivatization with a chiral reagent. *J. Chromatogr.* **315**, 279–285.

Buck, R.H. and K. Krummen (1987) High-performance liquid chromatographic determination of enantiomeric amino acids and alcohols after derivatization with *o*-phthaldialdehyde and various chiral mercaptans. *J. Chromatogr.* **387**, 255–265.

Cahill, W.R., E.P. Kroeff and D.J. Pietrzyk (1980) Applications of tert-butyloxycarbonyl-L-amino acid N-hydroxysuccinimide esters in the chromatographic separation and determination of D, L-amino acids and diastereomeric dipeptides. *J. Liq. Chromatogr.* **3**, 1319–1334.

Chen, T-M. and J.J. Contario (1984) High-performance liquid chromatographic resolution of enantiomers of γ-vinyl-γ-aminobutyric acid. *J. Chromatogr.* **314**, 495–498.

Christ, D.D. and T. Walle (1985) Stereoselective sulfate conjugation of 4-hydroxypropranolol in vitro by different species. *Drug Metab. Dispos.* **13**, 380–381.

Clark, C.R. and J.M. Barksdale (1984) Synthesis and liquid chromatographic evaluation of some chiral derivatizing agents for resolution of amine enantiomers. *Anal. Chem.* **56**, 958–962.

Corey, E.J. and S. Hashimoto (1981) A practical process for large-scale synthesis of (S)-5-hydroxy-6-trans-8, 11, 14-*cis*-eicosatetraenoic acid (5-HETE). *Tetrahedron Lett.* **22**, 299–302.

Darmon A. and J.P. Thernot (1986) Determination of betaxolol enantiomers by high-performance liquid chromatography. Application to pharmacokinetic studies. *J. Chromatogr.* **374**, 321–328.

Davies, J.S. and A.K.A. Mohammed (1984) Chiral analysis of the reaction stages in the Edman method for sequencing peptides. *J. Chem. Soc. Perkin Trans II*, 1723–1727.

Demian, J. and D.F. Gripshover (1987) High-performance liquid chromatographic determination of enantiomeric purity of 1-methyl-3-pyrrolidinol via derivatisation with (R, R)-O, O-dibenzoyltartaric acid anhydride. *J. Chromatogr.* **387**, 532–535.

Deyl, Z., J. Hyanek and M. Horakova (1986) Profiling of amino acids in body fluids and tissues by means of liquid chromatography. *J. Chromatogr.* **379**, 177–250.

Doolittle, R.E. and R.R. Heath (1984) (S)-Tetrahydro-5-oxo-2-furancarboxylic acid: a chiral derivatizing reagent for asymmetric alcohols. *J. Org. Chem.* **49**, 5041–5050.

Dunlop, D.S. and A. Niedle (1987) The separation of D/L amino acid pairs by high-performance liquid chromatography after precolumn derivatisation with optically active naphthylethyl isocyanate. *Anal. Biochem.* **165**, 38–44.

Edman, P. (1950) Method for determination of the amino acid sequence in peptides. *Acta. Chem. Scand.* **4**, 283–293.

Einarsson, S., S. Folestad and B. Josefsson (1987*a*) Separation of amino acid enantiomers using precolumn derivatization with *o*-phthalaldehyde and 2, 3, 4, 6-tetra-O-acetyl-1-thio-β-glucopyranoside. *J. Liq. Chromatogr.* **10**, 1589–1601.

Einarsson, S., B. Josefsson, P. Möller and D. Sanchez (1987*b*) Separation of amino acid enantiomers and chiral amines using precolumn derivatization with (+)-1-(9-fluorenyl)ethyl chloroformate and reversed-phase liquid chromatography. *Anal. Chem.* **59**, 1191–1195.

Falck, J.R., S. Manna, H.R. Jacobson, R.W. Estabrook, N. Chacos and J. Capdevila (1984) Absolute configuration of epoxyeicosatrienoic acids (EETs) formed during catalytic oxygenation of arachidonic acid by purified rat liver microsomal cytochrome P-450. *J. Am. Chem. Soc.* **106**, 3334–3336.

Foster, R.T. and F. Jamali (1987) High-performance liquid chromatographic assay of ketoprofen enantiomers in human plasma and urine. *J. Chromatogr.* **416**, 388–393.

Furukawa, H., E. Sakakibara, A. Kamei and K. Ito (1975) Separation of L- and D-amino acids as diastereomeric derivatives by high performance liquid chromatography. *Chem. Pharm. Bull.* **23**, 1625–1626.

Furukawa, H., Y. Mori, Y. Takeuchi and K. Ito (1977) Separation of L- and D-amino acids as diastereomeric derivatives by high-performance liquid chromatography. *J. Chromatogr.* **136**, 428–431.

Gal, J. (1984) Resolution to the enantiomers of ephedrine, norephedrine and pseudoephedrine by high-performance liquid chromatography. *J. Chromatogr.* **307**, 220–223.

Gal, J. and A.J. Sedman (1984) R-α-Methylbenzyl isothiocyanate, a new and convenient chiral derivatizing agent for the separation of enantiomeric amino compounds by high-performance liquid chromatography. *J. Chromatogr.* **314**, 275–281.

Gal, J. (1986) Determination of the enantiomeric composition of chiral aminoalcohols using derivatization and reversed-phase liquid chromatography. *J. Liq. Chromatogr.* **9**, 673–681.

Gal, J. and T.R. Brown (1986) Liquid chromatographic separation of enantiomers of adrenergic agonists. *J. Pharm. Meth.* **16**, 261–269.

Gal, J. (1987) Stereoisomer separations via derivatization with optically active reagents. Applications to compounds of pharmacological interest. *LC·GC* **5**, 106–126.

Gal, J. (1988) Indirect chromatographic methods for resolution of drug enantiomers. Synthesis and separation of diastereomeric derivatives. In *Drug Stereochemistry*, ed. I.W. Wainer and D.E. Drayer, Marcel Dekker, New York, chapter 4.

Gietl, Y., H. Spahn and E. Mutschler (1988) Simultaneous determination of R- and S-prenylamine in plasma and urine by reversed-phase high-performance liquid chromatography. *J. Chromatogr.* **426**, 305–314.

Görög, S., B. Herenyi and M. Löw (1986) Derivatization of N-protected amino acids for chiral separation by high-performance liquid chromatography. *J. Chromatogr.* **353**, 417–424.

Goto, J., M. Hasegawa, S. Nakamura, K. Shimada and T. Nambara (1977) New derivatization for liquid chromatographic resolution of amino acid enantiomers. *Chem. Pharm. Bull.* **25**, 847–849.

Goto, J., M. Hasegawa, S. Nakamura, K. Shimada and T. Nambara (1978) New derivatization reagents for the resolution of amino acid enantiomers by high-performance liquid chromatography. *J. Chromatogr.* **152**, 413–419.

Goto, J., N. Goto, A. Hikichi and T. Nambara (1979) Separation and determination of 2, 5-dimethoxy-4-methylamphetamine enantiomers in plasma by high-performance liquid chromatography. *J. Liq. Chromatogr.* **2**, 1179–1190.

Goto, J., N. Goto, A. Hikichi, T. Nishimaki and T. Nambara (1980) Sensitive derivatization reagents for the resolution of carboxylic acid enantiomers by high-performance liquid chromatography. *Anal. Chim. Acta.* **120**, 187–192.

Goto, J., N. Goto and T. Nambara (1982*a*) New type of derivatization reagents for liquid chromatographic resolution of enantiomeric hydroxyl compounds. *Chem. Pharm. Bull.* **30**, 4597–4599.

Goto, J., N. Goto and T. Nambara (1982*b*) Separation and determination of naproxen enantiomers in serum by high-performance liquid chromatography. *J. Chromatogr.* **239**, 559–564.

Goto, J., M. Ito, S. Katsuki, N. Saito and T. Nambara (1986) Sensitive derivatization reagents for optical resolution of carboxylic acids by high performance liquid chromatography with fluorescence detection. *J. Liq. Chromatogr.* **9**, 683–694.

Gulaid, A.A., G.W. Houghton and A.R. Boobis (1985) Separation of acebutolol and diacetolol diastereomers by reversed-phase high-performance liquid chromatography. *J. Chromatogr.* **318**, 393–397.

Gübitz, G. and S. Mihellyes (1984) Optical resolution of β-blocking agents by thin-layer chromatography and high performance liquid chromatography as diastereomeric R-( )-1-(1-naphthyl)ethylureas. *J. Chromatogr.* **314**, 462–466.

Halpern, B. (1977) Derivatives for chromatographic resolution of optically acitive compounds. In *Handbook of Derivatives for Chromatography*, ed. K. Blau, and G.S. King, Heyden, London, chapter 13.

Helmchen, G. and W. Strubert (1974) Determination of optical purity by high performance liquid chromatography. *Chromatographia* **7**, 713–715.

Helmchen, G., R. Ott and K. Sauber (1972) Gezielte trennung und absolute konfiguration von enantiomeren carbonsäuren und aminen. *Tetrahedron Lett.* 3873–3878.

Helmchen, G., H. Völter and W. Schühle (1977) Directed resolution of enantiomers via liquid chromatography of diastereomeric derivatives III. A convenient method to determine the absolute configuration of carboxylic acids $R^1R^2HCCOOH$. *Tetrahedron Lett.* 1417–1420.

Helmchen, G., G. Nill, D. Flockerzi, W. Schühle and M.S.K. Youssef (1979) Extreme liquid chromatographic separation effects in the case of diastereomeric amides containing polar substituents. *Angew. Chem. Int. Ed. Engl.* **18**, 62–63.

Hermansson, J. (1982) Separation and quantitation of (R)- and (S)-propranolol as their diastereomeric derivatives in human plasma by reversed-phase ion-pair chromatography. *Acta Pharm. Suec.* **19**, 11–24.

Hermansson, J. and C. von Bahr (1980) Simultaneous determination of d- and l-propranolol in human plasma by high-performance liquid chromatography. *J. Chromatogr.* **221**, 109–117.

Hermansson, J. and C. von Bahr (1982) Determination of (R)- and (S)-alprenolol and (R)- and (S)-metoprolol as their diastereomeric derivatives in human plasma by reversed-phase liquid chromatography. *J. Chromatogr.* **227**, 113–127.

Hirschmann, R., R.G. Strachan, H. Schwam, E.F. Schoenewaldt, H. Joshua, B. Barkemeyer, D.F. Veber, W.J. Paleveda, T.A. Jacob, T.E. Beesley and R.G. Denkewalter (1967) The controlled synthesis of peptides in aqueous medium. III. Use of Leuchs' anhydrides in the synthesis of dipeptides. Mechanism and control of side reactions. *J. Org. Chem.* **32**, 3415–3425.

Hoffmann, K.J., L. Renberg and C. Bäärnhielm (1984) Stereoselective disposition of RS-tocainide in man. *Eur. J. Drug. Metab. Pharmacokinet.* **9**, 215–222.

Hsyu, P-H. and K.M. Giacomini (1986) High performance liquid chromatographic determination of the enantiomers of β-adrenoceptor blocking agents in biological fluids I: Studies with pindolol. *J. Pharm. Sci.* **75**, 601–605.

Hutt, A.J., S. Fournel and J. Caldwell (1986) Application of a radial compression column to the high-performance liquid chromatographic separation of the enantiomers of some 2-arylpropionic acids as their diastereoisomeric S-(−)-1-(naphthen-1-yl)ethylamides. *J. Chromatogr.* **378**, 409–418.

Iwaki, K., S. Yoshida, N. Nimura, T. Kinoshita, K. Takeda and H. Ogura (1987) Activated carbamate reagent as chiral derivatizing agent for liquid chromatographic optical resolution of enantiomeric amino compounds. *Chromatographia* **23**, 899–902.

Jacques, J., A. Collet and S.H. Wilen (1981) *Enantiomers, Racemates, and Resolutions.* John Wiley & Sons, New York.

Jiang, M. and D.M. Soderlund (1982) Liquid chromatographic determination and resolution of the enantiomers of the acid moieties of pyrethroid insecticides as their (−)-1-(l-phenyl)ethylamide derivatives. *J. Chromatogr.* **248**, 143–149.

Johnson, D.M., A. Reuter, J.M. Collins and G.F. Thompson (1979) Enantiomeric purity of naproxen by liquid chromatographic analysis of its diastereomeric octyl esters. *J. Pharm. Sci.* **68**, 112–114.

Johnström, P., E. Ehrin, S. Stone-Elander and J.L.G. Nilsson (1984) Synthesis of ^{11}C-labelled D- and L-methionine for positron emission tomography. Investigation of the enantiomeric purity of the products. *Acta Pharm. Suec.* **21**, 189–194.

Joseph, M.H. and P. Davies (1983) Electrochemical activity of o-phthalaldehyde-mercaptoethanol derivatives of amino acids. *J. Chromatogr.* **277**, 125–136.

Kinoshita, T., Y. Kasahara and N. Nimura (1981) Reversed-phase high-performance liquid chromatographic resolution of non-esterified enantiomeric amino acids by derivatization with 2, 3, 4, 6-tetra-O-acetyl-β-D-glucopyranosyl isothiocyanate and 2, 3, 4-tri-O-acetyl-α-D-arabinopyranosyl isothiocyanate. *J. Chromatogr.* **210**, 77–81.

Kirchhoefer, R.D., G.M. Sullivan and J.F. Allgire (1985) Analysis of USP epinephrine injections for potency, impurities, degradation products, and d-enantiomer by liquid chromatography, using ultraviolet and electrochemical detectors. *J. Assoc. Off. Anal. Chem.* **68**, 163–165.

König, W.A. (1987) *The Practice of Enantiomer Separation by Capillary Gas Chromatography.* Dr Alfred Hüthig Verlag, Heidelberg.

Ladányi, L., I. Sztruhár, P. Slégel and G. Vereczekey-Donáth (1987) Determination of the enantiomeric composition of chiral carboxylic acids using chiral derivatization and HPLC. *Chromatographia* **24**, 477–481.

Lam, S. (1986) Resolution of D- and L-amino acids after precolumn derivatisation with *o*-phthalaldehyde by mixed chelation with Cu(II)-L-proline. *J. Chromatogr.* **355**, 157–164.

Lankamayr, E.P., K.W. Budna and F. Nachtmann (1980) Separation of enantiomeric iodinated thyronines by liquid chromatography of diastereomers. *J. Chromatogr.* **198**, 471–479.

Lee, E.J.D., K.M. Williams, G.G. Graham, R.O. Day and G.D. Champion (1984) Liquid chromatographic determination and plasma concentration profile of optical isomers of ibuprofen in humans. *J. Pharm. Sci.* **73**, 1542–1544.

Lee, H. and R.G. Harvey (1984) Absolute configuration of the 9,10-epoxides of 9, 10, 11, 12-tetrahydrobenzo[e]pyrene: Application of the exciton chirality rule to the *p*-methoxybenzoate of a bromohydrin. *J. Org. Chem.* **49**, 1114–1117.

Lindner, W., C. Leitner and G. Uray (1984) Liquid chromatographic separation of enantiomeric alkanolamines via diastereomeric tartaric acid monoesters. *J. Chromatogr.* **316**, 605–616.

Lindner, W. and C. Pettersson (1985) Resolution of optical isomers by liquid chromatographic techniques. In *Liquid Chromatography in Pharmaceutical Development*, ed. I. Wainer, Aster Publishing, Springfield, Part 1.

Lindner, W. (1987) Indirect separation of enantiomers by liquid chromatography. In *Chromatographic Chiral Separations*, ed. M. Zief and L.J. Crane, Marcel Dekker, New York, chapter 4.

Maibaum, J. (1988) Indirect high-performance liquid chromatographic resolution of racemic tertiary amines as their diastereomeric urea derivatives after N-dealkylation. *J. Chromatogr.* **436**, 269–278.

Maître, J.-M., G. Boss and B. Testa (1984) High-performance liquid chromatographic separation of the enantiomers of anti-inflammatory 2-arylpropionaties: Suitability of the method for in vitro metabolic studies. *J. Chromatogr.* **299**, 397–403.

Manning, J.M. and S. Moore (1968) Determination of D- and L-amino acids by ion exchange chromatography as L-D and L-L dipeptides. *J. Biol. Chem.* **243**, 5591–5597.

Marfey, P. (1984) Determination of D-amino acids. II. Use of a bifunctional reagent, 1, 5-difluoro-2, 4-dinitrobenzene. *Carlsberg Res. Commun.* **49**, 591–596.

Matzner, M., R.P. Kurkjy and R.J. Cotter (1964) The chemistry of chloroformates. *Chem. Rev.* **64**, 645–687.

McKay, S.W., D.N.B. Mallen, P.R. Shrubsall, B.P. Swann and W.R.N. Williamson (1979) Analysis of benoxaprofen and other α-methylarylacetic acids using high-performance liquid chromatography. *J. Chromatogr.* **170**, 482–485.

Michelsen, P., E. Aronsson, G. Odham and B. Åkesson (1985) Diastereomeric separations of natural glycero derivatives as their 1-(1-naphthyl)ethyl carbamates by high-performance liquid chromatography. *J. Chromatogr.* **350**, 417–426.

Miller, K.J., J. Gal and M.M. Ames (1984) High-performance liquid chromatographic resolution of enantiomers of 1-phenyl-2-aminopropanes (amphetamines) with four chiral reagents. *J. Chromatogr.* **307**, 335–342.

Mitchell, A.R., S.B.H. Kent, I.C. Chu and R.B. Merrifield (1978) Quantitative determination of D- and L-amino acids by reaction with tert-butyloxycarbonyl-L-leucine N-hydroxysuccinimide ester and chromatographic separation as L, D and L, L dipeptides. *Anal. Chem.* **50**, 637–640.

Nachtman, F. (1980) Determination of the L-isomer in D-penicillamine by derivatization liquid chromatography. *Int. J. Pharmaceutics* **4**, 337–345.

Nakazawa, H., Y. Kanamaru and A. Murano (1979) Determination of optical isomers of clinofibrate by high-performance liquid chromatography. *Chem. Pharm. Bull.* **27**, 1694–1696.

Nambara, T., S. Ikegawa, M. Hasegawa and J. Goto (1978) High-pressure liquid chromatographic resolution of amino acid enantiomers by derivatization with new chiral reagents. *Anal. Chim. Acta* **101**,111–116.

Newman, P. (1981–84) *Optical Resolution Procedures for Chemical Compounds.* Optical Resolution Information Center, New York, Vol. 1–3.

Nichols, D.E., C.F. Barfknecht, D.B. Rusterholz, F. Benington and R.D. Morin (1973) Asymmetric synthesis of psychotomimetic phenylisopropylamines. *J. Med. Chem.* **16**, 480–483.

Nimura, N., H. Ogura and T. Konoshita (1980) Reversed-phase liquid chromatographic resolution of amino acid enantiomers by derivatization with 2, 3, 4, 6-tetra-O-acetyl-β-D-glucopyranosyl isothiocyanate. *J. Chromatogr.* **202**, 375–379.

Nimura, N., Y. Kasahara and T. Kinoshita (1981) Resolution of enantiomers of norepinephrine and ephinephrine by reversed-phase high-performance liquid chromatography. *J. Chromatogr.* **213**, 327–330.

Nimura, N., A. Toyama and T. Kinoshita (1984) Optical resolution of amino acid enantiomers by high-performance liquid chromatography. *J. Chromatogr.* **316**, 547–552.

Nimura, N. and T. Kinoshita (1986) *o*-Phthalaldehyde-N-acetyl-L-cysteine as a chiral derivatization reagent for liquid chromatographic optical resolution of amino acid enantiomers and its application to conventional amino acid analysis. *J. Chromatogr.* **352**, 169–177.

Oatis, J.E., J.P. Baker, J.R. McCarthy and D.R. Knapp (1983) Synthesis and chromatographic separation of the glucuronides of (R)- and (S)-propranolol. *J. Med. Chem.* **26**, 1687–1691.

Palamareva, M.D. and S.R. Snyder (1984) Liquid-solid chromatography of diastereomers: Tetrasubstituted ethanes. *Chromatographia* **19**, 352–354.

Pedrazzini, S., W. Zanoboni-Muciaccia, C. Sacchi and A. Forgione (1987) Determination of flunoxaprofen enantiomers in biological fluids by high-performance liquid chromatography. *J. Chromatogr.* **415**, 214–220.

Pflügmann, G., H. Spahn and E. Mutschler (1987) Rapid determination of the enantiomers of metoprolol, oxprenolol and propranolol in urine. *J. Chromatogr.* **416**, 331–339.

Pirkle, W.H. and J. Finn (1983) Separation of enantiomers by liquid chromatographic methods. In *Asymmetric Synthesis*, ed. J, Morris, Academic Press, New York, **1**, chapter 6.

Pirkle, W.H. and J.R. Hauske (1977*a*) Broad spectrum methods for the resolution of optical isomers. A discussion of the reasons underlying the chromatographic separability of some diastereomeric carbamates. *J. Org. Chem.* **42**, 1839–1844.

Pirkle, W.H. and J.R. Hauske (1977*b*) Design of chiral derivatizing agents for the chromatographic resolution of optical isomers. Asymmetric synthesis of some chiral fluoroalkylated amines. *J. Org. Chem.* **42**, 2436–2439.

Pirkle, W.H. and M.S. Hoekstra (1974) An example of automated liquid chromatography. Synthesis of a broad-spectrum resolving agent and resolution of 1-(1-naphthyl)-2, 2, 2-trifluoroethanol. *J. Org. Chem.* **39**, 3904–3906.

Pirkle, W.H. and K.A. Simmons (1983) Improved chiral derivatizing agents for the chromatographic resolution of racemic primary amines. *J. Org. Chem.* **48**, 2520–2527.

Roth, M. (1971) Fluorescence reaction for amino acids. *Anal. Chem.* **43**, 880–882.

Ruterbories, K.J. and D. Nurok (1987) Thin layer chromatographic separation of diastereomeric amino acid derivatives prepared with Marfey's reagent. *Anal. Chem.* **59**, 2735–2736.

Ruud-Christensen, M. and B. Salvesen (1984) Separation of (R)- and (S)-proxyphylline as diastereoisomeric camphanates by reversed-phase liquid chromatography. *J. Chromatogr.* **303**, 433–435.

Schönenberger, B., A. Brossi, C. George and J.L. Flippen-Anderson (1986) Preparation of optically active secondary amines by thermal decomposition of (methylbenzyl)urea analogs: absolute configuration of (+)- and (−)-mecamylamine. *Helv. Chim. Acta* **69**, 283–287.

Scott, C.G., M.J. Petrin and T. McCorkle (1976) The liquid chromatographic separation of some acyclic isoprenoid acid enantiomers via diastereomer derivatization. *J. Chromatogr.* **125**, 157–161.

Sedman, A.J. and J. Gal (1983) Resolution of the enantiomers of propranolol and other beta-adrenergic antagonists by high-performance liquid chromatography. *J. Chromatogr.* **278**, 199–203.

Seeman, J.I., C.G. Chavdarian and H.V. Secor (1985) Synthesis of the enantiomers of nornicotine. *J. Org. Chem.* **50**, 5419–5421.

Shimada, K., E. Haniuda, T. Oe and T. Nambara (1987*a*) Ferrocene derivatization reagents for optical resolution of carboxylic acids by high-performance liquid chromatography with electrochemical detection. *J. Liq. Chromatogr.* **10**, 3161–3172.

Shimada, K., T. Oe and T. Nambara (1987*b*) Sensitive ferrocene reagents for derivatization of thiol compounds in high-performance liquid chromatography with dual-electrode coulometric detection. *J. Chromatogr.* **419**, 17–25.

Shimizu, R., K. Ishii, N. Tsumagari, M. Tanigawa and M. Matsumoto (1982) Determination of optical isomers in diltiazem hydrochloride by high-performance liquid chromatography. *J. Chromatogr.* **253**, 101–108.

Shimizu, R., T. Kakimoto, K. Ishii, Y. Fujimoto, H. Nishi and N. Tsumagari (1986) New derivatization reagent for the resolution of optical isomers in diltiazem hydrochloride by high-performance liquid chromatography. *J. Chromatogr.* **357**, 119–125.

Simons, S.S. and D.F. Johnson (1976) The structure of the fluorescent adduct formed in the reaction of *o*-phthalaldehyde and thiols with amines. *J. Am. Chem. Soc.* **98**, 7098–7099.

Sioufi, A., D. Colussi, F. Marfil and J.P. Dubois (1987) Determination of the (+)- and (−)-enantiomers of pirprofen in human plasma by high-performance liquid chromatography. *J. Chromatogr.* **414**, 131–137.

Slégel, P., G. Vereczkey-Donáth, L. Ladányi and M. Toth-Lauritz (1987) Enantiomeric separation of chiral carboxylic acids, as their diastereomeric carboxamides, by thin-layer chromatography. *J. Pharm. Biomed. Anal.* **5**, 665–673.

Snyder, L.R. and J.J. Kirkland (1979) *Introduction to Modern Liquid Chromatography*. John Wiley & Sons, New York.

Snyder, L.R. (1982) Theory for mobile phase effects in separations of isomers by liquid-solid chromatography. Application to the relative retention of certain diastereomers on silica. *J. Chromatogr.* **245**, 165–176.

Snyder, L.R., M.D. Palamareva, B.J. Kurtev, L.Z. Viteva and J.N. Stefanovsky (1986) Chromatographic behavior of diastereoisomers VIII. Solvent selectivity effects in separations on silica. *J. Chromatogr.* **354**, 107–118.

Souter, R.W. (1976) High-performance liquid chromatographic separation of enantiomeric amines. *Chromatographia* **9**, 635–637.

Souter, R.W. (1985) *Chromatographic Separations of Stereoisomers.* CRC Press, Boca Raton, Florida.

Spahn, H. (1987) Formation of diastereomeric derivatives of 2-arylpropionic acids using L-leucinamide. *J. Chromatogr.* **423**, 334–339.

Spahn, H., H. Weber, E. Mutschler and W. Möhrke (1984) α-Alkyl-α-arylacetic acid derivatives as fluorescence markers for thin-layer chromatographic and high-performance liquid chromatographic assay of amines and alcohols. *J. Chromatogr.* **310**, 167–178.

Sternson, L.A. (1981) General aspects of precolumn derivatization with emphasis on pharmaceutical analysis. In *Chemical Derivatisation in Analytical Chemistry*, ed. R.W. Frei and J.F. Lawrence, Plenum Press, New York, **1**, chapter 3.

Stoltenborg, J.K., C.V. Puglisi, F. Rubio and F.M. Vane (1981) High-performance liquid chromatographic determination of stereoselective disposition of carprofen in humans. *J. Pharm. Sci.* **70**, 1207–1212.

Takadate, A., M. Iwai, H. Fujino, K. Tahara and S. Goya (1983) Synthesis and properties of 2-dansylethyl chloroformate as a fluorescent labeling reagent for alcoholic hydroxyl groups. *Yakugaku Zasshi* **103**, 962–966.

Takaya, T., Y. Kishida and S. Sakakibara (1981) Determination of the optical purity of amino acids by high-performance liquid chromatography. Modification of the Manning and Moore procedure. *J. Chromatogr.* **215**, 279–287.

Thakker, D.R., H. Yagi, H. Akagi, M. Koreeda, A.Y.H. Lu, W. Levin, A.W. Wood, A.H. Conney and D.M. Jerina (1977) Metabolism of benzo[a]pyrene VI. Stereoselective metabolism of benzo[a]pyrene and benzo[a]pyrene 7, 8-dihydrodiol to epoxides. *Chem. Biol. Interactions* **16**, 281–300.

Thakker, D.R., H. Yagi, J.M. Sayer, U. Kapur, W. Levin, R.L. Chang, A.W. Wood, A.H. Conney and D.M. Jerina (1984) Effects of a 6-fluoro substituent on the metabolism of benzo[a]pyrene 7, 8-dihydrodiol to bay-region diol epoxides by rat liver enzymes. *J. Biol. Chem.* **259**, 11249–11256.

Thompson, J.A., J.L. Holtzman, M. Tsuru, C.L. Lerman and J.L. Holtzman (1982) Procedure for the chiral derivatization and chromatographic resolution of R-(+)- and S-(−)-propranolol. *J. Chromatogr.* **238**, 470–475.

Tung, R.D., M.K. Dhaon and D.H. Rich (1986) BOP-Cl mediated synthesis of the cycloproine A 8-11 tetrapeptide fragment. *J. Org. Chem.* **51**, 3350–3354.

Valentine, D., K.K. Chan, C.G. Scott, K.K. Johnson, K. Toth and G. Saucy (1976) Direct determinations of R/S enantiomers ratios of citronellic acid and related substances by nuclear magnetic resonance spectroscopy and high pressure liquid chromatography. *J. Org. Chem.* **41**, 62–65.

Wainer, I.W. (1980) Crown ether-catalyzed synthesis of diastereoisomeric esters. Enantiomeric purity of mandelic acid via high-performance liquid chromatographic analysis of its diastereoisomeric 2-octyl esters. *J. Chromatogr.* **202**, 478–482.

Walle, T., D.D. Christ, U.K. Walle and M.J. Wilson (1985) Separation of the enantiomers of intact sulfate conjugates of adrenergic drugs by high-performance liquid chromatography after chiral derivatization. *J. Chromatogr.* **341**, 213–216.

Weber, H., H. Spahn, E. Mutschler and W. Möhrke (1984) Activated α-alkyl-α-arylacetic acid enantiomers for stereoselective thin-layer chromatographic and high-performance liquid chromatographic determination of chiral amines. *J. Chromatogr.* **307**, 145–153.

Westley, J.W. and B. Halpern (1968) The use of (−)-menthyl chloroformate in the optical analysis of asymmetric amino and hydroxyl compounds by gas chromatography. *J. Org. Chem.* **33**, 3978–3980.

Wikby, A., A. Thalen and G. Oresten (1978) Separation of epimers of budesonide and related corticosteroids by high-performance liquid chromatography. A comparison between straight- and reversed-phase systems. *J. Chromatogr.* **157**, 65–74.

Williams, K.M. (1984) Kinetics of misonidazole enantiomers. *Clin. Pharmacol. Ther.* **36**, 817–823.

Wilson, M.J. and T. Walle (1984) Silica gel high-performance liquid chromatography for the simultaneous determination of propranolol and 4-hydroxypropranolol enantiomers after chiral derivatization *J. Chromatogr.* **310**, 424–430.

Wollard, P.M. (1986) Stereochemical difference between 12-hydroxy-5, 8, 10, 14-eicosatetraenoic acid in platelets and psoriatic lesions. *Biochem. Biophys. Res. Comm.* **136**, 169–176.

Wuis, E.W., E.W.J. Beneken Kolmer, S.E.C. van Beijsterveldt, R.C.M. Burgers, T.B. Vree and E. van der Kleyn (1987) Enantioselective high-performance liquid chromatographic determination of baclofen after derivatization with a chiral adduct of *o*-phthaldialdehyde. *J. Chromatogr.* **415**, 419–422.

Yagi, H., H. Akagi, D.R. Thakker, H.D. Mah, M. Koreeda and D.M. Jerina (1977) Absolute stereochemistry of the highly mutagenic 7,8-diol 9,10-epoxides derived from the potent carcinogen trans-7,8-dihydroxy-7,8-dihydrobenzo[a]-pyrene. *J. Am. Chem. Soc.* **99**, 2358–2359.

Yang, S.K. and P.P. Fu (1984) Stereoselective metabolism of 7-methylbenz[a]anthracene: absolute configuration of five dihydrodiol metabolites and the effect of dihydrodiol conformation on circular dichroism spectra. *Chem. Biol. Interactions* **49**, 71–88.

Yang, S.K., H.V. Gelboin, J.D. Weber, V. Sankaran, D.L. Fischer and J.F. Engel (1977) Resolution of optical isomers by high-pressure liquid chromatography. The separation of benzo[α]pyrene trans-diol derivatives. *Anal. Biochem.* **78**, 520–526.

Yoneda, M., M. Shiratsuchi, M. Yoshimura, Y. Ohkawa and T. Muramatsu (1984) Optical resolution and determination of absolute configuration of nipradilol. *Chem. Pharm. Bull.* **33**, 2735–2742.

References added in proof

Dieterle, W. and J.W. Faigle (1983) Multiple inverse isotope dilution assay for the stereospecific determination of R(+)- and S(−)-oxprenolol in biological fluids. *J. Chromatogr.* **259**, 313–318.

van den Oetelaar, P.J.M., J.R.C.M. van Beckhoven and H.J. Hoenders (1987) Analysis of aspartic acid racemization. Evaluation of a chiral capillary gas chromatographic method. *J. Chromatogr.* **388**, 441–447.

Part 3 Direct chiral resolution

5 Chiral ligand exchange chromatography

S. LAM

5.1 Introduction

Ligand exchange chromatography is separation effected by the preferential mixed chelate complexation of the analyte ligand with a metal ion and a second ligand introduced to the chromatographic system. Chromatography thus performed is highly selective for the differentiation of chelatable solutes. Stereoselective resolution of isomeric ligands can be achieved if the system ligand is optically active and forms a chiral complex in the presence of a metal ion. The chiral ligand can be incorporated into the chromatographic system by one of two general approaches: (a) by initially immobilizing a metal coordinating ligand on the stationary phase to create a specific solute sorbent interaction, and (b) by adding a chiral metal complex to the chromatographic eluent to effect isomeric separation on an achiral stationary phase, i.e. C_{18} bonded phase. Stereoselective separation of the enantiomeric amino acid pairs results when one of the isomers is differentially retained as the ternary complexes of the metal ion and the chiral ligand.

This chapter will deal primarily with the use of metal coordinating ligands as mobile-phase additives, as this type of system (a) is more flexible, (b) offers a wider range of variables, with which to manipulate retention and selectivity, and (c) is much more readily accessible to the practising chromatographer than systems employing immobilized metal-coordinating ligands. There is a limited range of commercially-available silica-based chiral stationary phases which employ an immobilized metal-coordinating ligand and they are invariably very expensive. Nevertheless it is worthwhile to consider immobilized metal-coordinating ligands as they have played an important role in the development of chiral ligand exchange chromatography.

5.2 Immobilized metal coordinating ligands

In 1968 Rogozhin and Davankov (1970) pioneered the chromatography of racemates on resins containing optically active bifunctional and trifunctional α-amino acids on to which a transition metal ion was loaded, and suggested ligand exchange was the separation mechanism. They bonded L-proline to a macroreticular isoporous polystyrene, charged the chelating resins with Cu(II) ion and succeeded the historical resolution of D-proline from the L with a mobile phase containing ammonia by column liquid chromatography (Davankov *et al.*, 1973). More recently, Davankov and co-workers bonded chiral ligands of amino acids on to the surface of microparticulate silica in an effort to improve the efficiency and performance of chromatography (Roumeliotis *et al.*, 1983; Kurganov *et al.*, 1983). The most significant of recent developments includes the work of Karger and co-workers (Shieh *et al.*, 1987) in synthesizing an L-proline bonded phase diluted with C_{18} alkyl groups. Such dilution improved resolution and was applied to the resolution of amino alcohol enantiomers as Schiff bases. In using micro-HPLC on an L-proline phase Takeuchi *et al.* (1987) achieved good resolution of racemates of dansyl amino acids but had difficulty in simultaneously resolving many pairs of enantiomers in a single chromatographic step. This problem was overcome by connecting chiral and hydrophobic columns in series. Saigo and co-workers have introduced new chiral stationary phases for ligand-exchange chromatography containing additional complexity and functionality in order to attempt to improve enantioselectivity. (1R, 2S)- and diastereomeric (1S, 2S)-2-carboxymethylamino-1, 2-diphenylethanol were prepared from (1R, 2S)- and (1S, 2S)-2-amino-1, 2-diphenylethanol respectively and were bound to silica gel pretreated with 3-glycidoxypropyltrimethoxysilane (Yuki *et al.*, 1987). A similar phase was later developed which utilized the sodium salt of ((1R, 2S)-2-hydroxy-1-methyl-2-phenylethylamino)acetic acid, conveniently prepared from (−)-norephedrine (Saigo *et al.*, 1988). Davankov and co-workers have demonstrated that chiral ligand exchange chromatography can even be extended to non-aqueous systems (Kurganov *et al.*, 1987). However, such systems are different in that they employ outer-sphere complexation and so far have only been applied to enantiomeric pairs of organometallic complexes.

Further research work (both chiral and achiral) on ligand exchange chromatography with bonded supports is described in several reviews (Davankov, 1979) and in a book (Davankov *et al.*, 1988).

5.3 Metal coordinating ligands as mobile phase additives

Although it was known that ligands of metal complexes interchange in solution long before the introduction of ligand exchange chromatography, more than a decade had elapsed after Davankov's initial work before it was realized that the optical isomers of an amino acid could exchange one of the two ligands of a bis[chiral ligand]-metal complex added in the eluent and the resulting ternary metal complexes could be differentially retained on an achiral stationary phase to facilitate chiral resolution of the racemates. Around the same time Karger and co-workers (LePage *et al.*, 1979; Lindner *et al.*, 1979) were working on mobile phase additives containing L-2-alkyl-4-octyldiethylene-triamine complexes of zinc and other metal ions for the separation of dansyl derivatives of amino acids on a reversed-phase column, Hare and Gil-Av also introduced optically active Cu(II)-proline eluents for the separation of free D and L amino acids on an ion exchange column (Hare and

Gil-Av, 1979; Gil-Av *et al.*, 1980). These works subsequently stimulated the interest of several research groups, and since then, a number of procedures have been developed for the resolution of optical isomers of amino acids by the addition of Cu(II) complexes of free amino acids or their alkyl derivatives to the eluents to facilitate chiral recognition.

Over the years, Lam and co-workers (1980–86) developed several stereoselective approaches for the resolution of dansyl derivatives as well as *o*-phthalaldehyde derivatives of amino acids on reversed-phase columns as the mixed chelate complexes with equimolar concentrations of Cu(II) and a second amino acid such as L-proline, L-arginine, L-histidine and L-histidine methyl ester in the mobile phases. The reagents for preparing these mobile phases are readily available from the commercial sources.

On the other hand, many research groups custom synthesized alkyl-amino acids which could form complexes with metal ions added to mobile phases for the ligand exchange chromatography of chiral ligands. Grushka and co-workers (Gilon *et al.*, 1979–81; Grushka *et al.*, 1983) successfully separated a number of free amino acids with the Cu(II) complexes of L-aspartate and detected the copper complexes at UV 235 nm. Later, they synthesized a series of L-aspartylalkylamides and used the copper(II) complexes for studying the mechanisms of chiral separation. Weinstein (1982) and Weinstein *et al.* (1982, 1984, 1984) synthesized N, N-dialkyl-L-amino acids and used these Cu(II) complexes for the resolution of both the underivatized amino acids and the dansyl derivatives. Nimura *et al.* (1981–82), using mobile phases containing Cu(II) complexes of N-(*p*-toluenesulphonyl-L-phenylalanine and N-(*p*-toluenesulphonyl-D-phenylglycine also separated the free amino acids on reversed phase columns.

Still by coating the C_{18} reversed phase columns with the long chain N-dodecyl L proline to which was subsequently charged with Cu(II) ions, Karger and co-workers (Tapuhli *et al.*, 1981) achieved the resolution of D and L dansyl amino acids. Davankov *et al.* (1981) also similarly treated the reversed phase columns with N-decylalkylamide and Cu(II) for the free amino acids. Recently, Marchelli *et al.* (1986) and Armani *et al.* (1988) also introduced the use of Cu(II) complexes of L-amino acylamides and tetradentate Cu(II) complexes of N, N′-bis(aminoacyl)ethane and propane diamines of L-phenylalanine, L-valine and L-alanine for the enantiomeric separation of D and L dansyl amino acids by reversed phase chromatography.

5.4 Ligand exchange mechanism

Mixed chelate complexes, which are common in biological systems, are exhibited in the ligand–metal complexes of amino acids and peptides, and enzyme–metal–substrate complexes.

The simplest model of a mixed chelate metal complex consists of a metal ion and two non-identical ligands. For illustration, we assume that the metal ion

M is divalent and the ligands, AX and AY are two different amino acids. Ignoring the charges, the metal ion is in equilibrium with the two ligands as follows:

$$M + 2AX \rightleftharpoons M(AX)_2$$
$$M + 2AY \rightleftharpoons M(AY)_2$$
$$M + AX + AY \rightleftharpoons M(AX)(AY)$$
$$M(AX)_2 + M(AY)_2 \rightleftharpoons 2M(AX)(AY)$$

The disproportionation constant K can be written as:

$$K = [M(AX)(AY)]^2/[M(AX)_2][M(AY)_2]$$

As there are two ways the mixed complex M(AX)(AY) can be formed but only one way the binary complexes $M(AX)_2$ and $M(AY)_2$ can be formed, statistically, the expected disproportionation constant is 4. When the stability constants for the mixed complex M(AX)(AY) and for the binary complexes $M(AX)_2$ and $M(AY)_2$ are determined, following Sigel (1973), it is possible to define the stability of the ternary complex as:

$$\log K = 2 \log M(AX)(AY) - \log M(AX)_2 - \log M(AY)_2$$

Thus the tendency towards ternary complex formation can be characterized by the sign and value of $\log K$. A value of K greater than 0.6 suggests stabilization of the ternary complex.

Ligand exchange is predominant in solution as suggested by the formation constants of L-valine-Cu(II)-histidine of 2.2, L-phenylalanine-Cu(II)-L- of 2.1 and L-tryptophan-Cu(II)-L-histidine of 2.4 (Brookes and Pettit, 1977). The $\log K$ values greater than 0.6 signify that histidine does indeed form ternary complexes with other amino acids. Chiral resolution of optical isomers was effected when the binary copper complex of histidine was added to the mobile phase (Lam *et al.*, 1980).

In the study by Marchelli *et al.* (1986) on the separation of D and L dansyl amino acids with an eluent containing bis(L-amino acid amidato) copper(II) complexes, two species of binary complexes responsible for chiral recognition were indeed found to exist in solution by potentiometric titration (Figure 5.1). The negatively charged amino acids are actually better resolved at the neutral pH where the binary Cu(II) complex is positively charged (Figure 5.2*a*), and the neutral amino acids are better resolved at higher pH in which the dianionic copper complex is formed (Figure 5.2*b*). Both the neutral $[Cu(II)L_2H_{-2}]$ and the positively charged $[Cu(II)L_2H^{+}_{-1}]$ binary complexes present in the mobile phase are responsible for the disproportionation reaction. On contact with an analyte ligand, the weaker protonated chelate is displaced from both the binary complexes and gives only the positively charged ternary complex D- or L-$AXCu(II)LH^{+}_{-1}$. Therefore only one chromatographic peak is observed for each isomer.

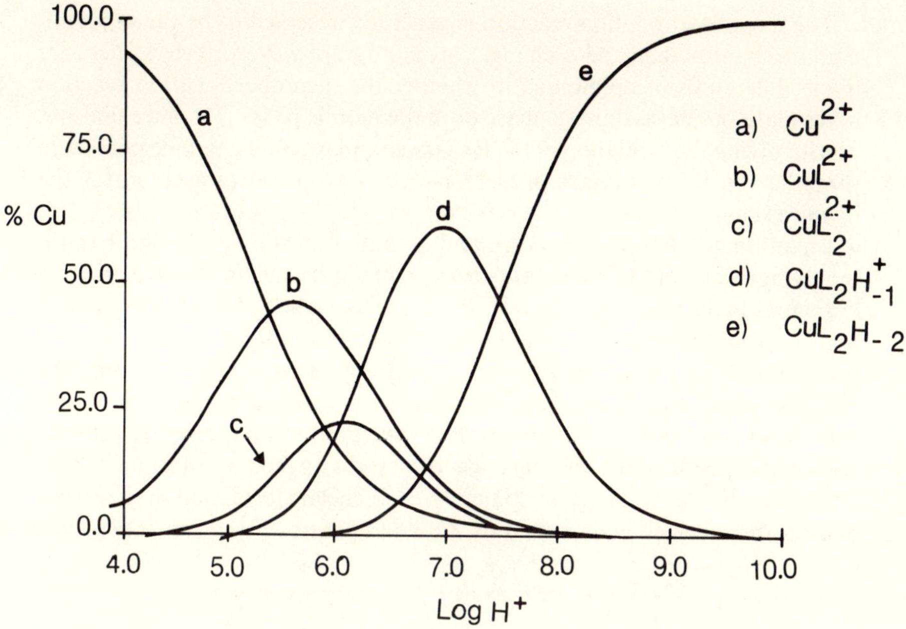

Figure 5.1 Species distribution for the L-phe-A/Cu(II) system as a function of pH, obtained by potentiometric titration. From Armani *et al.* (1988*a*).

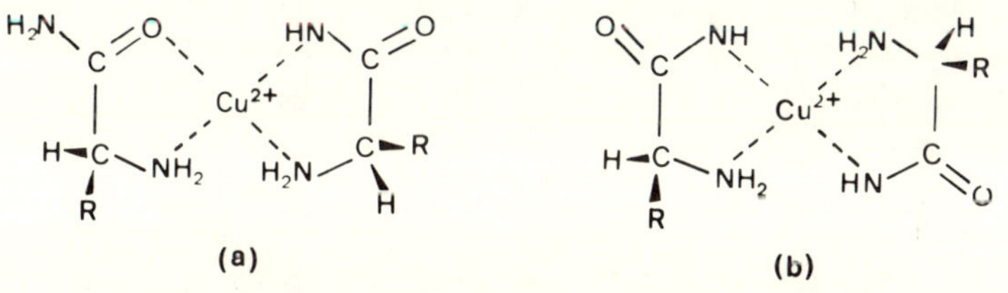

Figure 5.2 Structures of copper complexes: (*a*) $Cu_2LH^{+}_{-1}$; (*b*) C_2LH_{-2}.

Regardless of whether or not the chiral metal complex is immobilized on the stationary phase or added to the eluent, ligand exchange separation of D and L amino acid is effected on the formation of two diastereomeric ternary complexes from the parent complex according to the following equilibria:

$$[\text{L-AX}]\text{Cu(II)}[\text{L-AX}] + [\text{D-AY}] \rightleftharpoons [\text{L-AX}]\text{Cu(II)}[\text{D-AY}] + [\text{L-AX}]$$
$$[\text{L-AX}]\text{Cu(II)}[\text{L-AX}] + [\text{L-AY}] \rightleftharpoons [\text{L-AX}]\text{Cu(II)}[\text{L-AY}] + [\text{L-AX}]$$

The diastereomeric complex that is more stable and that interacts more strongly with the stationary phase will separate from the less stable complex.

The disproportionation reaction is generally accepted to be the predominant mechanism in ligand exchange chromatography. Nevertheless, there are many differences of opinion as to whether the disproportionation reaction takes place on the stationary phase or in the mobile phase. Two mechanisms, one involving the formation of the ternary complexes in the mobile phase and the other on the stationary phase, have been proposed to account for the ligand exchange reaction.

Enantiomeric resolution of D and L amino acid, by the mechanism involving the addition of a binary complex of Cu(II) and an amino acid or its derivative in the eluent, proceeds by the initial formation of the diastereomeric mixed chelate complexes in the mobile phase and follows by separation of these complexes on the hydrocarbon stationary phase (Lam, 1980–86). The diastereomeric metal complexes with different formation and dissociation rates display different chromatographic behaviour when interacting with the stationary phase and thus effect stereoselectivity and resolution. Chiral separation by means of diastereomeric mixed chelate formation in the eluent was unequivocally demonstrated by Zare and others (Gassmann *et al.*, 1985)

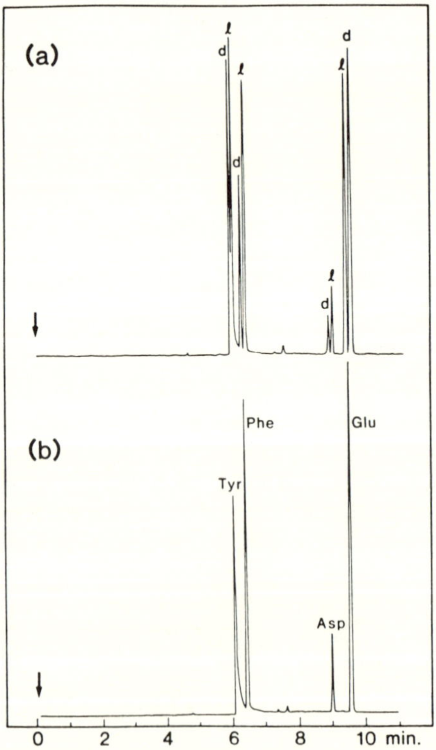

Figure 5.3 Electropherograms of DL-dansyl amino acids with Cu(II)-L-histidine electrolyte at pH 7.0. From Gassmann *et al.* (1985).

who separated D and L dansyl amino acids by zone electrophoresis in capillary columns with Cu(II)-L-histidine as the supporting electrolyte (Figure 5.3).

The second mechanism, involving the disproportionation of a stationary ligand from the binary metal complex on the stationary phase by an analyte ligand in the solution phase, is characteristic of ligand exchange chromatography with immobilized coordinating chelates or with long-chain alkylamide which is capable of forming metal complexes and which must first be coated on the reversed-phase column to saturation before the chromatographic analysis (Davankov *et al.*, 1981; Tapuhi *et al.*, 1981; Broge and Leussing, 1986). The number of moles of ligand adsorbed on the stationary phase was found to relate to the concentration of the ligand in the solution phase by the equation described for the Langmuir isotherm (Broge and Leussing, 1986). For very long chain alkylamides, such as N-decyl-L-histidine (Davankov *et al.*, 1981), a mobile phase containing trace quantities of the metal ion, and sometimes with ammonium chloride, was found adequate in eluting the chiral amino acids. Nevertheless, for long-term stability and reproducibility, a low concentration of the chiral alkylamide must also be added to the mobile phase in order to replenish leaching of the ligand from the stationary phase. Further, the small amounts of the chiral alkylamide metal complexes in the mobile phase also improves the kinetics of dissociation of the analyte from the stationary phase.

5.5 Enantiomeric resolution of D and L amino acids

As already summarized in the introduction, mobile phases containing copper complexes of various chiral ligands have been used to resolve optical isomers of amino acids. These ligands differ not only in selectivity, but also in availability. Some of the ligands were obtained by syntheses involving elaborate synthetic steps (LePage *et al.*, 1979; Grushka *et al.*, 1983), others by simpler steps (Weinstein, 1982; Nimura, 1981; Marchelli, 1988). Only a few of these are commercially available (Lam, 1986; Grushka, 1979; van der Haar, 1988). The ligands from the commercial source are favourable because of (i) inexpensive supply of highly pure resolving ligands, (ii) ease of use and implementation and (iii) reliable and reproducible high resolution separation of D and L amino acids. Therefore, several applications, proved to be easy to implement with ligands available through chemical suppliers or by simple organic syntheses, will be described.

5.5.1 *Free amino acids*

Chromatography of free amino acids usually requires post column derivatization of the analyte in the effluent with either *o*-phthalaldehyde for fluorescence detection or ninhydrin for spectrophotometric detection. In reality, resolution of enantiomers of free amino acids was only demonstrated with the approach

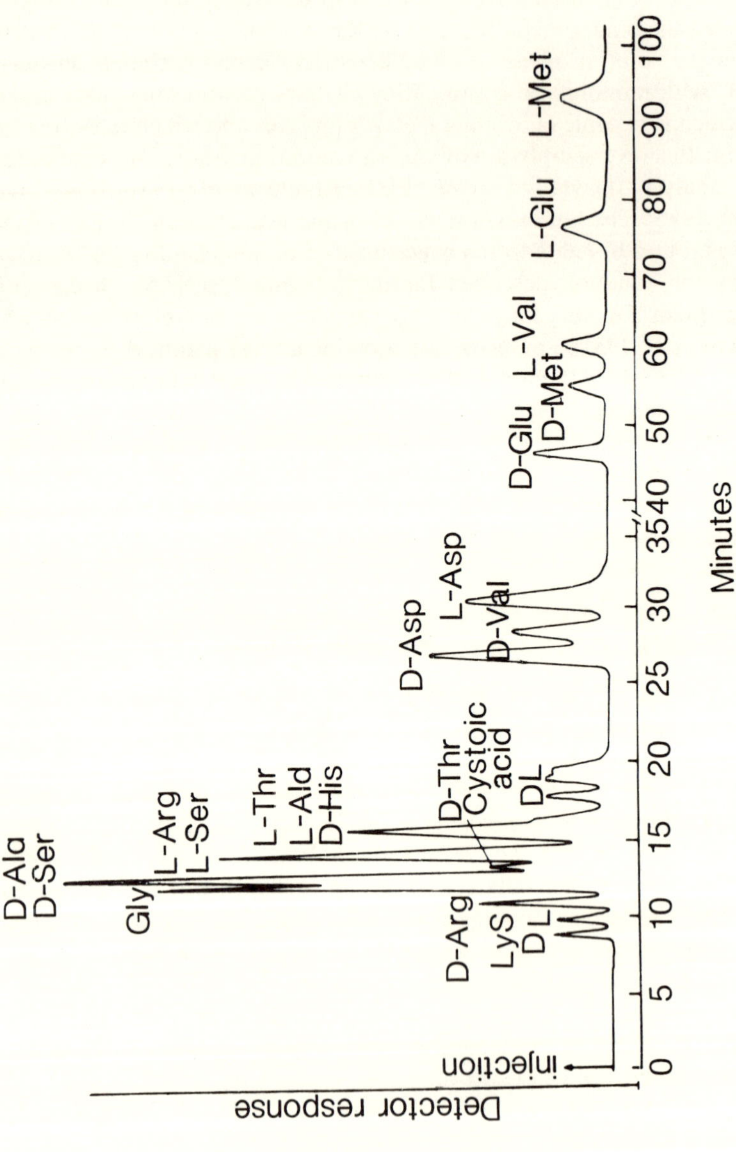

Figure 5.4 Separation of a mixture of amino acids into enantiomers on the reversed-phase column. Mobile phase: chiral additive of Cu(II)N, N-di-n-propyl-L-alanine in aqueous solution. Flow rate: 0.2 mL/min. Temperature: ambient. From Weinstein (1982), p. 375.

using imino acids and N-protected amino acids as the resolving ligands followed by post-column derivatization of the solutes with *o*-phthaladehyde for fluorometric detection. The procedure involving ninhydrin has not been explored probably because of its reaction with both the primary and secondary amines.

Hare and Gil-Av first used D- and L-proline for the resolution of D and L amino acids on an ion exchange column. Reversed-phase separation of enantiomers of free amino acid with N, N-di-n-propyl-L-alanine (Weinstein *et al.*, 1982) (Figure 5.4) and with N-(*p*-toluenesulphonyl)-L-phenylalanine (Nimura *et al.*, 1981) (Figure 5.5) were also reported.

5.5.2 *Dansyl amino acids*

Optical isomers of dansyl amino acids can be conveniently separated on reversed-phase columns with mobile phases containing chiral copper complexes. The simplest approach for resolving the dansyl amino acids is by using

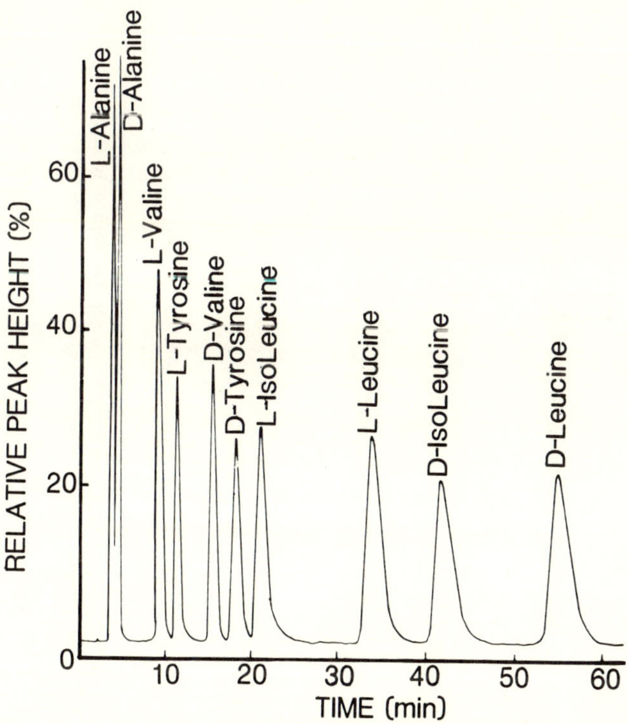

Figure 5.5 Separation of D, L amino acids with Cu(II) N-(*p*-toluenesulphonyl)-L-phenylalanine eluent. Mobile phase: 10% acetonitrile in an aqueous solutions containing 1 mM N-(*p*-toluene-sulphonyl)-L-phenylalanine and 0.5 mM $CuSO_4 \cdot 5H_2O$; pH 6.0. From Nimura *et al.* (1981).

Table 5.1 Capacity ratio k' and selectivity α of D-L-Dns-amino acids for four chiral eluents containing 2.5 mM Cu(II) complexes of the L amino acids shown in the headings*

	Proline			Histidine			Histidine methyl ester			Arginine		
	k'_L	k'_D	α	k'_L	k'_D	α	k'_D	k'_L	α	k'_D	k'_L	α
Ser	3.7	3.2	0.9	—	—	—	9.3	10.7	1.2	3.0	3.0	1.0
Thr	3.7	4.6	1.2	—	—	—	9.3	9.3	1.0	3.0	3.0	1.0
Ala	5.7	6.6	1.2	1.8	2.4	1.3	11.9	11.9	1.0	4.2	4.4.	1.1
αAB	7.3	9.2	1.2	2.2	3.8	1.7	13.6	15.0	1.1	5.7	6.2	1.1
Val	11.4	15.0	1.3	3.2	6.4	2.0	17.3	20.1	1.2	8.3	9.5	1.1
Met	11.4	14.8	1.3	4.4	7.8	1.8	27.2	30.7	1.1	10.3	11.7	1.1
N-Val	14.6	19.0	1.3	3.8	9.0	2.4	23.3	27.6	1.2	11.0	13.0	1.2
I-Leu	—	—	—	7.2	15.4	2.1	40.4	44.7	1.1	17.2	20.8	1.2
Leu	23.9	32.6	1.4	7.2	16.0	2.2	33.0	43.9	1.3	18.5	20.8	1.1
N-Leu	32.6	45.7	1.4	9.4	22.2	2.4	40.4	58.4	1.5	22.3	28.2	1.3
Phe	32.6	52.8	1.6	9.0	9.0	1.0	60.4	77.9	1.3	20.3	23.0	1.1
Trp	41.2	71.2	1.7	7.4	7.4	1.0	93.3	121.8	1.3	24.8	31.2	1.3

*Acetonitrile concentration was 15% for the proline and histidine systems and 20% for the histidine methyl ester and arginine systems.

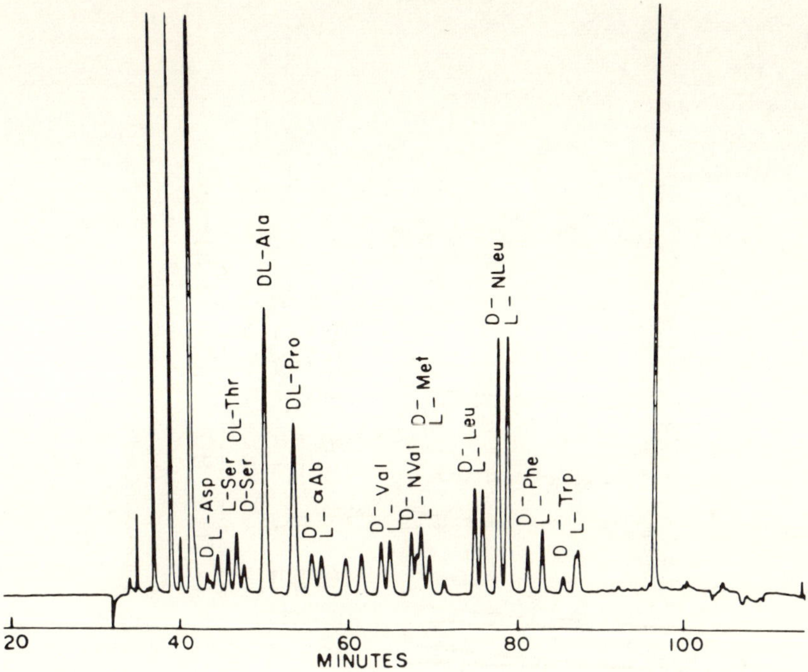

Figure 5.6 Separation of D, L-amino acid standards. Mobile phase 5.0 L-histidine methyl ester, 2.5 mM $CuSO_4{\cdot}5H_2O$, and 2.0 g of ammonium acetate, pH 5.5. A stepwise gradient was formed by blending the buffer with a 40% acetonitrile solution of the same buffer.

copper complexes of various amino acids and amino acid derivatives as reported by Lam *et al.* As shown in Table 5.1, these ligands all gave satisfactory separation of the amino acids. Among the various ligands, histidine methyl ester was unique in offering both good selectivity for enantiomeric pairs and excellent separation of the individual amino acids (Figure 5.6). L-phenylalanine acylamide introduced by Marchelli *et al.* (1986) was also demonstrated to be a selective ligand for the separation of amino acids and the enantiomeric pairs (Figure 5.7).

5.5.3 *Imino acids*

The simultaneous detection and resolution of enantiomers of secondary amino acids is a more complicated problem. N-dansylating provides highly fluorescent derivatives but the optical isomers are not resolved by the ligand exchange approach, possibly because the derivative, which contains a tertiary amino group, does not form the proper ternary complex with Cu(II) (Lindner *et al.*, 1979). Separation of enantiomers of amino acids by ligand exchange followed by post-column reaction with *o*-phthalaldehyde for detection is highly selective, but the imino acids, which do not form fluorescent

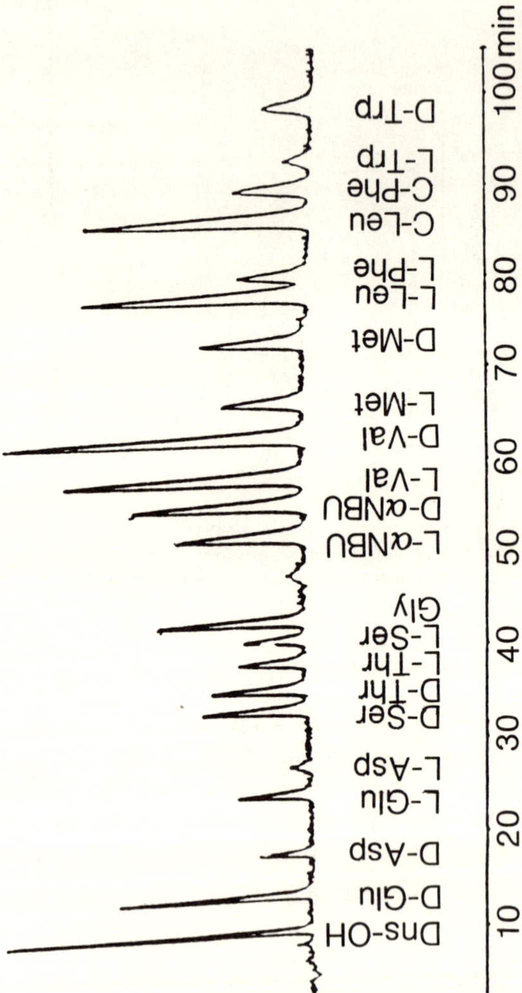

Figure 5.7 Enantiomeric separation of a synthetic mixture of D, L-Dns-amino acids with L-Phe-A/Cu(II). Conditions: 4 mML-Val-A, 2 mM copper acetate 0.3 M sodium acetate, pH 7; Column C18 Novapak. From Armani *et al.* (1988*b*).

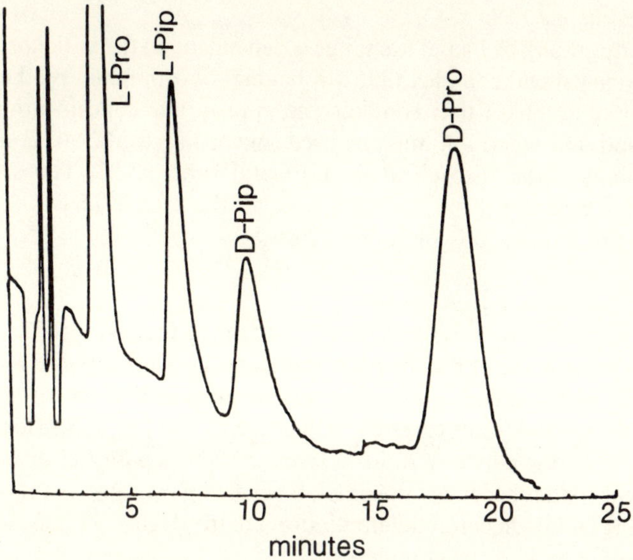

Figure 5.8 Separation of stereoisomers of proline and pipecolic acid. The Nucleosil 5 C18 column. 15 × 0.42 cm, was equilibrated with loading buffer containing 295 mg L-aspartame and 100 mg of copper sulphate in 1 L of water. Mobile phase: 100 mg of copper sulphate and 25 ml of loading buffer diluted to 1 L with water. Flow rate: 1.5 mL min^{-1}.

derivatives with *o*-phthalaldehyde, although enantiomerically resolved, are difficult to detect. A procedure using low concentration of Cu(II) L-aspartame (Grushka, 1980) as the resolving agent and detecting the copper complex by UV absorption, despite of lower sensitivity of detection, is a practical compromise. Both D, L-proline and D, L-pipecolic acid are resolved and detected (Figure 5.8).

5.6 Parameters affecting ligand exchange

Many of the reported ligand exchange procedures, although differing widely in the chiral metal complexes used, behave very much alike and share many common chemistries.

5.6.1 *Metal ion*

Copper(II) ions in the presence of the analyte ligands and a chiral ligand form highly stereoselective mixed chelate complexes that can be efficiently resolved by reversed-phase chromatography. Metal ions such as zinc, nickel and mercury, which form slightly less stable complexes, had also been used for ligand exchange chromatography of amino acid but were reported to give inferior resolution (Lindner *et al.*, 1979).

5.6.2 *Metal/ligand ratio*

As chromatography by ligand exchange is dependent on the initial presence of a chiral binary metal complex that can be disproportionated by the analyte ligand, a mobile phase that contains the appropriate concentration of the chiral ligand and metal ion must be used. According to the structure of the binary complex, the theoretical ligand/metal ratio is 2.1. The same was found experimentally to be the ideal ratio in the mobile phases for ligand exchange chromatography of amino acids.

5.6.3 *Concentration of metal complex*

As demonstrated with the mobile phase containing Cu(II)-L-proline, increasing the concentration of the ligand complex in the mobile phase resulted in more of the mixed chelate complex available for partition on the stationary phase, and an increase in capacity ratio k', selectivity α, and resolution R_S (Figure 5.9). As the capacity ratio k' levels off, the separation of individual amino acid pairs becomes maximal, suggesting at this point that the determining factors in chiral differentiation are the degree of dissociation and the hydrophobicity of the complexes.

5.6.4 *Eluent pH*

A change in the pH of the mobile phase would result in a change in the equilibrium distribution of the various metal species in the solution

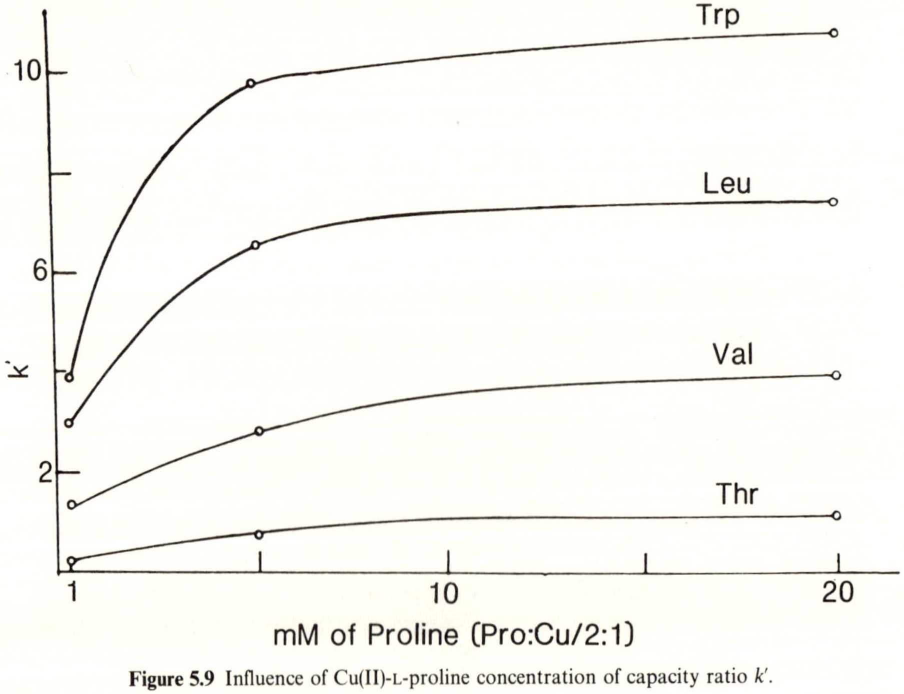

Figure 5.9 Influence of Cu(II)-L-proline concentration of capacity ratio k'.

(Figure 5.1). Bifunctional amino acids can add two protons. With the lowering of pH, the hydrogen ion would compete with the Cu(II) ion for the chelating sites; the resolution would diminish. A change in the pH also influences the interaction of polar amino acids with side chain with the charged metal species and affects chiral resolution. For practical applications, the pH of the mobile phase is in the range of 7.0–8.0. Higher pH causes not only precipitation of the copper complexes but also dissolution of the silica columns.

5.6.5 *Elution order*

The order of elution of amino acids is in general consistent with reversed-phase chromatography: the higher the carbon content, the bulkier the alkyl substituent on the α-carbon, the longer the retention (Table 5.1). The analyte amino acids with the highest carbon content form mixed chelate complexes that are more hydrophobic and partition favourably on the hydrocarbon stationary phase. For isomers with an equal number of carbon atoms, those pairs with a linear side chain, such as norvaline and norleucine, are retained more than those with a branched side chain, such as valine and leucine, and are better resolved. This behaviour can be explained by the stronger spatial interaction of the straight-chain isomers with the stationary phase. The order of elution between the individual amino acids and between the D- and L-isomers is also affected by the bulkiness of the alkyl substituents. The bulky substituents control the stability of the D,L amino acid-Cu(II)-chiral ligand complex as to whether the interaction between the alkyl groups of the analytes and the chiral ligand promotes or interferes with complex formation. With few exceptions, the greatest stereoselectivity and capacity ratio k', were always observed with the isomeric pairs of phenylalanine and tryptophan, which are both hydrophobic and have the largest alkyl substituent.

5.6.6 *Organic modifier*

The influence of the nature of organic modifier on the retention and stereoselectivity had been studied by van der Haar *et al.* (1988). Using a mobile phase containing Cu(II)-L-proline for the resolution of D and L dansyl amino acids, they observed similar retention behaviour with organic modifiers acetonitrile, methanol and 1,4-dioxane. The largest stereoselectivity α, with D isomer more retained, was observed with acetonitrile as the organic modifier. However, when tetrahydrofuran replaced acetonitrile as the organic modifier, a surprising retention behaviour with reversal of stereoselectivity with D isomer eluted before the L was noted. Tetrahydrofuran differs from the other organic solvents in that it possesses a pair of lone electrons on the oxygen atom. Like the oxygen atom of water, the oxygen atom of tetrahydrofuran can coordinate at the axial positions with the metal ion in the mixed chelate complex. As the bis(amino acid) ternary complex involving L-proline and the analyte coordinate around the Cu(II) ion in a trans conformation with two water molecules in the axial positions, stability enhancement results when the tetrahydrofuran molecule approaches the L-proline-Cu(II)-L-(dansyl AA)

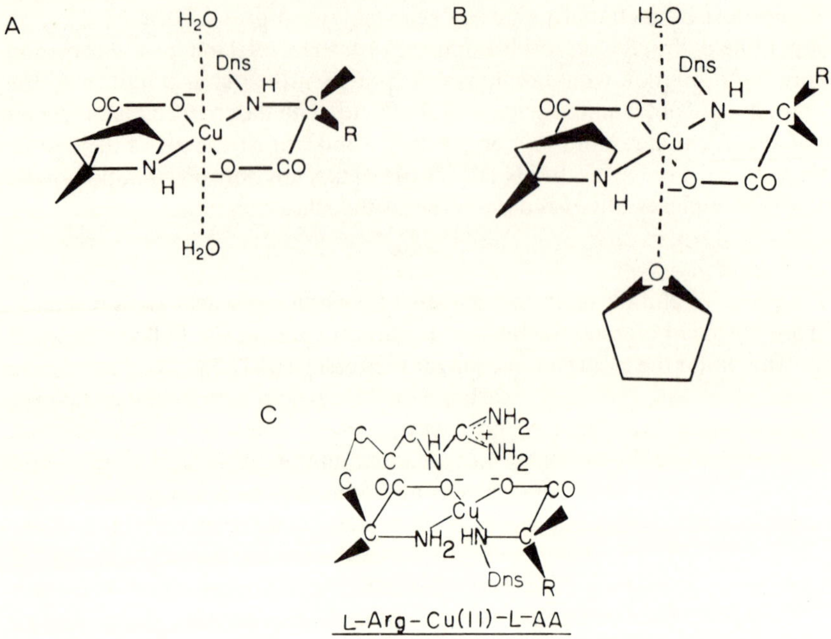

Figure 5.10 Ternary complexes of (A) D-Dns amino acid and L-proline with Cu(II) in aqueous acetonitrile; (B) L-Dns amino acid and L-proline with Cu(II) in aqueous tetrahydrofuran; and (C) L-Dns amino acid and L-arginine with Cu(II) in aqueous acetonitrile.

from the least sterically hindered plane and replaces the water molecule (Figure 5.10B).

Stereoselectivity of the amino acid pairs is also dependent in part on the hydrophobic retention of the diastereomeric metal complexes on the hydrocarbon stationary phase, the content of polar organic modifier such as acetonitrile and methanol would affect both the capacity ratio k' and the selectivity α. Lowering the organic modifier content of the mobile phase promotes hydrophobic interaction of the solute and sorbent, resulting in longer retention and improved resolution of the optical isomers.

5.6.7 *Stereoselectivity*

The stereoselectivity between the isomeric pairs depends primarily on the stability of the resulting ternary complexes formed with the chiral Cu(II) additive in the mobile phases. Ligand exchange chromatography of D and L dansyl amino acids with mobile phases containing Cu(II) complexes of chiral ligand of L-proline, L-arginine, L-histidine, and L-histidine methyl ester show that separation varies as a function of the chiral ligands by forming ternary mixed complexes of different conformation and stability.

In the Cu(II)-L-proline system, the bis(amino acids) coordinate around the metal in a trans conformation with two water molecules in the axial positions.

The lower stability of the L–L complex is apparently due to the interaction of the α-substituent of the dansyl amino acid, the pyrrole ring and the apically coordinated water molecules all in the same plane (Figure 5.10A).

With the exception of proline, the rest of the chiral ligands studied contain a chelatable side chain. These side chains are important in determining the conformation of the mixed chelate complexes by participation in coordination with the metal ion and electrostatic interaction of the analyte ligand. The contributions of these chelatable side chains is most dramatically illustrated in the reversal of stereoselectivity between the D and L pairs with the arginine complexes (Figure 5.10C). In the Cu(II)-L-arginine system, the bis(amino acid) coordinated around Cu(II) in the *cis* conformation, allowing charge interaction of the positive guanido group of arginine with the negatively charged carboxyl group of the neighbouring dansyl amino acid. Electrostatic ligand–ligand interactions within the complex molecule have been documented to be an important stabilizing factor (Yamauchi *et al.*, 1977). In the *cis* conformation, the L-arginine-Cu(II)-D-amino acid ternary complex is less stable because, as in the proline system, the alkyl group of the analyte amino acid, the guanido group of arginine, and the coordination water molecule are crowding in the same spatial coordinates.

Examination of the selectivity α and capacity factors k' of the four systems reveals that the Cu(II)-L-histidine system is the most stereoselective (Table 5.1). L-histidine differs from other amino acids in that it possesses on the α-carbon a bulky imidazole substituent, by which L-histidine functions as a terdentate ligand, which gives the mixed complex a rigid conformation and therefore high stereoselectivity. Proline has an intermediate selectivity because it has a ring system standing above the co-ordination plane, unlike the imidazole ring of histidine methyl ester which lies flat on the plane. Arginine, which does not have a rigid ring structure, gives selectivity similar to histidine methyl ester. Therefore, the three-dimensional conformation of the ternary complex is important in deciding stereoselectivity. Stronger spatial interaction with the metal complex results in greater differences in stability and better optical resolution.

5.7 Conclusion

Ligand exchange chromatography, by mixed chelate complexation of the analyte amino acids and a chiral ligand with a metal ion, is highly selective for the resolution of optical isomers of amino acid. Procedures for the separation of free amino acids and their dansylated derivatives, as well as imino acids, have been developed. These applications are simple in theory and implementation. They can be easily adopted in reversed-phase HPLC by conveniently replacing the mobile phase with one that contains a chiral metal complex. Ligand exchange chromatography with mobile phases containing chiral complexes on reversed-phase columns is attractive in that it has both the advantages of reversed phase chromatography for its high efficiency and ligand exchange chromatography for its excellent selectivity.

References

Armani, E., L. Barazzoni, A. Dossena and R. Marchelli (1988*a*) Bis(L-amino acid amidato) copper(II) complexes as chiral eluent in the enantiomeric separation of D, L-dansyl amino acids by HPLC (reversed phase) (unpublished).

Armani, E., A. Dossena, R. Marchelli and R. Virgili (1988*b*) Copper(II) complexes of diamino–diamido-type ligands as chiral eluent in the enantiomeric separation of D, L-dansyl amino acids by HPLC (reversed phase) (unpublished).

Broge, J.M. and D.L. Leussing (1986) Mechanism of the reversed phase high performance liquid chromatographic separation of D- and L-valine using copper(II) and L-aspartyl-L-phenylalanine methyl ester. *Anal. Chem.* **58**, 2237.

Brookes, G. and L.D. Pettit (1977) Complex formation and stereoselectivity in the ternary systems copper(II)-D/L-histidine-L-amino-acids. *J. Chem. Soc., Dalton* 1918.

Davankov, V.A., S.V. Rogozhin, A.V. Semechkin and T.P. Sachkova (1973) Ligand exchange chromatography of racemates: influence of the degree of saturation of the asymmetric resin by metal ions on ligand exchange. *J. Chromatogr.* **82**, 359.

Davankov, V.A. (1979) Resolution of racemates by ligand exchange chromatography. In *Advances in Chromatography*, ed. J.C. Giddings, E. Grushka, J. Cazes, and P.R. Brown, **17**, Marcel Dekker, New York.

Davankov, V.A., A.S. Bochkov and Y.P. Belov (1981) Ligand exchange chromatography of racemates: resolution of α-amino acids on reversed phase silica gels coated with N-decyl-L-histidine. *J. Chromatogr.* **218**, 547.

Davankov, V.A., J.D. Navratil, H.F. Walton (1988) *Ligand Exchange Chromatography*. CRC Press, Florida.

Gassmann, E., J.E. Kuo, and R.N. Zare (1985) Electrokinetic separation of chiral compounds. *Science* **230**, 813.

Gil-Av, E., A. Tishbee and P.E. Hare (1980) Resolution of underivatized amino acids by reversed phase chromatography. *J. Am. Chem. Soc.* **102**, 5115.

Gilon, C., R. Leshem, Y. Tapuhi and E. Grushka (1979) Reversed phase chromatographic resolution of amino acid enantiomers with metal aspartame eluents. *J. Am. Chem. Soc.* **101**, 7612.

Gilon, C., R. Leshem and E. Grushka (1980) Determination of enantiomers of amino acids by reversed phase high performance liquid chromatography. *Anal. Chem.* **52**, 1206.

Gilon, C., R. Leshem and E. Grushka (1981) Structure resolution relationship: the effect of the alkylamide side chain of aspartyl derivatives on the resolution of amino acid enantiomers. *J. Chromatogr.* **203**, 365.

Grushka, E., R. Leshem and C. Gilon (1983) Retention behaviour of amino acid enantiomers in reversed phase liquid chromatography. *J. Chromatogr.* **255**, 41.

Hare, P.E., and E. Gil-Av (1979) Separation of D and L amino acids by liquid chromatography: use of chiral eluents. *Science* **204**, 1226.

Kurganov, A.A., A.B. Tevlin and V.A. Davankov (1983) High performance ligand exchange chromatography of enantiomers. *J. Chromatogr.* **261**, 223.

Kurganov, A.A., C. Facklan and V.A. Davankov (1987) Enantiomeric resolution of cyclopentadienyl-rhodium–norbornadiene complexes using outer sphere ligand exchange HPLC. Abstracts *11th International Symposium of Column Liquid Chromatography*, Amsterdam Th-P-32, 51.

Lam, S. and F. Chow (1980) Resolution of D, L-dansyl amino acids by HPLC with a Cu(II)-L-proline eluent. *J. Liq. Chromatogr.* **3**, 1579.

Lam, S., F. Chow and A. Karmen (1980) Reversed phase high performance liquid chromatographic resolution of D- and L-dansyl amino acids by mixed chelate complexation. *J. Chromatogr.* **199**, 295.

Lam, S. and A. Karmen (1982) Resolution of optical isomers of dansyl amino acids by high performance liquid chromatography with L-histidine and its derivatives in the mobile phase. *J. Chromatogr.* **239**, 451.

Lam, S. and A. Karmen (1984) Stereoselective D- and L-amino acid analysis by high performance liquid chromatography. *J. Chromatogr.* **289**, 339.

Lam, S., H. Azumaya and A. Karmen (1984) High performance liquid chromatography of amino acids in urine and cerebrospinal fluid. *J. Chromatogr.* **302**, 21.

Lam, S. (1984) Stereoselective analysis of D- and L-dansyl amino acids as the mixed chelate copper(II) complexes by HPLC. *Chromatogr. Sci.* **22**, 416.

Lam, S. (1986) Resolution of D- and L-amino acids after precolumn derivatization with *o*-phthalaldehyde by mixed chelation with Cu(II)-L-proline. *J. Chromatogr.* **355**, 157.

LePage, J.N., W. Lindner, G. Davies, D.E. Seitz and B.L. Karger (1979) Resolution of optical isomers of dansyl amino acids by reversed phase liquid chromatography with optically active metal chelate additives. *Anal. Chem.*, **51**, 433.

Lindner, W., J.N. LePage, G. Davies, D.E. Seitz and B.L. Karger (1979) Reversed phase separation of optical isomers of dansyl amino acids and peptides using chiral metal chelate additives. *J. Chromatogr.* **185**, 323.

Marchelli, R., R. Virgili, E. Armani and A. Dossena (1986) Enantiomeric separation of D, L-dansyl amino acids by one- and two-dimensional thin layer chromatography. *J. Chromatogr.* **355**, 354.

Nimura, N., T. Suzuki, Y. Kasahara and T. Kinoshita (1981) Reversed phase liquid chromatographic resolution of amino acid enantiomers by mixed chelate complexation. *Anal. Chem.* **53**, 1380.

Nimura, N., A. Toyama, Y. Kasahara and T. Kinoshita (1982) Reversed phase liquid chromatographic resolution of underivatised D,L-amino acids using chiral eluents. *J. Chromatogr.* **23**, 671.

Rogozhin, S.V. and Davankov, V.A. (1970) *Dokl. Akad. Nauk SSR*, 192.

Roumeliotis, P., K.K. Unger, A.A. Kurganov, and V.A. Davankov (1983) High performance ligand exchange chromatography of α-amino acid enantiomers. *J. Chromatogr.* **255**, 51.

Saigo, K., Y. Yuki, H. Kimoto, T. Nishida and M. Hasegawa (1988) A novel chiral stationary phase for optical resolution of amino acids and their derivatives by ligand exchange of amino acids and their derivatives by ligand exchange high performance liquid chromatography. *Bull. Chem. Soc. Jpn* **61**, 322.

Shieh, C.H., B.L. Karger, L.R. Gelber and B. Feibusch (1987) Ligand exchange chromatography of amino alcohol enantiomers as Schiff bases. *J. Chromatogr.* **406**, 343.

Sigel, H. (1973) Structural aspects of mixed ligand complex formation in solution. In *Metal Ions in Biological Systems*, ed. H. Sigel, **2**, Marcel Dekker, New York.

Takeuchi, T., A. Homare and D. Ishii (1987) Enantiomeric separation of amino acids by micro high performance liquid chromatography on an L-proline-bonded stationary phase. *J. Chromatogr.* **407**, 151.

Tapuhi, Y., N. Miller and B.L. Karger (1981) Practical considerations in the chiral separation of dansyl amino acids by reversed phase liquid chromatography using metal chelate additives. *J. Chromatogr.* **205**, 325.

Van der Haar, J., J. Kip and J.C. Kraak (1988) Effect of the mobile phase composition and ligand structure on the separation of D- and L-dansyl amino acids, as mixed metal complexes, by reversed phase liquid chromatography (unpublished).

Weinstein, S. (1982) Resolution of D and L amino acids by HPLC using copper complexes of N, N-alkyl-α-amino acids as novel chiral additives. *Angew. Chem. Int. Ed. Eng.* **21**, 218.

Weinstein, S., M.H. Engel and P.E. Hare (1982) The enantiomeric analysis of a mixture of all common protein amino acids by high performance liquid chromatography using a new chiral mobile phase. *Anal. Biochem.* **121**, 370.

Weinstein, S. and S. Weiner (1984) Enantiomeric analysis of a mixture of the common protein amino acids as the dansyl derivatives. *J. Chromatogr.* **303**, 244.

Weinstein, S. and N. Grinberg (1984) Enantiomeric separation of dansyl-amino acids by reversed phase thin layer chromatography. *J. Chromatogr.* **303**, 251.

Yamauchi, O., T. Sakura, and A. Nakahara (1977) Stereoselectivity in mixed ligand copper(II) complexes with electrostatic ligand–ligand interactions. Application to optical resolution of α-amino acids with a charged side chain. *Bull. Chem. Soc. Jpn* **50**, 1776.

Yuki, Y., K. Saigo, H. Kimoto, K. Tachibana and M. Hasegawa (1987) Novel chiral stationary phases for optical resolution by ligand-exchange high performance liquid chromatography. *J. Chromatogr.* **400**, 65.

6 Synthetic multiple-interaction chiral bonded phases

T.D. DOYLE

6.1 Introduction

In the field of chiral liquid chromatography there is a large and important class of chiral stationary phases (CSPs), which may be termed synthetic multiple-interaction bonded phases. This class of CSP has arisen during the past ten years or so, nurtured by an eminently sensible idea: that the greater the number of specific, discrete, simultaneous interactions between chiral solute molecules and a chiral locus on the stationary phase, then the greater the likelihood of effective chiral discrimination, and thus of chromatographic resolution of enantiomeric solutes.

Starting from this premise, practitioners in this area have designed, constructed and bound to silica a variety of chiral organic molecules; these molecules, while relatively simple and well defined in structure, contain at least one each of three types of functional groups, each in proximity to the chiral centre:

(i) π-acidic or π-basic aromatic groups, capable of donor–acceptor interaction (charge–transfer complexation)
(ii) polar hydrogen-bond and/or dipole-stacking sites
(iii) bulky non-polar groups, providing steric repulsion, van der Waals interaction, and/or conformational control

CSPs bearing these three types of groups are highly effective in resolving enantiomeric compounds which contain similar but complementary groups. Further, chiral molecules lacking one or more of the required groups may often be derivatized with achiral reagents to introduce the necessary interaction sites.

This principle of complementary interaction has also led to an important corollary—reciprocity of CSPs and solutes. Chiral solutes which are resolved with high enantioselectivity are themselves excellent candidates for synthesis of new CSPs. These new phases in turn resolve solutes of the appropriate (complementary) type, often with improved selectivity compared to the original phase.

Thus, founded on rational design, success has bred success—this has been the history of developments in this area. At present, the literature describes nearly 100 structurally and configurationally distinct multiple-interaction CSPs; many are commercially available. This armoury of CSPs has been employed to resolve thousands of enantiomeric solutes, often with impressive selectivity and efficiency. Applications of these CSPs to the solution of synthetic, analytical and biomedical problems have begun to proliferate.

Additionally, one of the attractions of work in this area is its inherent rationality; there have been many fruitful investigations into details of the chromatographic process, designed to elucidate specific solute: CSP interaction mechanisms. An understanding of these factors is necessary for full utilization of this technique, especially for assignment of configuration based on elution order. However, these theoretical aspects have been well covered in a recent review Pirkle and Pochapsky (1987). Instead, the present chapter will concentrate on practical aspects of the selection (or preparation) and utilization of synthetic multiple-interaction chiral bonded phases.

6.2 Survey of multiple-interaction CSPs

Table 6.1 presents the structure, source and a brief description of 20 of the most representative multiple-interaction CSPs, taken from the much larger and continuously increasing number described in the literature. The selection and organization of CSPs for this table was made primarily in an attempt to show the great diversity which exists within the class. This variation arises from permutation among possible (i) aromatic donor:acceptor groups, (ii) hydrogen-bonding or dipolar groups, (iii) chiral environments, and (iv) modes of attachment to silica. However, in addition, commerical availability or ease of preparation was taken into consideration for inclusion of a CSP in the table, although, because of space limitations, not all such commercially available CSPs are included. The selection is also weighted in favour of those CSPs which have been most extensively studied, which have been shown to be most highly enantioselective, and/or which have demonstrated the greatest potential utility in practical applications.

6.2.1 *Historical development*

Although there are a number of antecedents in the field of gas-liquid chromatography, and although there were some related HPLC chiral stationary phases developed in the mid 1970s, notably the TAPA CSP of Gil-Av (Mikes *et al.*, 1976), the true impetus for the development of synthetic multiple-interaction bonded phases came, beginning in 1979, with the work of William H. Pirkle (Pirkle and House), his students and collaborators at the University of Illinois. Since then, the work of Pirkle's group has been the dominant influence in this area, not only for the variety of synthesized CSPs (nine are included in Table 6.1), but especially for their many and increasingly sophisticated studies of the underlying interaction mechanisms.

However, there have been other major contributors, notably from the group headed by Naobumi Oi, currently of the Sumika Chemical Analysis Service, Osaka, Japan; four of the CSPs listed in Table 6.1 are from this group. Many

Table 6.1 Chiral stationary phases

CSP	Structure	Comment	Reference
1	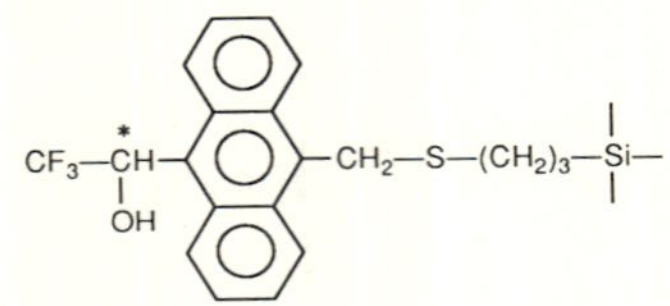	Trifluoroanthrylethanol CSP. 'First-generation' Pirkle CSP. Effective for wide variety of chiral solutes with π-acceptor groups.	Pirkle and House (1979)
π-Acceptor CSPs			
2	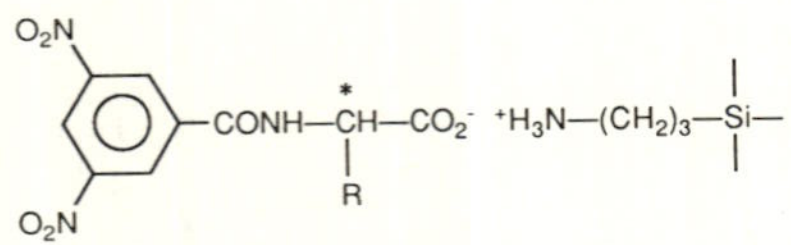 *2a*, R = Ph— *2b*, R = $(CH_3)_2CHCH_2$—	Ionic 3, 5-DNB-phenylglycine (*2a*) and leucine (*2b*) CSPs. 'Second-generation' Pirkle CSPs. Commercially available from various manufacturers.	Pirkle *et al.* (1981)
3	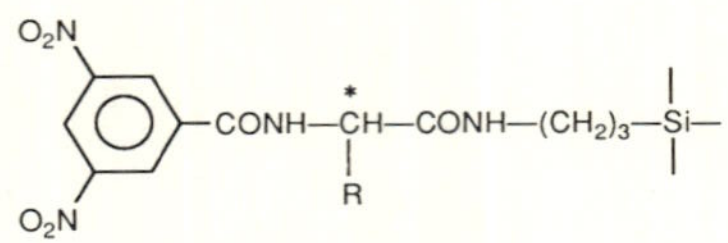 *3a*, R = Ph— *3b*, R = $(CH_3)_2CHCH_2$—	Covalent 3, 5-DNB-phenylglycine (*3a*) and leucine (*3b*) CSPs. Widely available commercially (3a in R, S, and RS configurations). Widely utilized.	Pirkle *et al.* (1980) Pirkle and Welch (1984)
4	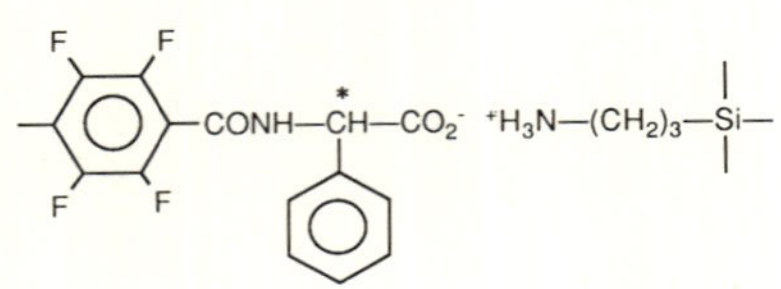	Ionic pentafluoro variant of CSP *2a*. Also prepared in mixture with CSP *2a* with interesting effects.	Kip *et al.* (1986)

π-Donor CSPs

5	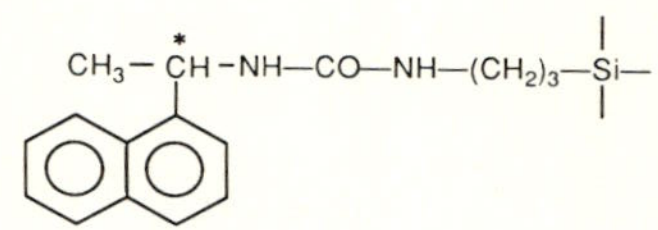	Naphthylethyl urea CSP. Readily prepared in-situ from commercially available, enantiomerically pure reagents.	Oi *et al.* (1983*a*)
6	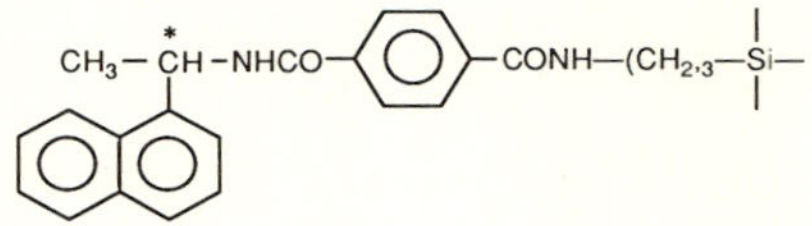	Amide-linked variant of CSP 5. Commercially available. Effective for resolution of chiral alcohols.	Oi *et al.* (1983*b*)
7	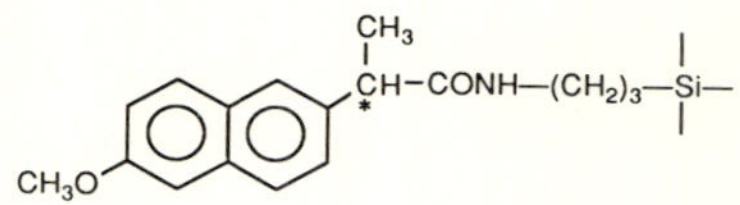	Readily prepared in-situ from enantiomerically pure pharmaceutical, naproxen. Effective for resolution of chiral acids.	Doyle (unpublished data)
8	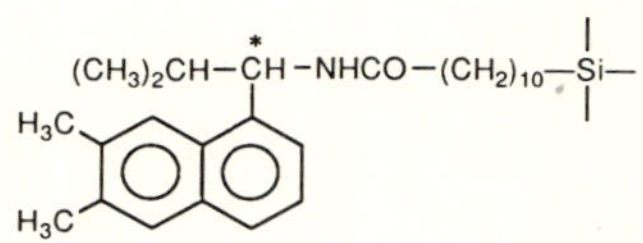	'Third-generation' Pirkle CSP, with structural features rationally chosen in attempt to maximize enantioselectivity.	Pirkle *et al.* (1986*a*)
9	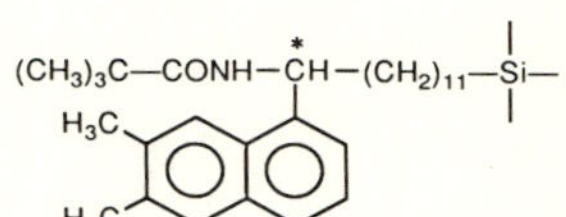	Pirkle's 'turned' CSP. Extremely high selectivities, but difficult to prepare.	Pirkle *et al.* (1984)
10	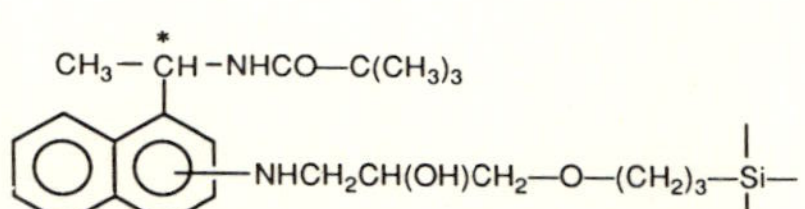	Unusual bonding mode to silica. Highly enantioselective for a variety of solutes. Commercially available.	Dappen *et al.* (1986)

Continued

Table 6.1 (*Contd.*)

CSP	Structure	Comment	Reference
11	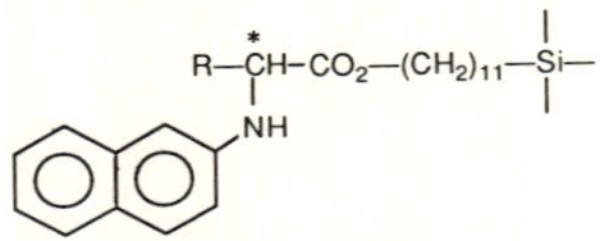 *11a*, R = CH_3— *11b*, R = $(CH_3)_2CH$—	N-naphthyl alanine (*11a*) and valine (*11b*) CSPs. Ester linkage to silica. Commercially available in R, S, and RS configurations.	Pirkle and Pochapsky (1986*a*)
CSPs with both π-acid and π-basic groups			
12	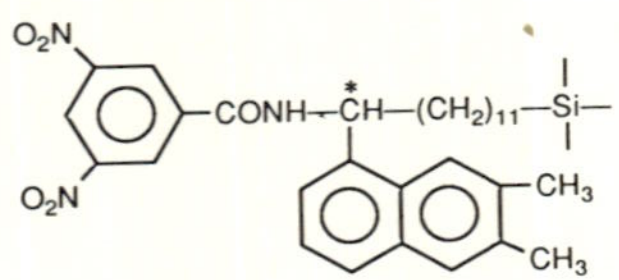	Has both π-acid and π-basic substituents; in limited studies was not dramatically superior to simpler types.	Hyun and Pirkle (1987)
13	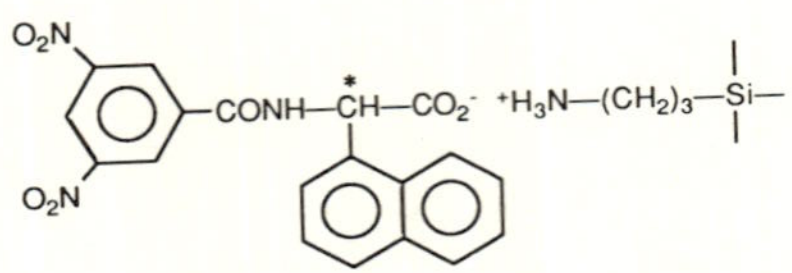	Ionic CSP with both π-acid and π-basic substituents. Superior to CSP *2a* for various solutes. Commercially available.	Oi *et al.* (unpublished data)
CSPs with multiple chiral centers			
14	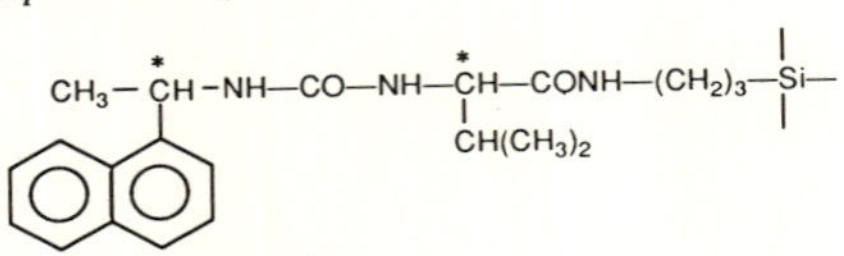 *14a* : S,S *14b* : S,R	Similar to naphthyl urea CSP *5*, but with addition of (S)-valine to system. S, S and S, R variations show different selectivities. Commercially available.	Oi and Kitahara (1986)

15	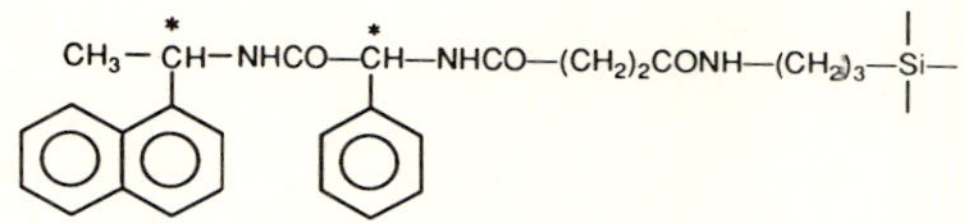 *15a*: R,S *15b*: R,R	Method of attachment to silica differs from CSP 14. Studied in R, S and R, R diastereomeric modifications.	Lloyd (1986)
16	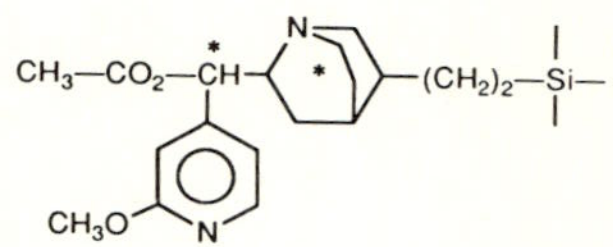	Modified quinine molecule bound to silica. Chiral bridged cyclic system. Moderately enantioselective for chiral acids without derivatization.	Pettersson and Gioeli (1987)
Miscellaneous CSPs			
17	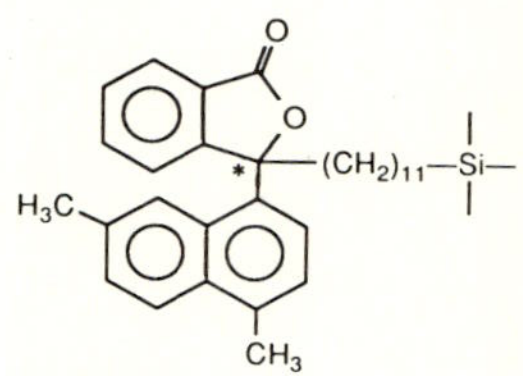	Phthalide-based CSP. Effective for resolution of chiral alcohols as ester derivatives.	Pirkle and Sowin (1987*b*)
18	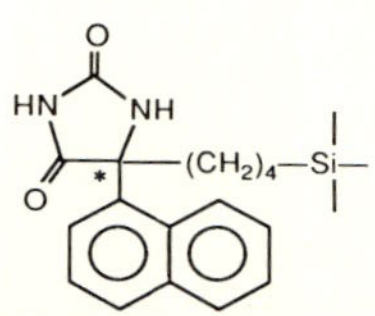	Hydantoin-based CSP. N-methyl analogues also synthesized. Highly enantioselective for a variety of solutes.	Pirkle and Hyun (1985)

Continued

Table 6.1 (*Contd.*)

CSP	Structure	Comment	Rèference
19	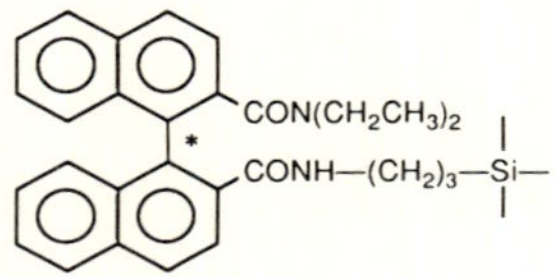	Simplest of several binaphthalene CSPs. Chirality dependent on hindered rotation. Resolves solutes similarly restricted.	Yamashita *et al.* (1987)
20	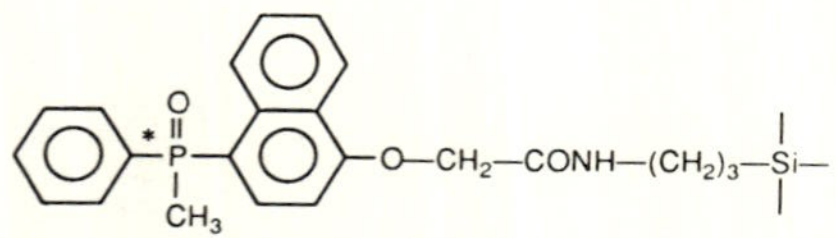	Phosphine oxide CSP. Resolves derivatized amino acids and amines.	Tambute *et al.* (1987)

other workers have also made significant contributions, not only to the design of new CSPs, but especially and increasingly in the area of utilization of these phases for the solution of practical problems in the analysis of chiral substances.

There is a closely related class of CSP which is outside the scope of this chapter, but which shares many of the features of multiple-interaction CSPs. These are the silica-bonded chiral diamide phases of Hara and co-workers (Hara and Dobashi, 1979). Although several of these phases contain phenyl groups, aromatic donor–acceptor interactions are probably of negligible or minor importance in these phases. Hydrogen-bond or dipole interaction processes are dominant and are sufficient to account for the observed enantioselectivity. Although these phases are highly effective in the resolution of chiral solutes, especially amino acids and their derivatives, they are not discussed further in this chapter. For the same reason, such phases as the *trans*-chrysanthemic acid CSP of Oi *et al.* (1983) are not discussed.

6.2.2 *Survey*

CSP 1 of Table 6.1, an anthryltrifluoromethanol phase, prepared by Pirkle and House (1979), is included chiefly for historical importance. This first-generation CSP resolved a variety of chiral solutes, especially those containing aromatic π-acid groups, which could interact effectively with the π-basic anthryl group. Among the solutes resolved, an impressive (for that time) separation factor of $\alpha = 1.78$ was observed for the n-butylamide of N-3, 5-dinitrobenzoylphenylglycine. This led, by the principle of reciprocity, to the design of second-generation CSPs, containing the 3, 5-dinitrobenzoyl (DNB) moiety. CSP 3a, a covalent DNB-phenylglycine phase, is the most direct exemplification of the group of π-acceptor CSPs in the table; differences within the group are minor, reflecting only the identity of the π-acid, or of the steric group attached to the chiral carbon, or of the mode (ionic or covalent) of bonding to silica. Because of the widespread commercial availability of CSPs 2 and 3, and because they have been available for some time, there is a considerable literature on their use. However, they are not necessarily the most effective, among those currently available.

6.2.2.1 *π-Donor CSPs.* In the third-generation CSPs of Pirkle, as well as in many of those developed by Oi and by others, the strongly electron donating (π-basic) naphthyl group is a key structural feature. However, the molecular architecture of this sub-group of π-donor CSPs is both distinct from, and considerably more varied than, the complementary π-acceptor group. Additionally, there is apparently no precise analogue in which a naphthoyl (rather than dinitrobenzoyl) group is attached to the N-terminus of an amino acid. Instead, in most cases (CSPs 5 to 10) a naphthyl substituent is directly bonded to the chiral carbon; in one example (CSP 11), it is attached to nitrogen. Attachment of naphthyl through both α- and β-positions has been utilized. In several cases, the π-basicity of the aromatic system has been enhanced by introduction of methyl (CSPs 8 and 9), methoxy (CSP 7) or amino (CSPs 10 and 11) substituents.

Although all the examples within this group involve covalent bonding to silica, the precise manner of attachment is diverse. Urea (CSP 5), ester (CSP 11), alkyl (CSP 9) and amide linkage (CSPs 7, 8 and 10) have been utilized. The orientation of the amide linkage has also been varied: compare CSP 7 with CSPs 8 and 9. A more drastic change of orientation is a feature of CSP 9 (Pirkle's 'turned CSP') and CSP 10; these are the only CSPs in the group where dipolar amide interaction sites are directed away from the strand leading to the silica support.

Another important feature incorporated within this sub-group is a lengthening of the alkyl chain leading to silica (CSPs 8, 9 and 11). Aminopropyl silanized silica is readily available in bulk and as pre-packed columns; it has therefore been most widely used. However, the CSPs with longer alkyl chains exhibit differing (sometimes improved) enantioselectivity towards various solutes.

6.2.2.2 *CSPs with both π-acid and π-basic groups.* CSPs 12 and 13 represent two different approaches to the incorporation of both π-acid and π-basic aromatic groups within the same molecule. Covalent CSP 12 is structurally analogous to CSP 9, while ionic CSP 13 is related to CSP 2.

Several CSPs incorporate two or more chiral centres within the molecule. CSPs 14 and 15 have each been studied in both diastereomeric modifications, whereas CSP 16 is a relatively complex molecule with multiple chiral centres which arise from the incorporation of the naturally occurring quinine molecular skeleton in the structure.

Two examples from Pirkle's group (CSP17 and 18) incorporate fused or cyclic dipolar ring systems. CSP 19 depends for its chirality on the restricted rotation of a binaphthalene system. Finally, CSP 20 is based on a chiral phosphorous atom.

6.2.3 *Preparation of CSPs*

The evolution of multiple-interaction CSPs over the past ten years has been shaped by an interplay between concept and feasibility. On the one hand there has been the goal of maximizing enantioselectivity by synthesis of the most conceptually promising chiral interaction molecules, as guided both by theory and by the accumulation of empirical results. This has led, for example, to CSP 9, which has produced some remarkable resolutions but which is difficult to prepare and is therefore, largely for economic reasons, not commercially available.

On the other hand there are a number of CSPs, perhaps of lower overall efficacy, which can be simply prepared, even with no chemical manipulation other than bonding to silica (CSPs 5 and 7, among others), or which can be synthesized with sufficient facility and economy so as to be commercially marketable (CSPs 2, 3, 10, 11, 13 and 14).

It is beyond the scope of this chapter to describe the synthesis of the more elaborate chiral molecules which have formed the basis of some of these CSPs; the details are generally well documented in the original literature. However, two general principles should be mentioned:

(i) Natural and synthetic amino acids, readily available in high enantiomeric purity, are convenient starting points for many CSPs. Although protection, modification and deprotection of these starting materials will normally be necessary, the literature of peptide synthesis abounds with model procedures which can be conducted efficiently, and, usually, without racemization.

(ii) Starting materials which are racemic (or which are racemized during modification) need not be excluded from consideration, because at an appropriate stage the intermediate product can be resolved by preparative scale HPLC using one of the many available CSPs, chosen for complementarity of aromatic interaction. For example, this approach has been used by Pirkle for preparation of CSPs 8 (Pirkle *et al.*, 1986) and 11 (Pirkle and Pochapsky, 1986).

6.2.3.1 *Attachment of the chiral component to silica.* Although it is beyond the scope of the survey to describe some of the more elaborate synthetic procedures for bonding the active chiral moiety to the silica support, there are several approaches of great simplicity and efficiency, that require no special synthetic expertise or equipment. These procedures are based on the ubiquitous commercial availability of n-aminopropylsilanized silica, both in bulk form and as pre-packed columns. Two examples are outlined:

(i) Chiral molecules with a terminal carboxylic acid group ($R^*\text{-}CO_2H$) may be activated in a variety of ways, most commonly with N-ethoxycarbonyl-2-ethoxy-1,2-dihydroquinoline (EEDQ), then bound through an amide

$$R^*\text{—}CO_2H + H_2N\text{—}(CH_2)_3\text{—}\overset{|}{\underset{|}{Si}}\text{—}$$

$$\xrightarrow{\text{EEDQ}} R^*\text{—}CONH\text{—}(CH_2)_3\text{—}\overset{|}{\underset{|}{Si}}\text{—}$$

linkage to the aminopropylsilanized silica. The reaction is normally conducted at room temperature, with reagents and silica dissolved or slurried in an inert solvent (Pirkle *et al.*, 1980; Yamashita *et al.*, 1987)

(ii) Similarly, chiral precursors with terminal amino groups may be converted to isocyanates (Pirkle and Hyun, 1985) or obtained as such from commercial sources (Doyle, unpublished data), then bound by way of formation of a urea linkage in a similarly gentle and non-racemizing reaction:

$$R^*\text{—NCO} + H_2N\text{—}(CH_2)_3\text{—}\overset{|}{\underset{|}{Si}}\text{—}\rightarrow R^*\text{—NH—CO—NH—}(CH_2)_3\text{—}\overset{|}{\underset{|}{Si}}\text{—}$$

The resulting covalently bonded chiral stationary phases may then be packed using conventional slurry-packing techniques and equipment. The majority of columns reported in the literature have been prepared by this technique, usually citing 'standard' packing procedures, although details have also been given (Lloyd, 1986).

6.2.3.2 *In-situ procedures.* A still more convenient method of CSP preparation, requiring no special equipment and therefore accessible to any analytical laboratory, consists of in-situ bonding of the chiral molecule to the N-terminus of pre-packed aminopropylsilanized silica columns. Both amide and urea linkages may be formed, utilizing the reactions described above. The chiral precursor and activator (if any), dissolved in methylene chloride or other inert solvent, is circulated for (typically) one hour through the column, at normal flow rates and with the detector disengaged. After brief washing, the columns are completely stable and ready for use.

Several examples of this exceedingly simple technique have been described (Pirkle and Welch, 1984; Pirkle and Hyun, 1985). Although it has been suggested that columns prepared in situ may exhibit slightly inferior chromatographic properties (Pirkle and Hyun, 1985), experience in the author's laboratory for the preparation of CSPs 5 and 7 (Doyle, unpublished data) indicate that columns of high selectivity, efficiency and durability are obtained, simply and economically. The chromatograms of Figure 6.1, to be discussed later, were obtained on a column prepared in situ.

Several phases (CSPs 2, 4 and 13) are ionically bound to the aminopropylsilanized support. These are also prepared by in-situ methods (Pirkle *et al.*, 1981; Kip *et al.*, 1986); however, the resulting CSP is typically less stable than those containing covalent attachments, and must be used with relatively non-polar mobile phases.

6.2.4 *General chromatographic considerations*

6.2.4.1 *Mobile phase selection.* Multiple-interaction bonded phases are most commonly utilized in normal phase mode. Most of the extensive chromatographic data reported by Pirkle and his co-workers, for their various CSPs, were obtained with mobile phases of isopropanol (0.2–20%) in hexane, and several studies have shown that isopropanol is often, if not always, the optimum active (i.e. polar) component of the mobile phase, if enantioselectivity (as measured by the separation factor α) is the primary criterion. However, tetrahydrofuran has been shown to be sometimes superior (Dappen *et al.*,

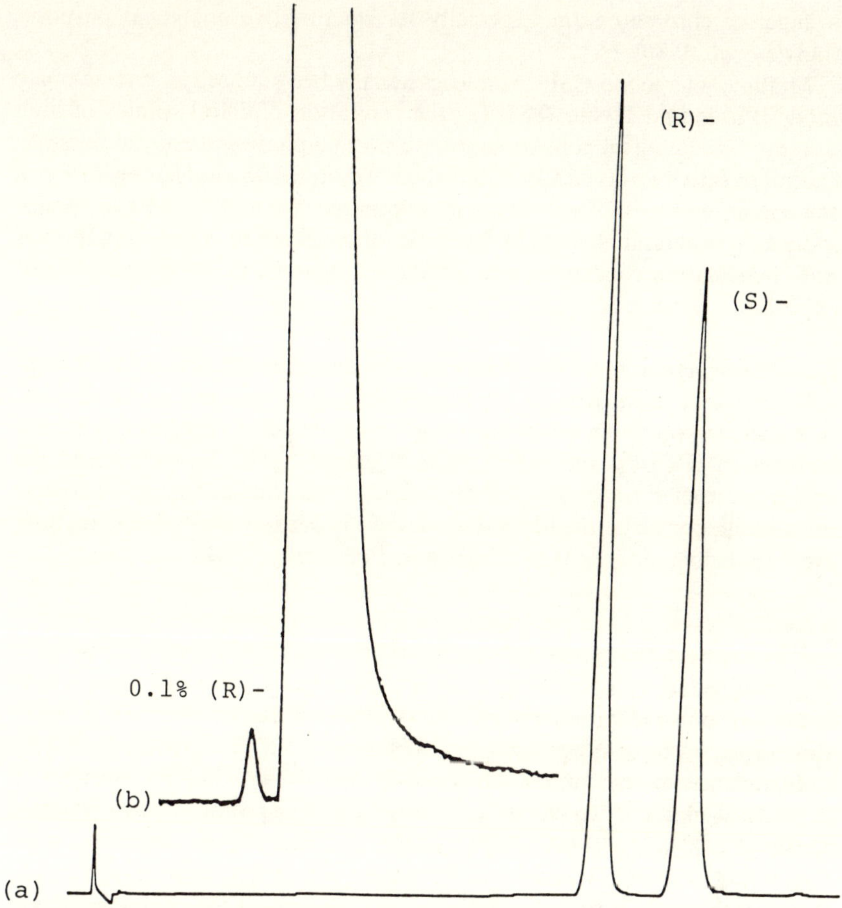

Figure 6.1 Resolution of amphetamine (Table 6.2, solute 7), as the 3,5-DNB amide derivative, on an (*S*)-1-(α-naphthyl)ethyl urea stationary phase (Table 6.1, CSP 5); (*a*) 50:50 *R*, *S* mixture; mobile phase = hexane: isopropanol:acetonitrile (97:1:2); $k'_1 = 22.0$, $k'_2 = 26.2$, $\alpha = 1.19$, $R_s = 4.08$, $N = 9800$ plates $25\,cm^{-1}$; (*b*) portion of chromatogram of 0.1:99.9 *R*, *S* mixture; mobile phase = hexane:isopropanol:acetonitrile (94:4:2), observed α (at this composition and mobile phase) = 1.12.

1986), and a variety of other hydrogen bond acceptors have been shown to be reasonable alternatives (Zief *et al.*, 1984). Much of the work of Oi has been with mixtures of hexane:dichloroethane:ethanol.

However, if the chromatographer's objective is to maximize overall resolution, then addition of small amounts of acetonitrile to isopropanol:hexane mixtures will often dramatically improve the observed R_s value. This is accomplished by improvement of peak symmetry (elimination of tailing) and by an increase in the observed number of theoretical plates (peak sharpness). Typically, the separation factor α will decrease, but the net result is a superior chromatogram, especially for quantitative analytical purposes (Doyle *et al.*, 1986).

Multiple-interaction CSPs have occasionally been utilized in reverse-phase mode (Pirkle and Hyun, 1985) for the resolution of chiral solutes of high polarity. Where direct comparisons with normal phase systems are possible, enantioselectivity is typically diminished. Additionally, the elution order of the enantiomers is often reversed, compared to normal phase results, indicating profound changes in the retention mechanism. At present, there is little information concerning the stability of these CSPs during continuous exposure to reverse-phase conditions.

6.2.4.2 *Temperature.* Enantioselectivity on multiple-interaction CSPs typically increases with decrease in temperature; however, below about 20°C (generally the optimum temperature), efficiency begins to degrade because of changes in viscosity and other physical effects. Pirkle has attributed the temperature effect to an expected dependence of the chromatographic process on the entropy factor, explicitly and inherently because multiple interactions are responsible for retention (Pirkle and Pochapsky, 1987).

6.2.4.3 *Efficiency.* Multiple-interaction CSPs, in common with other bonded silica stationary phases, exhibit characteristically high efficiency. This is one of the principal advantages of this class of chiral phases. Although there is little systematic documentation in the literature, these phases are clearly superior in this respect to most other types of CSPs.

Experience in the author's laboratory has shown that the number of theoretical plates N which can normally be realized varies over a range of 12 000–40 000 m^{-1} (3000–10 000 plates for the 25 cm columns most commonly available and utilized). Variations within this range arise not only from factors intrinsic to each column (such as nature of CSP, characteristics of the silica support), but also from the choice of mobile phase and from the number and kind of polar functional groups present in the enantiomeric solutes.

Given these high efficiencies and taking the criterion for baseline resolution as a resolution factor R_s of 1.5, a minimum separation factor α of 1.10 is a reasonable target for quantitative analytical-scale resolution of enantiomers on most of the available multiple-interaction CSPs. For trace analysis (1% or less of one enantiomer), greater selectivity is required, but an α of 1.20 will normally suffice. Such selectivities can in practice be achieved for a great number of chiral solutes, as will be seen in section 6.3.

Figure 6.1, an example from the author's laboratory, shows a previously unpublished resolution of amphetamine (as the 3, 5-dinitrobenzoyl amide) on a 1-(α-naphthyl)ethyl urea chiral stationary phase, CSP 5 (Doyle, unpublished data). Baseline resolutions of both the racemic (50:50) mixture and of trace (1 in 1000) enantiomeric contamination was readily achieved. Although the selectivity in this example ($\alpha = 1.19$) is not especially impressive, compared with others in section 6.3, the chromatogram is typical of the efficiency (9800 plates/25 cm) and peak symmetry which can be achieved by multiple-interaction bonded chiral stationary phases. The figure illustrates the quality of analytical results that can be routinely obtained.

6.3. Resolutions of enantiomeric solutes

6.3.1 *Survey*

Table 6.2 presents an overview of the variety of chiral solutes which can be resolved on multiple-interaction CSPs. The entries represent but a small fraction of the vast number of resolutions which have been reported in the past ten years, but an attempt was made to give representative examples of the principal structural types, the most appropriate CSPs for each and, in several instances, detailed comparisons of the selectivities which can be achieved for a given enantiomeric solute on various CSPs.

Most of the examples were chosen for simplicity of structure; in general, more complex analogues are also resolvable, often with enantioselectivity superior to that of the simpler types. The references in the table are the primary sources for locating such further examples.

As an illustration, solute 6, *sec*-butylamine, is structurally the simplest chiral aliphatic amine. It has been resolved by at least nine of the CSPs under discussion in this chapter and by a variety of strategies; this is an excellent demonstration of the power of the multiple-interaction approach. However, the higher aliphatic homologues of this amine will usually exhibit superior enantioselectivities (barring cross-over effects (Pirkle *et al.*, 1984) as will structures with branching, attached aromatic and alicyclic groups, and other introduced functionalities.

However, selectivity does not increase without limit. A levelling off to a maximum α value with increasing molecular weight can be expected. Additionally, incorporation of additional aromatic or polar functionalities in the molecule may diminish resolution, by contributing alternative, competitive, perhaps non-chiral interactions to the overall solute: CSP interaction process.

Many of the examples in the table are from the area of pharmaceutical chemistry, where application of the techniques of chiral chromatography has been fruitful.

6.3.2 *Derivatization of solutes*

Although many of the solutes listed in Table 6.2 are resolvable directly, most require preliminary derivatization; this subject is therefore an unavoidable and integral part of any discussion of multiple-interaction CSPs. Three objectives are attained by derivatization; the first two are often essential for enantiomeric resolution; the third is non-essential but often desirable:

Table 6.2 Resolutions of enantiomeric solutes

No.	Type and structure	Derivative	CSP	α	Reference
Alcohols					
1	$CH_3(CH_2)_7$—CH(CH_3)—OH	DNAn	6	1.10	Oi and Kitahara (1983)
2	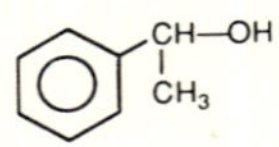	(direct)	13	1.06	Oi *et al.* (1988)
		DNAn	11a	1.15	Pirkle *et al.* (1986*a*)
		DNB	10	1.10	Dappen *et al.* (1986)
		DNB	17	1.19	Pirkle and Sowin (1987)
3	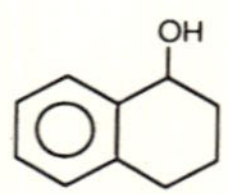	DNAn	11a	1.46	Pirkle *et al.* (1986*a*)
4	CH_3—CH(OH)—CH_2OH	bis-DNAn	11a	1.19	Pirkle *et al.* (1987)
		bis-DNAn	9	1.35	Pirkle *et al.* (1987)
Thiols					
5	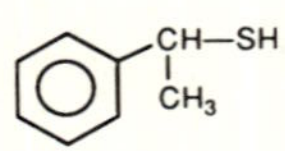	DNAn	11a	1.20	Pirkle *et al.* (1986*a*)
Amines, primary					
6	CH_3—CH_2CH(CH_3)—NH_2	Amide, α-naphthoyl	3b	1.03	Pirkle and Welch (1984)
		Amide, DNB	6	1.04	Oi *et al.* (1983*b*)
		Amide, DNB	5	1.10	Pirkle and Hyun (1985*a*)
		Amide, DNB	11b	1.19	Pirkle *et al.* (1986*a*)
		Amide, DNB	10	1.23	Dappen *et al.* (1986)
		Amide, DNB	8	1.26	Pirkle *et al.* (1984)
		Amide, DNB	9	1.43	Pirkle *et al.* (1984)
		Urea, DNAn	11a	1.15	Pirkle *et al.* (1986*a*)
		Carbamate, β-naphthoyl	3a	1.06	Doyle *et al.* (1986)

7	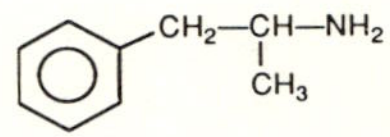	Amide, β-naphthyl	3a	1.09	Wainer and Doyle (1983)
		Carbamate, β-naphthyl	3a	1.13	Pirkle and Hyun (1985*a*)
		Amide, DNB	9	1.39	Pirkle and Hyun (1984)
		Amide, DNB	5	1.19	Doyle (unpublished data)
		Urea, DNAn	5	1.39	Doyle (unpublished data)
8	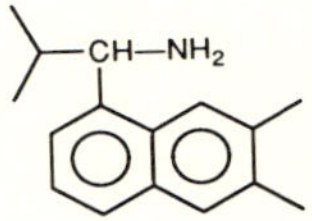	Amide, acetyl	3a	2.20	Pirkle *et al.* (1983)
9	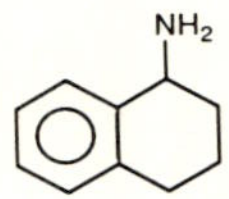	Urea, DNAn	11	2.58	Pirkle *et al.* (1986*a*)
Amines, secondary					
10	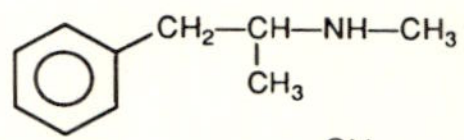	Carbamate, β-naphthoyl	3a	1.14	Doyle *et al.* (1986)
Amines, tertiary					
11	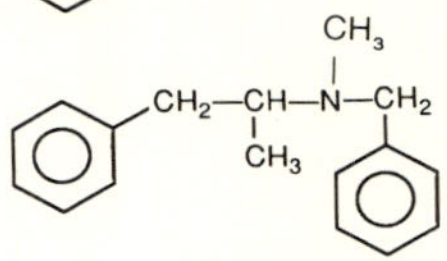	Carbamate, β-naphthoyl (by cleavage)	3a	1.14	Doyle *et al.* (1986)
Aminoalcohols					
12	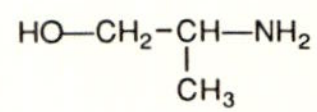	Urea, DNAn	11b	4.09	Pirkle *et al.* (1986*a*)
13	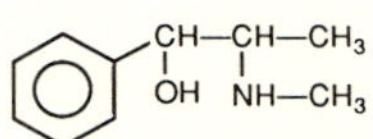	Oxazolidine	2a	1.36	Wainer *et al.* (1986)
14	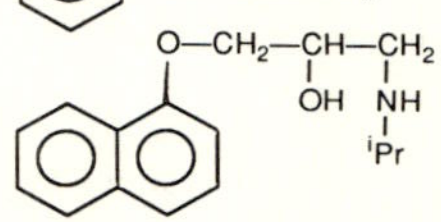	Oxazolidone	3a	1.09	Wainer *et al.* (1984)
		Amide, DNB	10	1.09	Dappen *et al.* (1986)

Continued

Table 6.2 (*Contd.*)

No.	Type and structure	Derivative		CSP	α	Reference
Amino acids						
15	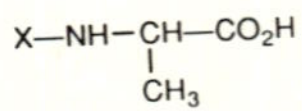	X = DNB (reverse phase)		8	1.28	Pirkle and Hyun (1985)
16	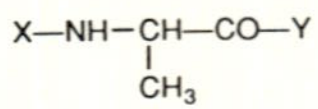	X = DNB, Y = —OCH_3		20	1.18	Tambute *et al.* (1987)
				15b	1.26	Lloyd (1986)
				18	1.38	Pirkle and Hyun (1985*b*)
				6	1.43	Oi and Kitahara (1983)
				15a	1.48	Lloyd (1986)
				12	1.57	Hyun and Pirkle (1987)
				14b	1.73	Oi and Kitahara (1986)
				14a	1.86	Oi and Kitahara (1986)
				9	3.34	Pirkle and Hyun (1984)
17	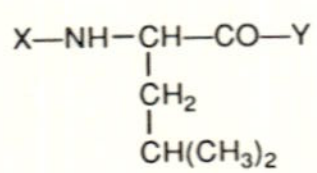	X = DNB, Y = —NH—$(CH_2)_3CH_3$		8	1.57	Pirkle and Hyun (1985*a*)
				5	1.74	Pirkle and Hyun (1985*b*)
				11b	17.66	Pirkle *et al.* (1986*a*)
18	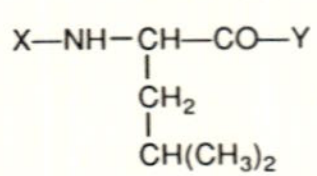	X = DNB, Y = 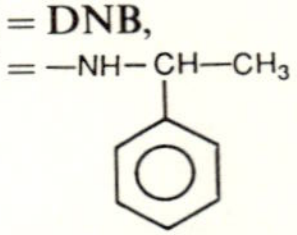	RR, SS	11b	11.86	Pirkle *et al.* (1986*a*)
			RS, SR	11b	18.00	Pirkle *et al.* (1986*a*)
19	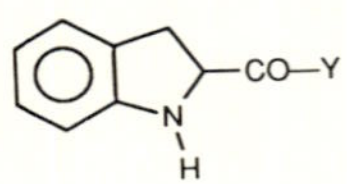	Y = —O—CH_2CH_3		3a	1.50	Pirkle *et al.* (1985)
Cyclic imidides						
20	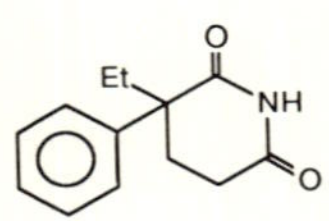	(direct)		26	1.12	Yang *et al.* (1985)

Benzodiazepenones					
21	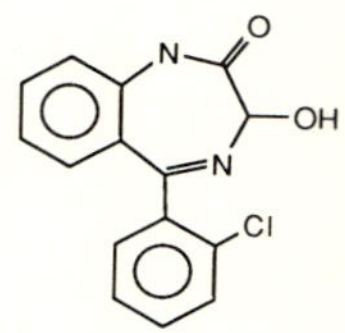	(direct)	3a	1.21	Pirkle and Tsipouras (1984)
Carboxylic acids					
22	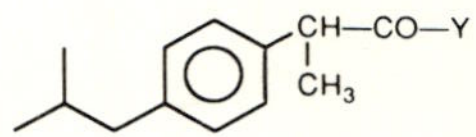	Y = —NH—CH_2—(α-naph.)	3a	1.12	Wainer and Doyle (1984)
		Y = —NH—(α-naphthyl)	3a	1.23	Nicholl-Griffith (1987)
		Y = DNAn	11a	1.33	Pirkle *et al.* (1986*a*)
		Y = DNAn	7	1.45	Doyle (unpublished data)
		Y = DNAn	5	3.81	Doyle (unpublished data)
23	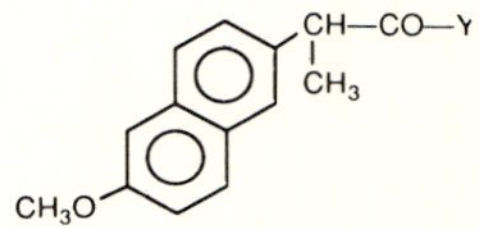	Y = OH (direct)	16	1.04	Pettersson and Gioeli (1987)
		Y = —NH—CH_2—(α-naph.)	3a	1.23	Wainer and Doyle (1984)
		Y = DNAn	7	1.26	Doyle (unpublished data)
Phthalides					
24	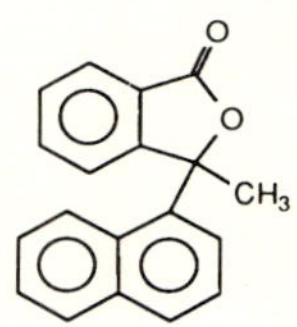	(direct)	3a	1.57	Pirkle and Sowin (1987*a*)
Sulphoxides					
25	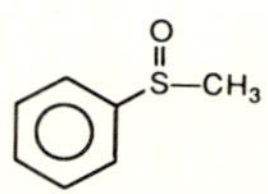	(direct)	2a	1.05	Pirkle *et al.* (1981)

Continued

Table 6.2 (*Contd.*)

No.	Type and structure	Derivative	CSP	α	Reference
Phosphine oxides					
26	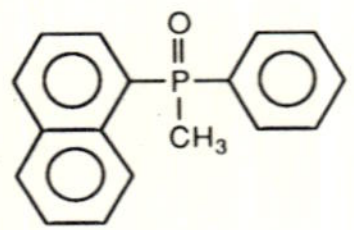	(direct)	3a	1.30	Tambute *et al.* (1986)
Binaphthols					
27	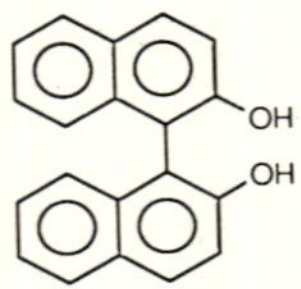	(direct)	19	1.07	Yamashita *et al.* (1987)
		(direct)	4	1.10	Kip *et al.* (1986)
		(direct)	2a	1.33	Oi *et al.* (1983*c*)

(i) attenuation (but not cancellation) of the polarity of functional groups of the solute, so that individual polar interactions are not of a magnitude to overwhelm the influence of other interactions
(ii) introduction of an aromatic functionality with π-basicity or acidity, complementary to the aromatic function in the CSP
(iii) enhancement of the ultraviolet/visible or fluorescence detectability of the solutes.

6.3.2.1 *Amines* are conveniently converted to amides with α- or β- naphthoyl chloride (for π-acceptor CSPs) or with 3,5-dinitrobenzoyl chloride (for π-donor CSPs). Other acid chlorides, possessing less strongly interacting donor/acceptor groups, may sometimes be effective, depending on the overall importance of the aromatic interaction in the chiral recognition mechanism.

Amines may also be derivatized with aryl chloroformates (Doyle *et al.*, 1986) to afford carbamates. The resolutions of secondary amines are often superior following this reaction than other derivatization products. Additionally, tertiary amines may often be cleaved to afford resolvable carbamates—virtually the only strategy for resolution of tertiary amines on these CSPs (Doyle, unpublished data).

Amines and alcohols react readily with isocyanates, to form ureas and carbamates respectively. Many isocyanate reagents are stable and commercially available; a convenient analytical scale preparation of 3,5-dinitrophenylisocyanate has been described by Pirkle *et al.* (1986).

6.3.2.2 *Aminoalcohols* undergo the reactions described above, to produce either mono- or bis-derivatives, or alternatively, can be cyclized to produce oxazolidones or oxazolidines (Wainer *et al.*, 1984, 1986).

Chiral alcohols are resolved well as esters, provided an appropriate CSP is selected. However, resolution of chiral acids as esters has seldom been reported; the author's experience is that this approach is often ineffective. Chiral acids can usually be resolved with high selectivity as amides, prepared by the action of the acid chloride (or otherwise activated carboxyl) on an aryl- or alkylamine.

Most of the above reactions are straightforward, well established, and experimentally convenient; details can usually be found in the individual references (Table 6.2). However, certain important considerations are relevant to chiral analysis and apply to all of these reactions.

In the simplest and most common cases, the substance of analytical interest is enantiomeric (one chiral centre) and is treated with an achiral reagent. The product is therefore also enantiomeric, rather than diastereomeric. There are numerous desirable consequences of this:

(i) chiral purity of the reagent is not a factor
(ii) reaction rates are identical for both enantiomers

(iii) spectrophotometric properties are identical for both
(iv) the original enantiomeric composition is retained provided the product is not recrystallized—an important procedural point
(v) quantitative reaction is not a prerequisite.

All these consequences greatly increase the range of reaction type and increase the number of reagents that can be utilized effectively for chiral analysis.

There are special cases where either the reagent is deliberately chosen to be chiral, or the analyte is initially diastereomeric. The results of chiral chromatography of such diastereomeric solutes also provide valuable analytical information and are of increasing theoretical importance in studies of interaction mechanisms (Pirkle *et al.*, 1987a).

In certain cases, an achiral reagent may have a pro-chiral locus which generates a new chiral centre in the product molecule, in unequal diastereomeric proportion. Of the reactions discussed in this chapter, only oxazolidine formation (Wainer *et al.*, 1986) normally presents this potential complication.

6.3.2.3 *Racemization.* The question of racemization of the chiral analyte during derivatization also needs careful consideration. Up to the present, there has been little consideration of this possibility, at least in the literature directly associated with chiral chromatography, although the classical reactions involved have been thoroughly studied in other contexts. However, remarkably, until the advent of chiral CSPs, there were few, if any, analytical techniques that could give unambiguous evidence of chiral purity at the trace level (1% or less). In the author's view, many aspects of this subject should be re-examined.

In the interim, the present best evidence is that most of the reactions suggested herein for amines proceed with no detectable racemization (i.e. $< 0.1\%$). Data for alcohols is scant, but here also no racemization has been detected. Reactions of carboxylic acids and, especially, amino acids are more suspect; low-level racemization has been detected for certain methods of carboxyl activation (Pirkle *et al.*, 1987*b*). Studies to evaluate and to minimize this are required. Fortunately, multiple-interaction CSPs are now available as the ideal analytical tools for such investigations.

6.3.3 *Limitations*

Multiple-interaction CSPs have perhaps a wider demonstrated applicability than virtually any other class of CSP, but there are nevertheless some limitations and/or restrictions. Chief among these are:

(i) the frequent and, in some respects, conceptually inherent need for derivatization
(ii) the requirement for relatively lipophilic solutes, and for normal-phase elution, although this requirement is not absolute

(iii) the requirement that at least one of the interacting groups be proximate (usually, adjacent) to the chiral centre. Resolutions in which the chiral centre is more remote are exceptional.

6.4 Special applications

6.4.1 *Preparative separations*

Semi-preparative and preparative scale versions of several of the more popular CSPs are now commercially available. To date, the most important application has been in the chiral purification of precursors for the synthesis of other CSPs (Pirkle *et al.*, 1986).

The requirement of derivatization for resolution of many of the chiral solutes constitutes a serious limitation to the utility of preparative work, especially as some of the more common derivatives, e.g. amides, are difficult to cleave. However, other derivatives (e.g. esters of alcohols, carbamates) are more practical candidates, as are the many solutes which can be directly resolvable. As should be apparent, multiple-interaction CSPs have more-than-adequate efficiency and selectivity for preparative work; further, the use of normal phase systems facilitates solute recovery by vacuum stripping of mobile phases.

6.4.2 *Enantiomeric trace analysis*

Many multiple-interaction CSPs are available as either (R)- or (S)-modifications. Accordingly, for the detection of trace enantiomeric contamination of chiral substances, the analyst can readily select the CSP in which the trace component elutes first, avoiding overlap problems from tailing of the peak of the preponderant component. The chromatogram in Figure 6.1 is an excellent illustration of this approach. Support for the identity of the trace peak can be achieved by obtaining a comparison chromatogram on the CSP with opposite configuration (resulting in reversal of elution order) or by use of a racemic CSP (a few of these are commercially available; enantiodiscrimination is lost on these CSPs). Although this approach has been studied in some detail (Perry *et al.*, 1987), absolute identification of trace peaks, such as by HPLC/MS, is preferable, and has been accomplished (Lee *et al.*, 1986).

6.4.3 *Elution order and configuration*

The most interesting and controversial aspect of resolutions on multiple-interaction CSPs concerns the consistency and predictability of enantiomeric elution order, and the uses to which this regularity may be put. Extensive compilations of chromatographic data, especially by Pirkle and his co-workers, have firmly established the empirical validity of this regularity.

Where exceptions occur, these are readily explicable in terms of:

(i) gross structural differences such as introduction of strongly polar

functional groups (possibly, but not necessarily involving change in the Cahn–Ingold–Preleg priorities for stereochemical nomenclature (Wainer and Alembik, 1986))
(ii) change from a π-acidic to π-basic aryl substituent or vice versa, with consequent fundamental change of orientation of solute to CSP (Doyle and Wainer 1984; Wainer *et al.*, 1986)
(iii) gradual progression in the relative contributions of competing interaction mechanisms within a homologous series, resulting in cross-over of elution order at a point characteristic of the particular combination of solute series, derivative, and CSP (Pirkle *et al.*, 1984).

Because the causes of these exceptions are understood and can, with proper caution, be eliminated or at least accounted for in experimental work, the present status of the use of elution order as a probe for absolute configuration is that:

> Given unambiguous data for the elution order of a sufficient number of model compounds of known configuration, this order serves as a reliable criterion for *probable* assignment of absolute configuration for other, closely related molecules.

The criteria for 'unambiguous data', 'sufficient number' and 'closely related' are flexible, depending on, among other factors, the simplicity or complexity of the structural variation in the series, the magnitude of the separation factors α, the variation in α between members of the series, the likelihood of and evidence for cross-over phenomena, the position of the unknown within the series, and the extent to which the interaction mechanism for the given class of solutes with the specific CSP is understood.

In other words, configurational assignments based on elution orders are still very much dependent on experience and judgement; notwithstanding this, such assignments have been made and will continue to be made (Pirkle and Tsipouras, 1984).

The situation is, however, clouded by remaining uncertainties in the details of specific solute–CSP interaction mechanisms. Although many highly detailed interaction mechanisms have been proposed, and although there is an encouraging trend towards the generation of independent spectrophotometric evidence for such interaction models (Pirkle and Pochapsky, 1987*b*) many of the currently proffered mechanisms have an unfortunate flavour of *post-hoc* rationalization. Subtle intramolecular conformational and intermolecular orientational preferences are invoked with little evidence for their validity, other than that suggested by the observed elution orders.

Regardless, the empirical evidence for regularity and uniformity is impressive, is demonstrably of great practical utility, and amply justifies the continued vigorous investigation of underlying mechanisms.

6.5 Conclusion

Developments in HPLC chiral separations in the 1980s have resulted in a dramatic improvement in chemists' ability to resolve enantiomers and, equally, in their perception of what represents an acceptable resolution. At the beginning of this period, there were few successful resolutions in the literature, except for certain classes of easily resolvable compounds such as amino acids. The initial laboratory objective was thus to generate any observable enantioselectivity. The prevailing attitude was that chromatographic discrimination between enantiomers would in most cases be too slight to produce measurable (much less useful) separations. The basis for this attitude was apparently a consensus that (again, except in special cases) interactions between the stationary phase and solutes that could be eluted in reasonable time would be inherently too weak, too non-specific, or, at least, too fundamentally achiral to provide a practical basis for resolution. This situation was perceived to be in contrast to that with respect to pre-chromatographic formation of diastereomeric derivatives, where covalent bond formation would assure both persistence and sufficient magnitude of dissimilar chiral environments.

The scientists responsible for the development of multiple-interaction CSPs, especially Pirkle and his co-workers, proceeded in a manner directly and deliberately counter to this prevailing pessimistic viewpoint. They recognized that, by rational design, chiral stationary phases could provide specific, discrete and stereochemically dependent interactions, which could be in principle, and which proved to be in fact, highly efficacious for enantioselection. These interactions depend for their success on multiplicity (classically, three interactions (Dalgliesh, 1952)), simultaneity, comparability of magnitude, and non-degeneracy. They depend also on complementary interaction sites on the solute molecules, if necessary introduced by derivatization. Proximity of the interaction sites to the chiral loci of both CSP and solute is also required.

The record of the past ten years demonstrates that, when these conditions are met, chiral resolution is virtually assured. It has been shown that in many cases the conditions are rather easily met, by a variety of strategies, especially when an appropriate facile achiral derivatization can be carried out. Most of the multiple-interaction CSPs described in this chapter have proved to be broadly applicable; the majority of solutes with the appropriate structural features can be resolved with more than sufficient enantioselectivity for analytical purposes, and on any of a variety of readily accessible CSPs. Indeed, resolutions are often 'too-good' to be practical; methods of attenuation of resolution (not difficult to devise) must be employed.

Thus the current situation is that there is now an entirely different perspective from which chemists view chiral separations. The focus is now less on raw feasibility than on a more refined search for such features as simplicity, efficiency and ruggedness; in short, on all the criteria for practicality that apply to other, achiral, separations.

This great progress in the area of HPLC chiral separations is only one of several developments in the field of stereochemistry that are leading to a fundamental and profound change in how chemists view enantiomers and racemates. There is a growing appreciation (of what should be a truism) that each enantiomer of a pair is a chemically distinct species, in a real and practical sense, rather than identical except merely for such esoteric phenomena as the ability to rotate plane polarized light. Daily observation in the analytical laboratory of the direct, facile HPLC separation of enantiomeric substances is a forceful demonstration of this fundamental non-identity of enantiomers; the conclusion is driven home when retention times differ by factors of two, ten, even, in extreme cases (Pirkle and Pochapsky, 1986*b*) of over 100.

All that is required for effective differentiation of enantiomers is a suitable chiral environment. Such an environment has always been a feature of such natural matrices as the human body. For many chiral molecules it can now also be provided, readily, with great utility, and through logical design, by multiple-interaction chiral bonded phases.

References

Dalgliesh, C.E. (1952) Optical resolution of aromatic amino acids on paper chromatograms. *J. Chem. Soc.* **47**, 3940–3942.

Dappen, R., V.R. Meyer and H. Arm (1986) New chiral, covalently bonded, π-donor stationary phases for high-performance liquid chromatography, based on derivatives of optically active 1-(α-naphthyl)ethylamine. *J. Chromatogr.* **361**, 93–105.

Doyle, T.D., W.M. Adams, F.S. Fry, Jr. and I.W. Wainer (1986) The application of HPLC chiral stationary phases to stereochemical problems of pharmaceutical interest: a general method for the resolution of enantiomeric amines as α-naphthylcarbamate derivatives. *J. Liq. Chromatogr.* **9**, 455–471.

Doyle, T.D. and I.W. Wainer (1984) A unique reversal of elution order during direct enantiomeric resolution of amide derivatives of 1-phenyl-2-aminopropane by high performance liquid chromatography on chiral stationary phases. *J. High Resolut. Chromatogr., Chromatogr. Commun.* **7**, 38–40.

Hara, S. and A. Dobashi (1979) Liquid chromatographic resolution of enantiomers on normal-phase chiral amide-bonded silica gel. Retentions of optically active α-amino acid derivatives on N-acyl homologues of L-valylamino-propylsilanised silica phases. *J. Chromatogr.* **186**, 543–552.

Hyun, M.H. and W.H. Pirkle (1987) Preparation and evaluation of a chiral stationary phase bearing both π-acidic and π-basic sites. *J. Chromatogr.* **393**, 357–365.

Kip, J., P. Van Haperen and J.C. Kraak (1986) R-N-(pentafluorobenzoyl)phenylglycine as a chiral stationary phase for the separation of enantiomers by high-performance liquid chromatography. *J. Chromatogr.* **356**, 423–427.

Lee, E.D., J.D. Henion, C.A. Brunner, I.W. Wainer, T.D. Doyle and J. Gal (1986) High-performance liquid chromatographic chiral stationary phase separation with filament-on thermospray mass spectrometric identification of the enantiomer contaminant (S)-(+)-methamphetamine. *Anal. Chem.* **58**, 1349–1352.

Lloyd, M.J.B. (1986) Preparation and evaluation of new chiral stationary phases for high-performance liquid chromatographic separation of enantiomers. *J. Chromatogr.* **351**, 219–229.

Mikes, F., G. Boshart and E. Gil-Av (1976) Resolution of optical isomers by high-performance liquid chromatography, using coated and bonded chiral charge-transfer complexing agents as stationary phases. *J. Chromatogr.* **122**, 205–221.

Nicoll-Griffith, D.A. (1987) Stereoelectronic model to explain the resolution of enantiomeric ibuprofen amides on the Pirkle chiral stationary phase. *J. Chromatogr.* **402**, 179–187.
Oi, N. and H. Kitahara (1983) High-performance liquid chromatographic separation of chiral alcohols on chiral stationary phases. *J. Chromatogr.* **265**, 117–120.
Oi, N. and H. Kitahara (1986) Enantiomer separation by HPLC with some urea derivatives of L-valine as novel chiral stationary phases. *J. Liq. Chromatogr.* **9**, 511–517.
Oi, N., T. Kitahara, T. Doi and S. Yamamoto (1983*a*) Urea derivatives of chiral amino acids and amines as novel stationary phases for enantiomer separation by high performance liquid chromatography. *Bunseki Kagaku* (*Jap. Anal.*) **32**, 345–346.
Oi, N., H. Kitahara, Y. Matasumoto, H. Nakajima and Y. Horikawa (1988) Enantiomer separation by HPLC with some novel chiral stationary phases (unpublished).
Oi, N., M. Nagase and T. Doi (1983*b*) High-performance liquid chromatographic separation of enantiomers on (S)-1-(α-naphthyl)ethylamine bonded to silica gel. *J. Chromatogr.* **257**, 111–117.
Oi, N., M. Nagase, Y. Inda and T. Doi (1983*c*) High-performance liquid chromatographic separation of enantiomers on (1R, 3R)-trans-chrysanthemic acid and its amide derivatives bonded to silica gel. *J. Chromatogr.* **259**, 487–493.
Perry, J.A., J.D. Rateike and T.J. Szczerba (1987) Eluting trace components before major constituents. *J. Chromatogr.* **389**, 57–64.
Pettersson, C. and C. Gioeli (1987) Separation of enantiomeric acids using immobilized acetylquinine as a chiral stationary phase. *J. Chromatogr.* **398**, 247–254.
Pirkle, W.H., D.M. Alessi, M.H. Hyun and T.C. Pochapsky (1987*a*) Separation of some enantiomeric di- and tripeptides on chiral stationary phases. *J. Chromatogr.* **398**, 293–309.
Pirkle, W.H., R. Dappen and D.S. Reno (1987*b*) Applications for racemic versions of chiral stationary phases. *J. Chromatogr.* **407**, 211–216.
Pirkle, W.H., J.M. Finn, J.L. Schreiner and B.C. Hamper (1981) A widely useful chiral stationary phase for the high-performance liquid chromatography separation of enantiomers. *J. Chem. Soc.* **103**, 3964–3966.
Pirkle, W.H. and D.W. House (1979) Chiral high-pressure liquid chromatographic stationary phases. 1. Separation of the enantiomers of sulphoxides, amines, amino acids, alcohols, hydroxy acids, lactones, and mercaptans. *J. Org. Chem.* **44**, 1957–1960.
Pirkle, W.H., D.W. House and J.M. Finn (1980) Broad spectrum resolution of optical isomers using chiral high-performance liquid chromatographic bonded phases. *J. Chromatogr.* **192**, 143–158.
Pirkle, W.H. and M.H. Hyun (1984) A chiral stationary phase for the facile resolution of amino acids, amino alcohols, and amines as the N-3, 5-dinitrobenzoyl derivatives. *J. Org. Chem.* **49**, 3043–3046
Pirkle, W.H. and M.H. Hyun (1985) Reversed-phase chromatographic resolution of N-(3, 5-dinitrobenzoyl)-α-amino acids on chiral stationary phases. *J. Chromatogr.* **322**, 287–293.
Pirkle, W.H. and M.H. Hyun (1985*a*) α-Arylalkylamine-derived chiral stationary phases. Evaluation of urea linkages. *J. Chromatogr.* **322**, 295–307.
Pirkle, W.H. and M.H. Hyun (1985*b*) Preparation and use of hydantoin-based chiral stationary phases. *J. Chromatogr.* **322**, 309–320.
Pirkle, W.J., M.H. Hyun and B. Bank (1984) A rational approach to the design of highly-effective chiral stationary phases. *J. Chromatogr.* **316**, 585–604.
Pirkle, W.H., G. Mahler, and M.H. Hyun (1986*b*) Separation of the enantiomers of 3, 5-dinitrophenyl carbamates and 3, 5-dinitrophenyl ureas. *J. Liq. Chromatogr.* **9**, 443–453.
Pirkle, W.H., G.S. Mahler, T.C. Pochapsky and M.H. Hyun (1987) Direct chromatographic separation of enantiomeric diol derivatives. *J. Chromatogr.* **388**, 307–314.
Pirkle, W.H. and M.H. Hyun (1985*b*) Preparation and use of hydantoin-based chiral stationary phase for the chromatographic separation of enantiomers. *J. Am. Chem. Soc.* **108**, 352–354.
Pirkle, W.H. and T.C. Pochapsky (1986*b*) Generation of extreme selectivity in chiral recognition. *J. Chromatogr.* **369**, 175–177.
Pirkle, W.H. and T.H. Pochapsky (1987*a*) Chiral stationary phases for the direct LC separation of enantiomers. In *Advances in Chromatography*, ed. J.C. Giddings, Marcel Dekker, New York.
Pirkle, W.H. and T.C. Pochapsky (1987*b*) Chiral molecular recognition in small bimolecular systems: a spectroscopic investigation into the nature of diastereomeric complexes. *J. Chem. Soc.* **109**, 5975–5982.
Pirkle, W.H., T.C. Pochapsky, G.S. Mahler, D.E. Corey, D.S. Reno, and D.M. Alessi (1986*a*)

Useful and easily prepared chiral stationary phases for the direct chromatographic separation of the enantiomers of a variety of derivatized amines, amino acids, alcohols, and related compounds. *J. Org. Chem.* **51**, 4991–5000.

Pirkle, W.H., T.C. Pochapsky, G.S. Mahler and R.E. Field (1985) Chromatographic separation of the enantiomers of 2-carboalkoxyindolines and N-aryl-α-amino esters on chiral stationary phases derived from N-(3, 5-dinitrobenzoyl)-α-amino acids. *J. Chromatogr.* **348**, 89–96.

Pirkle, W.H. and J.L. Schreiner (1981) Chiral high-pressure liquid chromatographic stationary phases. 4. Separation of the enantiomers of bi-β-naphthols and analogues. *J. Org. Chem.* **46**, 4988–4991.

Pirkle, W.H. and T.J. Sowin (1987*a*) Direct liquid chromatographic separation of phthalide enantiomers. *J. Chromatogr.* **387**, 313–321.

Pirkle, W.H. and T.J. Sowin (1987*b*) Design, preparation and performance of a phthalide-based chiral stationary phase. *J. Chromatogr.* **396**, 83–92.

Pirkle, W.H. and A. Tsipouras (1984) Direct liquid chromatographic separation of benzodiazepinone enantiomers. *J. Chromatogr.* **291**, 291–298.

Pirkle, W.H. and C.J. Welch (1984) Chromatographic separation of the enantiomers of acylated amines on chiral stationary phases. *J. Org. Chem.* **49**, 138–140.

Pirkle, W.H., C.J. Welch, M.H. Hyun (1983) A chiral recognition model for the chromatographic resolution of N-acylated 1-aryl-1-aminoalkanes. *J. Org. Chem.* **48**, 5022–5026.

Tambute, A., A. Begos, M. Lienne, M. Caupe and R. Rosset (1987) New chiral stationary phases containing a phosphorus atom as an asymmetric centre. 1. Synthesis and first chromatographic results. *J. Chromatogr.* **396**, 65–81.

Tambute, A., P. Gareil, M. Caude and R. Rosset (1986) Preparative separation of a racemic tertiary phosphine oxides by chiral high-performance liquid chromatography. *J. Chromatogr.* **363**, 81–93.

Wainer, I.W. and M.C. Alembik (1986) Steric and electronic effects in the resolution of enantiomeric amides on a commercially available Pirkle-type high-performance liquid chromatographic chiral stationary phase. *J. Chromatogr.* **367**, 59–68.

Wainer, I.W. and T.D. Doyle (1983) Application of high-performance liquid chromatographic chiral stationary phases to pharmaceutical analysis. Direct enantiomeric resolution of amide derivatives of 1-phenyl-2-aminopropane. *J. Chromatogr.* **259**, 465–472.

Wainer, I.W. and T.D. Doyle (1984) Application of high-performance liquid chromatographic chiral stationary phases to pharmaceutical analysis: structural and conformational effects in the direct enantiomeric resolution of α-methylarylacetic acid anti-inflammatory agents. *J. Chromatogr.* **284**, 117–124.

Wainer, I.W., T.D. Doyle, K.H. Donn and J.R. Powell (1984) The direct enantiomeric determination of (−)- and (+)-propranolol in human serum by high-performance liquid chromatography on a chiral stationary phase. *J. Chromatogr.* **306**, 405–411.

Wainer, I.W., T.D. Doyle, F.S. Fry, Jr. and Z. Hamidzadeh (1986) Chiral recognition model for the resolution of ephedrine and related α, β-aminoalcohols as enantiomeric oxazolidine derivatives. *J. Chromatogr.* **355**, 149–156.

Yamashita, J., T. Numakura, H. Kita, T. Suzuki, S. Oi, S. Miyano, H. Hashimoto and T. Nobuhara (1987) High-performance liquid chromatographic separation of enantiomers on axially chiral binaphthalene derivatives bonded to silica gel. *J. Chromatogr.* **403**, 275–279.

Yang, Z.Y., S. Barkan, C. Brunner, J.D. Weber, T.D. Doyle and I.W. Wainer (1985) Resolution of enantiomeric barbiturates, succinimides and related molecules on four commercially available chiral stationary phases. *J. Chromatogr.* **324**, 444–449.

Zief, M., L.J. Crane and J. Horvath (1984) Selection of the mobile phase for enantiomeric resolution via chiral stationary phase columns. *J. Liq. Chromatogr.* **7**, 709–730.

7 Immobilized proteins as HPLC chiral stationary phases

I.W. WAINER

7.1 Introduction

Proteins are high-molecular-weight polymers composed of chiral subunits (L-amino acids). These polymers play a number of different roles in a biological system including the complexation (or binding) of xenobiotic materials. As proteins are chiral polymers, it is not surprising that the binding of small enantiomeric molecules is often stereospecific. This is especially true for the serum proteins α_1-acid glycoprotein (AGP) and albumin (SA).

Human AGP appears to contain a single drug-binding site which has some stereoselectivity (Muller and Wollert, 1983; Muller *et al.*, 1983). Differences in binding have been found for the enantiomers of propranolol (Sjoholm, 1988; Walle *et al.*, 1983), disopyramide (Lima *et al.*, 1984; Hermansson, 1983), verapamil (Vogelgesang and Echizen, 1985), prilocaine, mepivacaine and bupivacaine (Hermansson, 1983). For example, Walle *et al.* (1983) have demonstrated that the l-propranolol is bound to AGP to a greater extent than the d-isomer.

Human serum albumin (HSA) appears to contain two major drug-binding sites: the warfarin–azapropazone-binding area and the indole and benzodiazepine binding site (Sjoholm, 1988; Trillement *et al.*, 1984; Muller and Wollert, 1979). Both of these sites display some stereoselectivity.

The stereoselectivity of the warfarin–azapropazone-binding area has been investigated by a number of laboratories. Sellers and Koch-Weser (1975) have demonstrated that S-warfarin is more highly bound to HSA than R-warfarin although there is only a slight difference between the isomers (S, 99.47% bound; R, 99.15% bound; Brown *et al.* 1977). A much higher stereoselectivity has been found for phenprocoumon with an almost two-fold higher affinity of S-phenprocoumon over R-phenprocoumon (Otagiri *et al.*, 1980).

The indole and benzodiazepine binding site has also been the subject of a number of investigations and appears to be more stereospecific than the warfarin–azapropazone-binding area. For example, the affinity of (+)-oxazepam hemisuccinate is 49.5 times greater than the (−)-isomer (Muller and Wollert, 1975) while the ratios of the affinities of the S-enantiomers of N-desmethyl-3-methyldiazepam and N-desmethyl-3-methylmedazepam to the corresponding R-isomers are 135 and 28, respectively (Alebic-Kolbah *et al.*, 1979; Gratton *et al.*, 1985). A less dramatic example is the difference between the affinities of the enantiomers of ketoprofen where the affinity of the (+)-enantiomer is three times greater than that of the (−)-isomer (Konowal *et al.*, 1978).

The ability of AGP and SA to stereoselectively bind small molecules has been utilized in the development of two HPLC chiral stationary phases (HPLC CSPs) which are commercially available. The AGP-CSP was synthesized by Hermansson (1983) and Allenmark *et al.* used bovine serum albumin (BSA) to create the BSA-CSP (1983). Both of these phases can be used to resolve a wide variety of enantiomeric compounds. Each of these phases will be discussed in this chapter.

7.2 The AGP-CSP

7.2.1 *Preparation of the CSP*

The commercially available form of AGP-CSP is prepared by first ionically binding the AGP to diethylaminoethyl silica and then crosslinking the proteins using a process which involves the oxidation of the terminal alcohol groups to aldehydes, Schiff base formation and reduction of the resulting enamines to secondary amines (Lindgren, 1985). This process yields a stable immobilized-protein coating of approximately 180 mg of protein per gram of silica or 4.4×10^{-6} moles of protein per gram of silica. The resulting column is a useful CSP with a great deal of applications but also with a very low capacity.

7.2.2 *Solute selectivity*

Human AGP has a molecular weight of about 41 000 and is composed of a single 181-unit peptide chain and five carbohydrate units. The protein has an isoelectric point of 2.7 in phosphate buffer and is one of the key sites for the binding of cationic molecules (Foster, 1975).

The ability of human AGP to bind cationic molecules and the stereoselectivity of this binding is the basis for the chiral resolutions which have been achieved on the AGP-CSP. A wide range of enantiomeric cationic molecules have been resolved on this CSP. Table 7.1 lists some of the. reported resolutions.

The magnitude of the stereochemical resolution of a solute appears to be highly dependent upon the molecular structure. The results from the initial studies with the AGP-CSP indicate that the solute should contain at least two bonding groups, for example, an ammonium ion and a hydrogen bonding moiety, and a bulky or rigid structure at or near one of the binding sites. Structural features which tend to increase the magnitude of the stereoselectivity include steric bulk and aromatic groups at the chiral carbon and the chiral centre as part of a ring system.

The effect on the stereoselectivity of the steric bulk at the ammonium ion

Table 7.1 Enantiomeric cationic drugs resolved on AGP-CSP

Solute	Solute
Alprenolol*	Methylphenidate
Atropine	Metoprolol
Bromodiphenhydramine	Nadolol A
Brompheniramine	Nadolol B
Bupivicaine	Oxyphencyclimine
Butorphanol	Oxprenolol*
Carbinoxamine	Pentazocine
Chlorpheniramine	Phenmetrazine
Clidinium	Phenoxybenzamine
Cocaine	Phenyramidol
Cyclopentolate	Pindolol*
Dimethindene	Promethazine
Diperodone	Pronethalol
Disopyramide	Propoxyphene
Dobutamine	Propranolol*
Doxylamine	Pseudoephedrine
Ephedrine	Ranolazine
Homatropine	Salbutamol
Ketamine	Sotalol
Labetalol A	
Labetalol B	Tazifylline
Mepenzolate	Terbutaline
Mepivicaine	Tetrahydrazoline
Methadone	Tocainide
Methorphan	Tridihexethyl
Methylatropine	Tropicamide
Methylhomatropine	Verapamil

*Resolved as the oxazolidone derivative

Source: Wainer, I.W. and M.C. Alembik (1988). The enantiomeric resolution of biologically active molecules on commercially available liquid chromatographic chiral stationary phases. In *Chromatographic Chiral Separations* ed. M. Zief and L.J. Crane, Marcel Dekker, New York.

is illustrated by the chromatographic results obtained with terbutaline and metaproterenol (Schill *et al.* 1986*a*). Terbutaline and metaproterenol differ only in the size of the alkyl moiety on the amine function, tert-butyl and 2-propyl moieties, respectively (Figure 7.1). When these compounds are chromatographed on the AGP-CSP, terbutaline is resolved with a stereochemical selectivity α of 1.22, while metaproterenol is unresolved.

The distance between the ammonium and hydrogen bonding moieties also effects the stereoselectivity. This is illustrated by the results from the chromatography of a series of compounds related to metoprolol (Table 7.2) (Schill *et al.* 1986*a*). When the number of methylene groups in the chain between the amine and the alcohol groups is increased from 1 to 3, α is reduced from 1.49 to 1.00.

Chiral molecules which contain an amine function can be resolved on the AGP-CSP without precolumn derivatization, and unlike a number of other

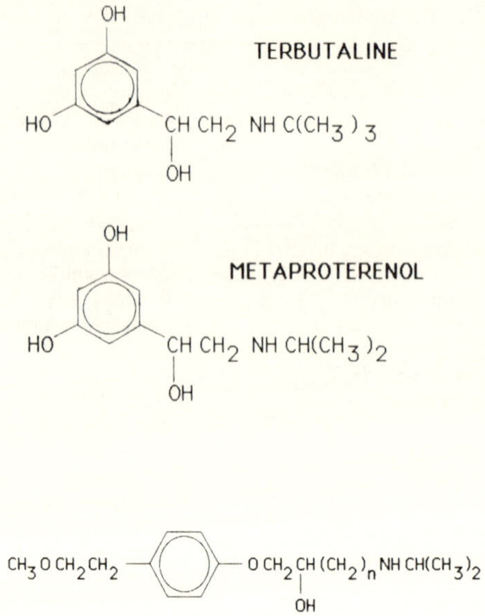

Figure 7.1 The structure of (*A*) terbutaline; (*B*) metaproterenol.

CSPs, the amine function does not have to be at the chiral centre. Enantiomeric α, β-aminoalcohols such as ephedrine, propranolol and labetalol can also be resolved without precolumn derivatization (Schill *et al.*, 1986*a*, 1986*b*). However, the efficiency and stereoselectivity of the resolution can be improved by converting the compounds to oxazolidone derivatives (Hermansson, 1985).

Although AGP primarily binds cationic compounds, a number of anionic chiral molecules can be resolved on the AGP-CSP. These resolutions can be obtained when the mobile phase contains an ion pairing agent such as N, N-dimethyloctylamine or tetrabutylammonium bromide (Schill *et al.*, 1986*a*, 1986*b*; Hermansson and Eriksson, 1986). The acidic compound resolved on the AGP-CSP include α-methylarylacetic acids such as ibuprofen and naproxen and barbiturates such as hexobarbital. A list of some of the anionic compounds which have been resolved on the CSP are presented in Table 7.3.

Chiral carboxylic acid derivatives such as amides and esters can also be resolved on the AGP-CSP, often with an increase in stereoselectivity. For example, mandelic acid is not resolved on this CSP but its methyl and ethyl esters are resolved with $\alpha = 1.27$ and 1.93, respectively (Hermansson, 1985).

7.2.3 *Mobile phase effects*

Most mobile phases used with the AGP-CSP are composed of phosphate buffer and an organic or inorganic modifier. The retention k' and stereoselectivity α of a solute can often be dramatically altered by changing the pH and/or

Table 7.2 Effect on stereoselectivity of distance between ammonium and hydrogen-bonding sites.

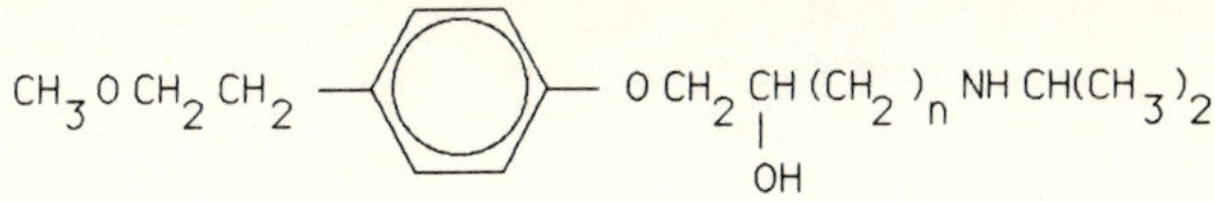

n	α
1	1.49
2	1.14
3	1.0

the composition of the mobile phase. Each of these parameters will be discussed below.

7.2.4 *Effect of pH*

The pH of the mobile phase should be kept between 3.0 and 7.5 as the silica based column is unstable outside this region (Schill *et al.*, 1986*b*). Within this range, altering the pH of the mobile phase is an excellent method for adjusting k'. For cationic solutes, a decrease in pH usually results in a drop in k' (Table 7.4). The magnitude of the effect is dependent on both the structure of the solute and the composition of the mobile phase. For anionic solutes, a decrease in pH often results in an increase in k'.

The stereoselectivity of the AGP-CSP can also be changed by modifying the pH of the mobile phase. This effect is extremely dependent on the composition of the mobile phase. For example, when cyclopentolate and doxylamine are chromatographed using a mobile phase composed of 0.02 M phosphate buffer modified with 0.003 M tetrabutylammonium bromide, a decrease in the pH of

Table 7.3 Enantiomeric anionic drugs resolved on AGP-CSP

Solute	References
Ethotoin	Hermansson and Eriksson (1986)
Fenoprofen	Schill *et al.* (1986)
Hexobarbital	Hermansson and Eriksson (1986)
Ibuprofen	Schill *et al.* (1986*b*) Hermansson and Eriksson (1986)
Ketoproxen	Hermansson and Eriksson (1986)
Naproxen	Schill *et al.* (1986*b*) Hermansson and Eriksson (1986)
2-phenylbutyric acid	Schill *et al.* (1986*b*)
3-phenylbutyric acid	Schill *et al.* (1986*b*)
2-phenoxypropionic acid	Schill *et al.* (1986*b*) Hermansson and Eriksson (1986)
2-phenylpropionic acid	Schill *et al.* (1986*b*)

Table 7.4 Effect of pH on retention and stereoselectivity*

Solute	pH	k'*	α
Cyclopentolate	7.0	18.0	1.70
	6.0	4.8	2.09
Doxylamine	7.0	9.5	1.23
	6.0	6.1	1.37

*Capacity factor of the first eluted enantiomer.

the mobile phase from 7.0 to 6.0 results in an increase in α (Table 7.4). However, if the mobile phase is changed to 0.02 M phosphate buffer modified with 0.1 M sodium chloride and 0.33 M 2-propanol, a decrease in the pH from 7.5 to 6.5 reduces the α for cyclopentolate from 1.96 to 1.79 whereas the α for doxylamine is virtually unchanged: 1.15 as against 1.16 (Schill *et al.*, 1986*b*).

7.2.5 *Effect of mobile phase modifiers*

While the alteration of the pH of the mobile phase can be used to change k' and α, the most dramatic changes can be brought about through the use of mobile phase modifiers. Both uncharged and charged modifiers have been studied and their effects often differ (Schill *et al.*, 1986*a*, 1986*b*; Hermansson, 1985; Hermansson and Eriksson, 1986). The compounds which have been used as mobile phase modifiers for the AGP-CSP are presented in Figure 7.2.

7.2.5.1 *Uncharged modifiers.* Uncharged modifiers such as monovalent alcohols are primarily used to reduce k'. For example, the k' for (R)- and (S)-disopyramide are 8.51 and 31.48, respectively, when the mobile phase is composed of 0.02 M phosphate buffer modified with 4% 2-propanol. When the amount of 2-propanol is increased to 8%, the k's are reduced by about 80% to 1.77 and 5.66, respectively (Hermansson, 1985). However, these modifiers also tend to reduce α. In this case, the stereoselectivity for disopyramide was reduced from $\alpha = 3.70$ to $\alpha = 3.20$ (Hermansson, 1985).

Other uncharged modifiers such as divalent alcohols and uncharged amino acids also tend to reduce k' but do not have as great an effect on α (Hermansson, 1985; Schill *et al.*, 1986*a*, 1986*b*). For example, a ten-fold increase in the concentration of aminohexanoic acid from 0.01 M to 0.10 M decreases the k' for most cationic solutes by about 33% but has little effect on α (Schill *et al.*, 1986*b*).

7.2.5.2 *Charged modifiers.* A variety of cationic and anionic compounds have been used as mobile phase modifiers for the AGP-CSP (Figure 7.2). The effect of these modifiers on k' and α varies according to the structures of the solute and the modifier and the pH of the mobile phase. There is in fact a highly

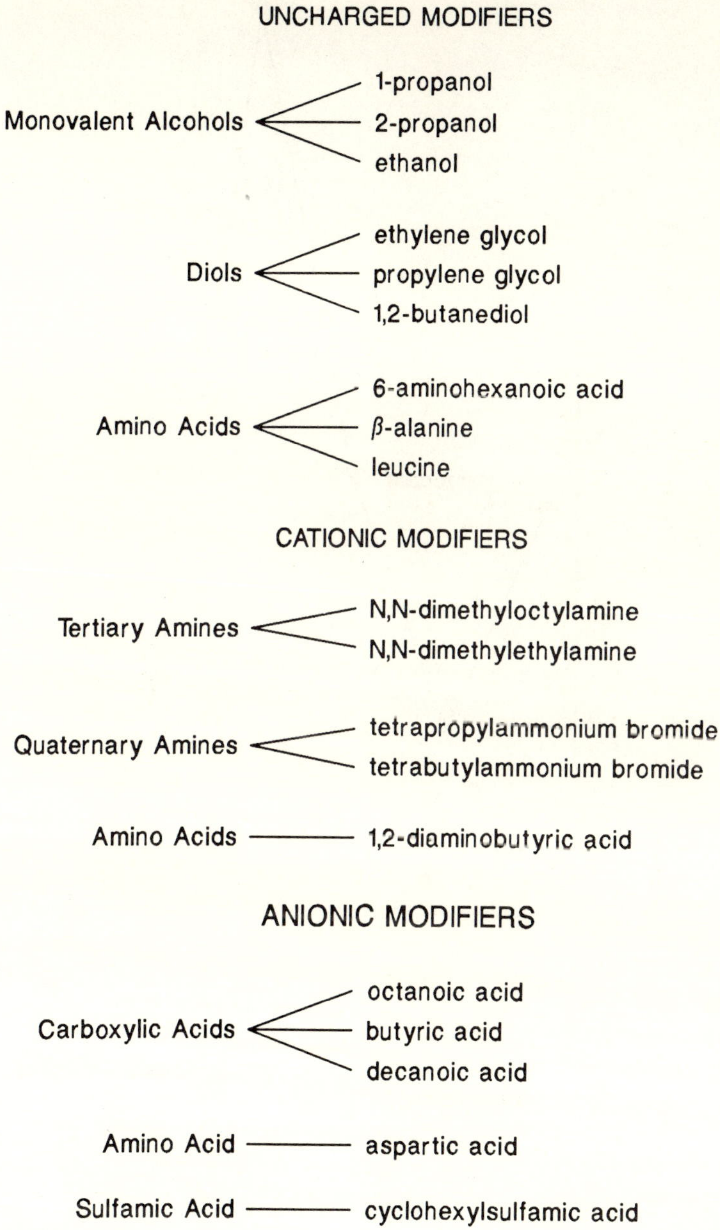

Figure 7.2 Mobile phase modifiers used with the AGP-CSP. From I.W. Wainer, S.A. Barkan, and G. Schill, *L.C./G.C.* **4** (1986) 422.

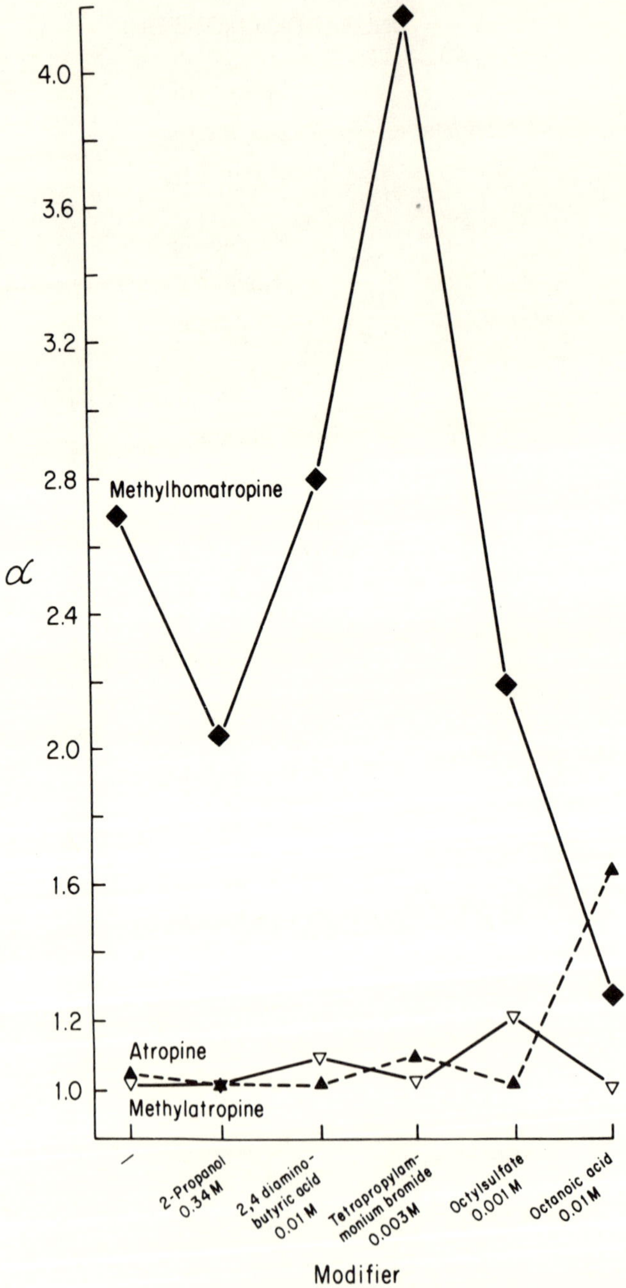

Figure 7.3 Effect of mobile phase modifiers on the stereoselectivity (α) of atropine, methylatropine and methylhematropine. From I.W. Wainer, S.A. Barkan, and G. Schill, *L.C./G.C.* **4** (1986) 422.

complex relationship between these variables which, however, is still not fully understood.

In general, the addition of a cationic mobile phase modifier will decrease the retention of a cationic solute and will increase the k' for anionic solutes (Schill *et al.*, 1986*a*, 1986*b*) whereas an anionic modifier will have the opposite effect. Cationic mobile phase modifiers usually increase α for anionic solutes whereas very low values for α or no chiral resolution at all is observed with anionic modifiers. The effect on the stereoselectivity of cationic solutes is more variable. This is exemplified by the effect of different mobile phase modifiers on α for the solutes atropine, methylatropine and methylhomatropine, (Figure 7.3) (Schill *et al.*, 1986*b*).

Methylatropine is the N-methyl derivative of atropine and these molecules differ only in the charge and steric bulk at the bridgehead nitrogen. When cationic modifiers are added to the mobile phase, there is virtually no difference in the stereoselectivity for both of these compounds. However, a dramatic difference is observed with anionic modifiers. The addition of octylsulphate to the mobile phase results in the chiral resolution of methylatropine ($\alpha = 1.3$) but not of atropine. When octanoic acid is the modifier, atropine is resolved ($\alpha = 1.6$) but not methylatropine.

The structures of atropine and methylhomatropine differ in two respects: a tertiary amine as against a quaternary ammonium ion at the bridgehead; atropine has a methylene group between the hydroxyl moiety and the asymmetric centre which is absent in homatropine. These small structural differences result in large chromatographic differences. Homatropine is easily resolved without a modifier in the mobile phase. When cationic modifiers are added the stereoselectivity increases relative to the results obtained with the unmodified mobile phase. The addition of anionic modifiers yields the opposite results.

7.2.6 *Applications of the AGP-CSP to pharmacological studies*

The AGP-CSP is an excellent analytical column which is finding a number of uses in clinical and pharmacological studies. Two of the recent examples of this application are the work by Lee *et al.* (1987) on the assay of bupivicaine enantiomers and Tan and Soldin (1987) on the disposition kinetics of salbutamol enantiomers in man.

7.2.6.1 *Bupivacaine.* Bupivacaine is a local anaesthetic which is administered as a racemic mixture. While R-(+)-bupivicaine and S-(−)-bupivicaine are equipotent as *in vitro* nerve blockers, the R-enantiomer is more toxic than its enantiomorph. Lee *et al.* (1987) have developed a chromatographic method to study the serum concentration of the two isomers after the administration of the racemate. The stereochemical resolution of the enantiomers is accomplished on an AGP column using a mobile phase composed of sodium phosphate buffer (0.008 M):2-propanol (9%):sodium chloride (0.1 M). Some chromatograms obtained using this procedure are presented in Figure 7.4.

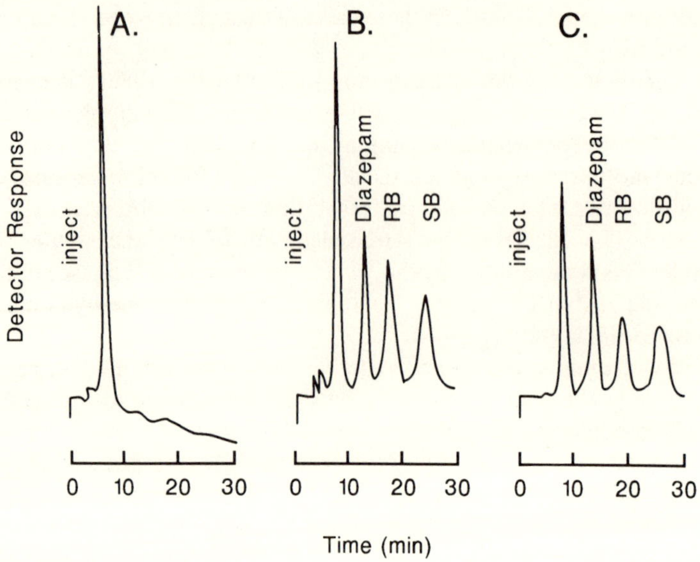

Figure 7.4 Representative chromatograms of (*A*) blank serum; (*B*) serum spiked with 1 $\mu g\,mL^{-1}$ diazepam and 2 $\mu g\,mL^{-1}$ racemic bupivacaine. From Lee *et al.* (1987).

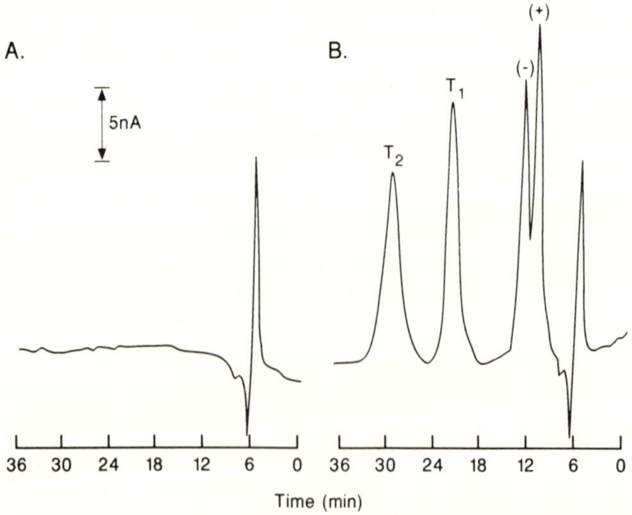

Figure 7.5 Representative chromatograms of (*A*) blank urine; and (*B*) urine containing 2 $\mu g\,mL^{-1}$ racemic salbutanol where (+) = *S*-(+)-salbutanol; (−) = *R*-(−)-salbutanol; T_1, T_2 = (+/−)-terbutaline, the internal standard. From Tan and Soldin (1987).

7.2.6.2 *Salbutamol.* Salbutamol is a therapeutic agent used in the treatment of bronchial asthma. Tan and Soldin (1987) have developed an enantioselective HPLC assay for the determination of the urinary excretion of R-(–)-and S-(+)-salbutamol which is presented in Figure 7.5. The assay uses an AGP column and a mobile phase composed of 5.3 mM citrate buffer (pH 7.2) modified with triethylamine (0.1% v/v). A column temperature of 0°C and a flow rate of 0.2 mL. min^{-1} were maintained throughout the assay. The salbutamol was quantified using electrochemical detection.

7.3 The bovine serum albumin (BSA) CSP

7.3.1 *Preparation of the CSP*

In 1973, Stewart and Doherty (1973) immobilized BSA on succinoylaminoethyl-Sepharose and used the resulting CSP to resolve stereochemically D, L-tryptophan. Allenmark *et al.* (1983) reported the covalent immobilization of BSA on 10-μm silica. The work of Allenmark was used as the basis for the development of the commercially available BSA-CSP. However, it appears that the actual immobilization procedure has not been reported.

In addition to the BSA-CSP developed by Allenmark, two other silica-based BSA columns have been reported. Erlandsson *et al.* (1986) prepared a chromatographic stationary phase by adsorbing BSA on silica. Both analytical and preparative columns were prepared and used to resolve stereochemically a variety of solutes. A glutaraldehyde cross-linked BSA-silica CSP has also been reported (Hsu *et al.*, 1987; Aubel and Rogers, 1987). This column appears to be stable and to have a stereoselectivity comparable to the commercially available CSP.

7.3.2 *Solute selectivity*

Bovine serum albumin is a globular, hydrophobic protein with a molecular weight of about 66 000. The protein contains a single 581 amino acid chain with 17 intrachain disulphide bridges forming nine double loops. BSA has an isoelectric point of 4.7 and a net charge of – 18 at pH 7.0. The protein is able to bind uncharged hydrophobic molecules as well as anionic compounds.

These binding properties and the observed stereoselectivity of the native BSA are the basis for the enantioselectivity of the BSA-CSP. A wide variety of anionic and neutral enantiomeric molecules can be resolved on the BSA-CSP but not as cationic compounds. Some representative classes of solutes are presented in Table 7.5. From initial studies, it appears that enantioselectivity on the BSA-CSP requires that the solute contain aromatic and polar moieties (Allenmark, 1986). Steric effects also appear to be highly important but are, at this point, unpredictable (Allenmark, 1986).

Some aromatic amino acids such as kynurenine can be stereochemically

Table 7.5 Compounds resolved on BSA-CSP

Class	Example	Reference
Aromatic amino acids	Kynurenine	Allenmark (1986)
Amino acid derivatives	N-Benzoyl	Allenmark *et al.* (1984)
	Dansyl	Allenmark and Bomgren (1982)
	N-(2,4-dinitrophenyl)	Allenmark *et al.* (1984) Allenmark and Bomgren (1982)
Aromatic hydroxy ketones	Benzoin	Allenmark (1986) Allenmark *et al.* (1984)
Benzodiazepinones	Oxazepam	Allenmark (1986)
Coumarine derivatives	Warfarin	Allenmark (1986)
Lactams	4-Amino-3-(*p*-chlorophenyl)butyric acid lactam	Allenmark (1986)
Reduced folates	Leucovorin	Wainer and Stiffin (1988)
Sulphoxides	Omeprazole	Allenmark (1986)

resolved on the BSA-CSP without derivatization. However, most amino acids require precolumn derivatization of the amine moiety. The N-derivatives which have been used include: acetyl, benzenesulphonyl, phthalimido, DANSYL, 2,4-dinitrophenyl and 2,4,6-trinitrophenyl (Allenmark and Bomgren, 1982; Allenmark *et al.*, 1984; Allenmark, 1986). The retention of the N-nitrobenzoyl- and N-nitrophenyl-amino acids tends to increase with the degree of nitro substitution which may be due to π–π interaction between the aromatic moiety on the amino acids and two tryptophanyl residues in the protein (Allenmark, 1986).

The BSA-CSP has been used to resolve stereochemically enantiomeric molecules containing asymmetric sulphoxide and sulphoximine moieties and an aromatic group near or at the chiral centre (Allenmark and Bomgren, 1982; Allenmark, 1986). The coumarin derivatives phenprocoumon and warfarin and a number of benzodiazepine derivatives can be resolved without pre-column derivatization (Allenmark, 1986). The BSA-CSP has also been used to resolve stereochemically 2-hydroxy-2-phenylacetophenone (benzoin) and the reduced folates leucovorin and 5-methyltetrahydrofolate (Wainer and Stiffin, 1988).

7.3.3 *Mobile phase effects*

The standard mobile phase used with the BSA-CSP is composed of phosphate buffer modified with 1-propanol although other modifiers have been investigated. Mobile phases containing methanol and acetonitrile should not be used with this CSP. The buffer concentration can range from 0.01 M to 0.20 M and the pH from 4.5 to 8.0. The resolution and retention of solutes can be regulated by changing these parameters.

7.3.4 *Effect of pH*

In general, for N-derivatized amino acids, an increase in pH will result in a decrease in retention and stereochemical resolution (Allenmark and Bomgren, 1982; Allenmark *et al.*, 1984). This effect is believed to be due to the fact that a decrease in the pH will result in a corresponding decrease in the net negative charge of the BSA. This means that the Coulomb interaction between the BSA and the N-derivatized amino acids or other uncharged carboxylic acids will increase resulting in an increase in k' (Allenmark *et al.*, 1984).

A similar effect of pH on k' has been found for the enantiomers of warfarin but the effect on α was opposite to that found with the N-derivatized amino acids (Wainer and Chu, 1988). An increase in the pH of the mobile phase from 5.0 to 8.0 resulted in a decrease in k' for S-warfarin from 33.6 to 3.4 (a 90% drop) and a decrease in k' for R-warfarin from 38.6 to 4.6 (an 88% drop), the net result being an increase in α from 1.15 (at pH 5.0) to 1.35 (at pH 8.0). This result could be due to the small difference in the binding of S- and R-warfarin to BSA (S-warfarin, 99.47% bound; R-warfarin, 99.15% bound (Brown, 1970)). The increase in pH improved the efficiency of the column.

7.3.5 *Effect of buffer concentration*

For solutes containing a carboxylic acid moiety, the general rule appears to be that the lower the buffer concentration the higher the retention (Allenmark and Bomgren, 1982; Allenmark *et al.*, 1984; Allenmark, 1986). This holds for buffer concentrations ranging from 0 to 100 mM. However, when N-benzoyl-amino acids were chromatographed with buffer concentrations of 200 mM and above, an increase in the buffer concentration resulted in an increase in k' (Allenmark *et al.*, 1984). These results have been interpreted as a reflection of the two types of binding forces between the solute and the CSP, i.e. Coulomb attraction and hydrophobic interaction. The rapid decrease in k' with increasing buffer strength up to 100 mM should mainly reflect the decrease in Coulomb attraction, whereas the increase above 200 mM should be due to the increasing role of binding by hydrophobic interaction (Allenmark *et al.*, 1984).

7.3.6 *Effect of mobile-phase modifiers*

7.3.6.1 *1-Propanol as mobile-phase modifier.* To date, 1-propanol has been the most commonly used mobile-phase modifier with the BSA-CSP. As with the AGP-CSP, the addition of an alcohol to the mobile phase appears to reduce the hydrophobic interactions between the solute and the CSP, which results in lower k'-values and a reduction in α. This effect has been demonstrated for N-benzoyl-amino acids (Allenmark *et al.*, 1984), N-(2,4-dinitrophenyl)- and dansyl-amino acids (Allenmark and Bomgren, 1982) and warfarin (Wainer and Chu, 1988). In the latter case, an increase in 1-propanol concentration from 1% to 5% resulted in a decrease in k' for S-warfarin from 41.1 to 12.0

and for R-warfarin from 54.6 to 13.8. However, there was also a decrease in α from 1.33 to 1.15.

7.3.6.2 *Other mobile-phase modifiers.* In an attempt to reduce the retention on the BSA-CSP without reducing the stereoselectivity, Wainer and Chu (1988) have investigated the use of a number of compounds as mobile-phase modifiers using warfarin as the initial solute. The compounds chosen as mobile-phase modifiers were known to compete with warfarin for binding to BSA.

The utility of this approach is demonstrated by the effect of the mobile-phase additive trichloroacetic acid. When S- and R-warfarin are chromatographed using a mobile phase composed of phosphate buffer (0.2 M, pH 7.5):1-propanol (97:3 v/v) the k' for S- and R-warfarin are 18.08 and 21.83, respectively, and $\alpha = 1.21$. The addition of trichloroacetic acid (5 mM) reduces the k's by almost 50% (9.42, S-warfarin; 10.96, R-warfarin) whereas the stereoselectivity falls by only 4% to $\alpha = 1.16$. The effect on trichloroacetic acid on the chromatography of R, S-warfarin is also presented in Figure 7.6.

7.3.7 *Applications of BSA-CSP to pharmacological studies*

The BSA-CSP is an excellent analytical column which can be adapted for use in pharmacological studies. One of the inviting properties of this CSP as well as the AGP-CSP is their use of aqueous mobile phases and the fact that many solutes do not require pre-column derivatization. This allows for the development of chromatographic systems where the CSP is coupled to an achiral

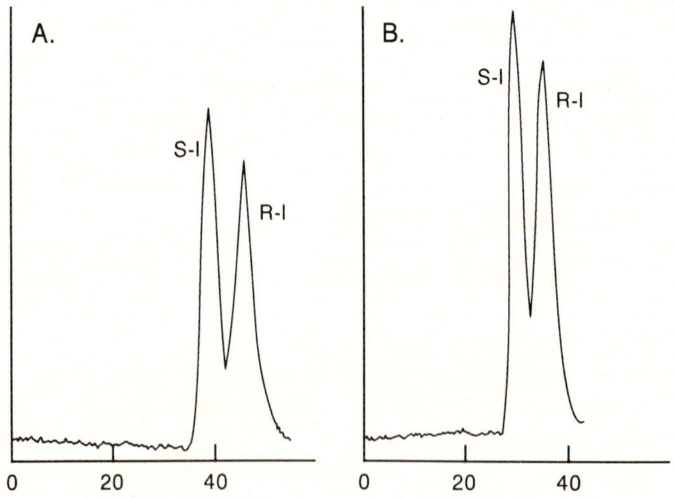

Figure 7.6 The effect on the retention and stereoselectivity of warfarin on the BSA-CSP when trichloroacetic acid is added as a mobile phase modifier. (*A*) mobile phase – phosphate buffer (0.2 M, pH 7.5):1-propanol (97:3, v/v); (*B*) mobile phase – phosphate buffer (0.2 M, pH 7.5):1-propanol (97:3, v/v) with 3 mM trichloroacetic acid. *S-I* = *S*-Warfarin; *R-I* = *R*-Warfarin.

pre-column. In this manner, biological samples can be processed with a minimum of initial workup and the analytical systems can be easily automated. This approach is exemplified by the methods developed for the analysis of the stereoisomeric composition of leucovorin and warfarin in serum.

7.3.7.1 *Leucovorin.* Leucovorin (LV) is a reduced folate which is used to treat or prevent host toxicity in cancer patients due to the administration of supralethal doses of methotrexate and in combination with 5-fluorouracil in the treatment of colorectal cancer. LV is administered as a diastereomeric mixture where the isomers differ in the configuration about the 6-carbon of the tetrahydropteridine ring.

In order to facilitate the analysis of the stereoisomeric composition of LV in serum after the clinical administration of the diastereomeric mixture, Wainer and Stiffin (1988) developed an HPLC method for the resolution of the LV diastereomers using the BSA-CSP. The system was adapted for the analysis of serum samples by coupling the BSA-CSP to an achiral HPLC column packed with a phenyl support. The LV fractions can be isolated from interfering serum peaks and quantified using the phenyl column and then selectively switched to the BSA-CSP for the determination of the stereoisomeric composition. The

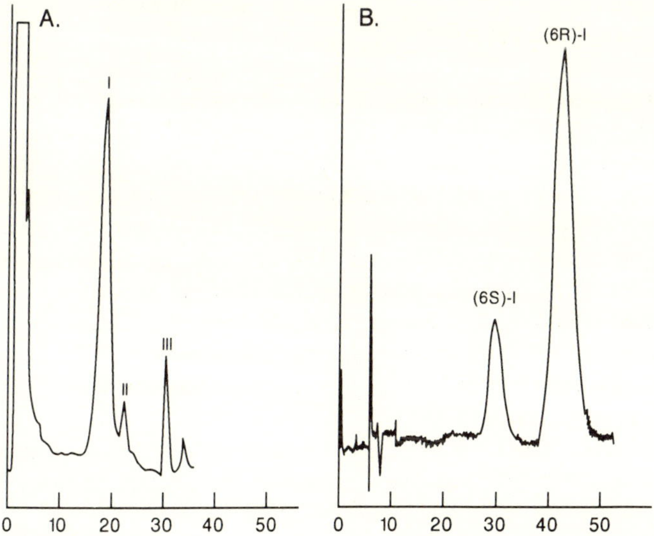

Figure 7.7 The determination of the stereoisomeric composition of leucovorin in serum using achiral/chiral coupled column chromatography. (*A*) A representative chromatogram from the chromatography of a serum sample on the achiral phenyl support where 1 = (R, S)-leucovorin; 2 = (R, S)-5-methyltetrahydrofolate; 3 = methotrexate (internal standard). (*B*) A representative chromatogram from the chromatography on the BSA-CSP where (6S)-1 = (6S)-leucovorin; (6R)-1 = (6R)-leucovorin.

system has been automated and is currently in clinical use. Representative chromatograms are presented in Figure 7.7.

7.3.7.2 *Warfarin.* Warfarin is an anticoagulant which is administered as a racemic mixture. The enantiomorphs of warfarin, (−)-(S)-warfarin (S-War) and (+)-(R)-warfarin (R-War), differ in their potency, rate of elimination and plasma protein binding. The pharmacological and pharmacokinetic differences between the enantiomers of warfarin raises the issue of the rapid and accurate determination of the serum concentration of these isomers.

One approach to this problem is the stereochemical resolution of S-War and R-War on a BSA-CSP. This resolution was initially reported by Allenmark (1986). However, while this separation was adequate for the analysis of warfarin standards, it could not be directly used with serum samples due to interference from serum components and warfarin metabolites. To overcome these problems, Chu and Wainer (1988) coupled the BSA-CSP to an achiral pre-column.

In this system, the achiral pre-column was used to separate R, S-War from the serum components and warfarin metabolites and to quantitate the total warfarin concentration. The eluent containing R, S-War was then selectively transferred to the BSA-CSP where R-War and S-War were stereochemically resolved ($\alpha = 1.19$) and the enantiomeric composition was determined. This system is rapid, accurate, does not require extensive pre-column manipulation and can be automated for use in large-scale clinical studies.

7.4 Chiral phases based on enzymes and other biological polymers

7.4.1 *Immobilized ovomucoid*

Ovomucoid is a relatively abundant protein which can be easily purified from chicken egg whites. Miwa *et al.* (1987*a*, 1987*b*) have immobilized this protein on aminopropyl silica gel. The resulting chiral stationary phase (OV-CSP) was capable of resolving stereochemically a number of different compounds including amines and carboxylic acids. The cationic compounds resolved on the OV-CSP include chloropheniramine ($\alpha = 2.28$) and pindolol ($\alpha = 1.48$) and the anionic compounds include flurbiprofen ($\alpha = 1.28$) and ketoprofen ($\alpha = 1.21$).

The effect on k' and α of the pH of the mobile phase has been investigated (Miwa *et al.*, 1987*b*). For anionic compounds, an increase in pH results in a decrease in k' with very little change in α. For example, when the pH of the mobile phase (0.2 M phosphate buffer) was raised from 5.0 to 6.0, the k' of the first eluted enantiomer of flurbiprofen fell from 13.12 to 7.28 whereas the stereoselectivity was only slightly reduced from $\alpha = 1.28$ to 1.24. An increase in pH has the opposite effect on the cationic solute chloropheniramine. When the pH of the mobile phase (0.2 M phosphate buffer) was raised from 5.5 to 6.0, the k' of the first eluted enantiomer rose from 3.20 to 9.50 and α also increased from 2.28 to 3.14. However, the effect is not consistent. When pindolol was studied under the same conditions, the k' of the first eluted enantiomer increased from 0.46 (pH 5.5) to 1.31 (pH 6.0) but α decreased from 1.48 to 1.44.

Cationic and anionic mobile phase modifiers can also be used to alter k' and α (Miwa *et al.*, 1987*b*). For example, when a cationic molecule such as tetra-n-butylammonium hydrogen sulphate (TBS-HSO_4) was added to the mobile phase, there was a moderate effect on the k' and α of the anionic solutes and a dramatic effect on the chromatography of the cationic solutes. For example, with 0.5 M TBS-HSO_4 in the mobile phase (0.5 M phosphate buffer, pH 5.0) the k' of the first eluted enantiomer of flurbiprofen fell from 13.12 to 11.32 and α rose from 1.28 to 1.30. For chloropheniramine, the addition of 0.5 M TBS-HSO_4 to the mobile phase (0.5 M phosphate buffer, pH 5.5) reduced the k' of the first eluted enantiomer from 3.20 to 0.46 and the stereoselectivity was lost.

In addition to the composition of the mobile phase, k' and α also depend on the silica support (Miwa *et al.*, 1987*b*). When the ovomucoid is immobilized on LiChrosorb NH_2, the k's for the enantiomers of flurbiprofen were 52.26 and 79.31 and $\alpha = 1.52$ (mobile phase: 0.2 M phosphate buffer, pH 5.0). Using the same mobile phase on an OV-CSP based upon Unisil Q NH_2, the k's for these solutes were 13.12 and 16.73 and $\alpha = 1.28$.

7.4.2 *Immobilized α-chymotrypsin*

Enzymes are biopolymers which often have high substrate stereoselectivity. For example, α-chymotrypsin (ACHT) has a greater affinity for L-amino acids than for D-amino acids (Neurath and Schwert, 1950). Wainer, *et al.* (1988) have used this property of ACHT to develop a new type of CSP based on immobilized enzymes.

In this work, a hydrophylic polymer was bound to 15 μm silica with 300 Å pores. Gluteraldehyde was then covalently attached to the polymer and ACHT was covalently bound to this moiety. The resulting ACHT-CSP is stable and capable of stereochemically resolving a number of amino acids, amino acid derivatives and dipeptides. For example, using a mobile phase of 0.1 M phosphate buffer (pH 6.5), the ACHT-CSP can stereochemically resolve: D,L-tryptophan ($\alpha = 1.10$), D,L-N-acetyltryptophan ($\alpha = 1.14$), D,L-N-acetylphenylalanine ($\alpha = 1.36$), D,L-N-benzoylleucine ($\alpha = 1.18$), D,L-tryptophan amide ($\alpha = 1.87$) and D,L-tryptophan methylester ($\alpha = 1.71$). The effect of solute structure and mobile phase composition on retention and stereoselectivity are currently under investigation.

7.5 Conclusion

The AGP- and BSA-CSPs are important advances in the science of enantiomeric separations. They are rapidly becoming important analytical tools in chemical and pharmacological research. However, their usefulness has only been partially explored. Recent results in our laboratory indicate that these columns can be used as rapid probes of the *in vivo* biological activity of AGP and BSA. In addition, these CSPs have opened the door for the development of other HPLC phases created from the immobilization of other proteins and enzymes. The analytical and biological applications of the protein-based stationary phase should be evident in the near future.

References

Alebic-Kolbah, T., F. Kajfez, S. Rendic, V. Sunic, A, Konowal and G. Snatzke (1979) Circular dichroism and gel filtration study of binding of prochiral and chiral 1,4-benzodiazepin-2-ones to human serum albumin. *Biochem. Pharmacol.* **28**, 2457–2464.

Allenmark, S., and B. Bomgren(1982) Direct liquid chromatographic separation of enantiomers on immobilized protein stationary phases II. Optical resolution of a sulphoxide, a sulphoximine and a benzoyl-amino acid. *J. Chromatogr.* **252**, 297–300.

Allenmark, S., B. Bomgren and H. Boren (1983) Direct LC separation of enantiomers on immobilised protein stationary phases. III. Optical resolution of a series of N-aroyl-D,L-aminoacids by HPLC on bovine serum albumin covalently bound to silica. *J. Chromatogr.* **365**, 63–68.

Allenmark, S., B. Bomgren and H. Boren, (1984) Direct liquid chromatographic separation of enantiomers on immobilised protein stationary phase IV. Molecular interaction forces and retention behavior in chromatography on bovine serum albumin as a stationary phase. *J. Chromatogr.* **316**, 617–624.

Allenmark, S. (1986) Optical resolution by liquid chromatography on immobilised bovine serum albumin. *J. Liq. Chromatogr.* **9**, 425-442.

Aubel, M. and L. B. Rogers (1987) Effects of pretreatment on enantioselectivity of silica-bound bovine serum albumin used as high-performance liquid chromatographic stationary phase. *J. Chromatogr.* **392**, 415–420.

Brown, N. A., E. Jahnchen, W. E. Muller and U. Wollert (1977) Optical studies on the mechanism of the interaction of the enantiomers of the anticoagulant drugs phenprocoumon and warfarin with human serum albumin. *Mol. Pharmacol.* **13**, 70–78.

Chu, Y.-Q. and I.W. Wainer (1988) The measurement of warfarin enantiomers in serum using coupled achiral/chiral HPLC. (unpublished).

Delée, E., L. LeGarrec, I. Jullien, S. Beranger, J.C. Pascal and H. Pinhas (1978) Direct HPLC resolution of β-aminoalcohol (tazifylline, ranolazine, sotalol) enantiomers. *Abstracts 11th Int. Symposium on Column Liquid Chromatography*, Amsterdam, Abs. No. FR-P-12.

Erlandsson, P., L. Hansson and R. Isakkson (1986) Direct analytical and preparative resolution of enantiomers using albumin adsorbed to silica as a stationary phase. *J. Chromatogr.* **370**, 475–483.

Foster, J.F. (1975) Plasma albumins. In *The Plasma Proteins*, ed. K. Schmid. Academic Press, New York.

Gratton, G., E. Decorte, F. Moimas, C. Angeli and V. Sunjic (1985) Enantioselecitivity of the binding of S- and R-7-chloro-1,3-dihydro-3-methyl-5-phenyl-2H-1,4-benzodiazepines to human serum albumin. *Il Farmaco Ed. Sci.* **40**, 209–214.

Hermansson, J. (1983) Direct liquid chromatographic resolution of racemic drugs using α_1-acid glycoprotein as the stationary phase. *J. Chromatogr.* **269**, 71–80.

Hermansson, J. (1984) Liquid chromatographic resolution of racemic drugs using a chiral α_1-acid glycoprotein column. *J. Chromatogr.* **298**, 67–78.

Hermansson, J. (1985) Resolution of racemic aminoalcohols (β-blockers), amines and acids as enantiomeric derivatives using a chiral α_1-acid glycoprotein column. *J. Chromatogr.* **325**, 379–384.

Hermansson, J. and M. Eriksson (1986) Direct liquid chromatographic resolutions of acidic drugs using a chiral α_1-acid glycoprotein column (Enantiopac). *J. Liq. Chromatogr.* **9**, 621–639.

Hsu, T.-B., P.A. Shah and L.B. Rogers (1987) Synthesis and characterisation of chiral stationary phases from amino acids and small peptides for liquid chromatography fractionations of racemic alcohols. *J. Chromatogr.* **391**, 145–160.

Konowal, A., G. Snatzke, T. Alebic-Lolbach, F. Kajfez, S. Rendic and V. Sunjic (1978) General approach to chiroptical characterisation of binding of prochiral and chiral 1,4-benzodiazepin-2-ones to human serum albumin. *Biochem. Pharmacol.*, **28**, 3109–3113.

Lee, E.D.J., S.B. Ang and T.L. Lee (1987) Stereoselective high-performance liquid chromatographic assay for bupivacaine enantiomers. *J. Chromatogr.* **420**, 203–206.

Lima, J.J., G.L. Jungbluth, T. Devine and L.W. Robertson (1984) Stereoselective binding of diisopyramide to human plasma protein. *Life Sci.* **35**, 835–837.

G. Lindgren (1985) (LKB Produkter AB, Bromma, Sweden), personal communication.

Miwa, T., M. Ichikawa, M. Tsuno, T. Hattori, T. Miyakawa, M. Kayano and Y. Miyake (1987*a*) Direct liquid chromatographic resolution of racemic compounds. Use of ovomucoid as a column ligand. *Chem. Pharm. Bull.* **35**, 682–686.

Miwa, T., T. Miyakawa, M. Kayano and Y. Miyake (1987*b*) Application of an ovomucoid-conjugated column for the optical resolution of some pharmaceutically important compounds, *J. Chromatogr.* **408**, 316–322.

Muller, W.E. and U. Wollert (1975) High stereospecificity of the benzodiazepine binding site on human serum albumin. *Mol. Pharmacol.* **11**, 52–60.

Muller, W.E. and U. Wollert (1979) Human serum albumin as a 'silent receptor' for drugs and endogenous substances. *Pharmacology* **19**, 59–68.

Muller, W.E. and A.E. Stillbauer (1983) Characterisation of a common binding site for basic drugs on human α_1-acid glycöprotein (orosomucoid). *Nauyn-Schmiedeberg's Arch. Pharmacol.* **322**, 170–178.

Muller, W.E., A.E. Stillbauer and S. El-Gamal (1983) Psychotropic drug competition for [^{3}H] imipramine binding further indicates the presence of only one high-affinity drug binding site on human α_1-acid glycoprotein (orosomucoid). *J. Pharm. Pharmacol.* **35**, 684–693.

Neurath, H. and Schwart, G.W. (1950) The mode of action of the crystalline pancreatic proteolytic enzymes. *Chem. Rev.* **46**, 69–153.

Otagiri, M., J.S. Fleitman and J.H. Perrin (1980) Investigations into the binding of phenprocoumon to albumin using fluorescence spectroscopy. *J. Pharm. Pharmacol.* **32**, 478–484.

Schill, G., I.W. Wainer and S.A. Barkan (1986*a*) Chiral separation of cationic drugs on an α_1-acid glycoprotein bonded stationary phase. *J. Liq. Chromatogr.* **9**, 641–699.

Schill, G., I.W. Wainer and S.A. Barkan (1986*b*) Chiral separations of cationic and anionic drugs on an α_1-acid glycoprotein-bonded stationary phase (Enantiopac). *J. Chromatogr.* **365**, 73–88.

Seller, E.M. and J. Koch-Weser (1975) Interaction of warfarin stereoisomers with human albumin. *Pharmacol. Res. Comm.* **7**, 331–341.

Sjoholm, I. (1988) Specificity of binding sites on plasma proteins. In *Drug Protein Binding*. Ed. M.M. Reidenberg and S. Erill, Prager Publishers, Philadelphia.

Stewart, K.K. and R.F. Doherty (1973) Resolution of D,L-tryptophan by affinity chromatography on bovine serum albumin-agarose columns. *Proc. Natl. Acad. Sci. USA* **70**, 2850–2852.

Tan, Y.K. and S.J. Soldin (1987) Analysis of salbutanol enantiomers in human urine by chiral high-performance liquid chromatography and preliminary studies related to the stereoselective disposition kinetics in man. *J. Chromatogr.* **422**, 187–195.

Tillement, J.P., G. Houin, R. Zini, S. Urien, E. Albengres, J. Barre, M. Lacomte, P. D'Athis and B. Sebille (1984) The binding of drugs to blood plasma macromolecules: Recent advances and therapeutic significance. *Adv. Drug Res.* **13**, 59–86.

Vogelgesang, B. and H. Echizen (1985) Stereoselective protein binding of verapamil isomers *Nauyn-Schmiedeberg's Arch. Pharamcol.*, R 98.

Wainer, I.W. and R.M. Stiffin (1988) Direct resolution of the stereoisomers of leucovorin and 5-methyltetrahydrofolate using a bovine serum albumin high-performance liquid chromatographic chiral stationary phase coupled to an achiral phenyl column *J. Chromatogr.* **424**, 158–162.

Wainer, I.W. and Y.-Q. Chu (1988) The use of mobile phase modifiers to alter retention and stereoselectivity on a bovine serum albumin HPLC chiral stationary phase. (unpublished).

Wainer, I.W., P. Jadaud, G.R. Schonbaum, S.V. Kakodkar and M.P. Henry. Enzymes as HPLC supports for chiral resolutions: Initial investigations with α-chymotrypsin. *Chromatographia*, (unpublished).

Walle, U.K., T. Walle, S.A. Bai and L.S. Olanoff (1983) Stereoselective binding of propranolol to human plasma, α_1-acid glycoprotein and albumin. *Clin. Pharmacol. Therap.* **34**, 718–723.

8 Cyclodextrin inclusion complexation

L. COVENTRY

8.1 Structure of cyclodextrin

The type of bonded phase for HPLC based on the class of oligosaccharides known as cyclodextrins employs inclusion complexing to achieve chiral selectivity. How inclusion complexing accomplishes this is best understood by examining the physical structure of cyclodextrins. Cyclodextrins are toroidally shaped molecules containing from six to twelve glucose units bonded through α-(1, 4) linkages (Figure 8.1). The physical shape of the molecule is that of a

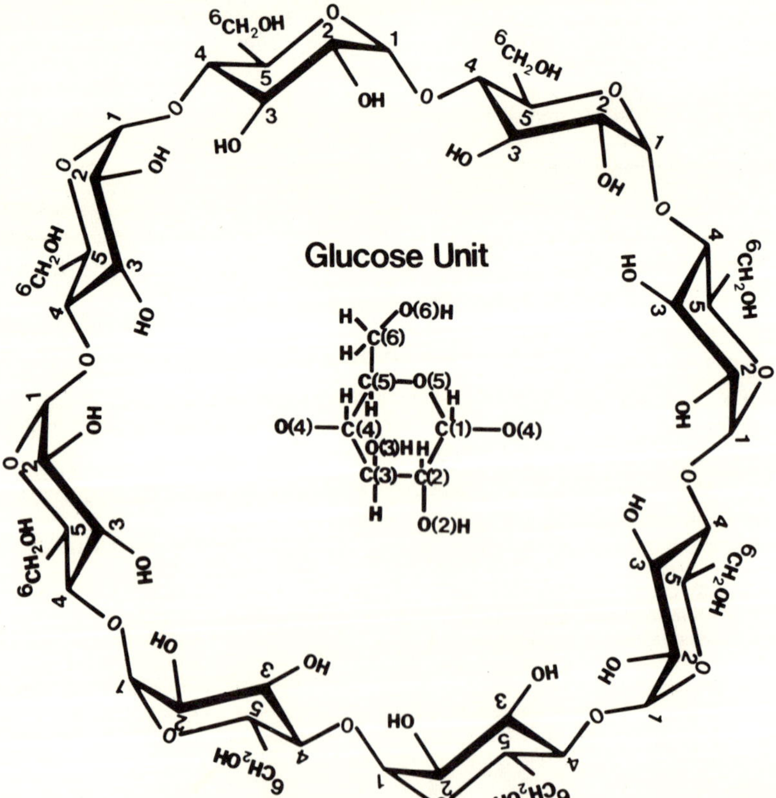

Figure 8.1 Cyclodextrin molecule.

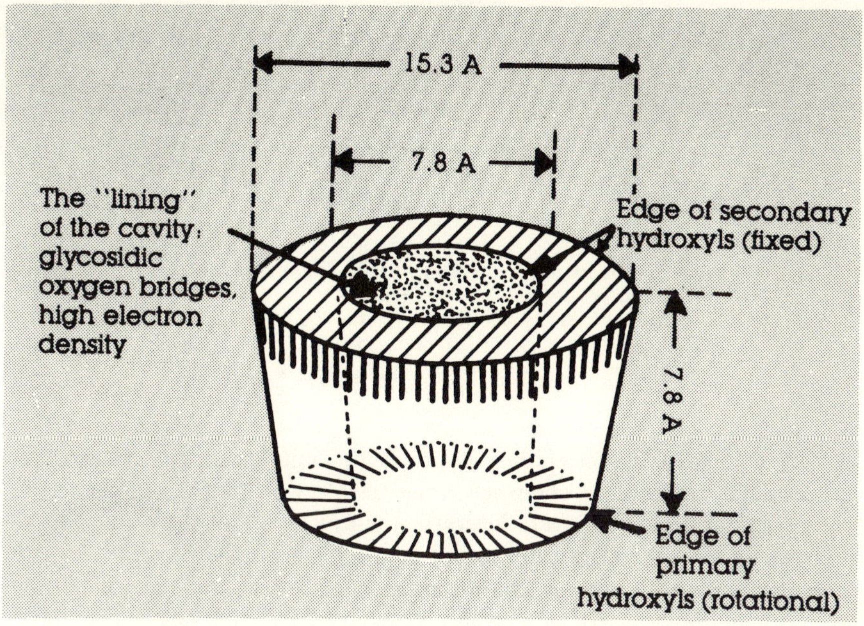

Figure 8.2 Geometry of β-cyclodextrin. From *Cyclobond Handbook.*

truncated cone, the internal diameter of which is proportional to the number of glucose units. Due to the orientation of the glucose units, there are no hydroxyls on the interior cavity and it is therefore relatively hydrophobic. Each glucose unit contributes five chiral centres to the molecule, and the 2-hydroxyl groups at the entrance of the cavity are orientated in a clockwise direction. This can best be appreciated by studying Figure 8.2 which shows the geometry of β-cyclodextrin.

8.2 Mechanism of chiral separation

Chiral separations require the solute molecule to enter the hydrophobic cavity in such a way as to place the centre of asymmetry in association with the polar hydroxyl groups at the edge of the cavity. Where there is no association between these polar groups and the groups attached to or near the solute's chiral centre, separation is minimal or nil. Generally it is not the degree or nature of the penetration into the cyclodextrin cavity that is the main criterion for resolution to occur, but the existence of interaction between the secondary hydroxyls and the guest molecule.

For a racemic molecule to be resolved into its enantiomers there must be a difference in stability of the inclusion complex formed for each isomer. This inclusion complex and the subsequent interaction is shown simply in Figure 8.3.

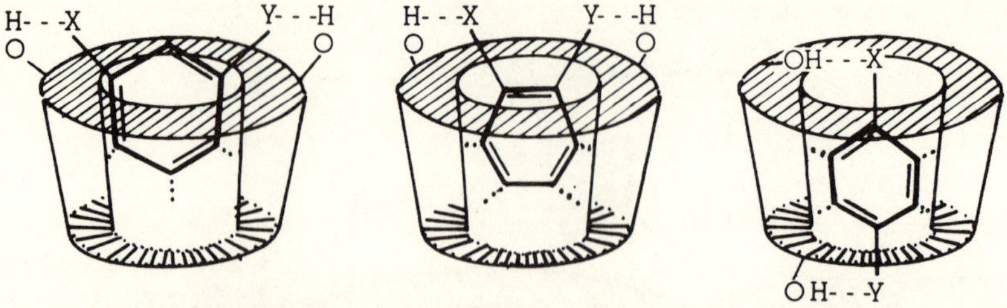

Figure 8.3 Inclusion complex schematic. From *Cyclobond Handbook*.
--- Hydrogen bonding
... Electrostatic association
X Hydrophobic groups that can enter cavity
Y Polar groups that can hydrogen bond.

Different cavity diameters may be selected to suit the size and structure of the solute molecule by choosing cyclodextrins with differing numbers of glucose units in their structure. However, these all have the same cavity depth. To extend this a form of cyclodextrin can be used where the 2-hydroxyls have been derivatized by acetylation. This derivatized form of cyclodextrin allows the interaction to take place where previously the molecule would have projected well beyond the edge of the underivatized cyclodextrin. An example of this will be discussed in section 8.4.3. Commercially there are three types available with varying numbers of glucose units. Two of these are available in the acetylated form. The differences between these columns are summarized in Table 8.1.

Ideally it would be desirable to be able to look at the structure of the molecule to be resolved and logically be able to choose the correct column and mobile phase to effect the separation. This is not often the case, but it is possible to make decisions from a knowledge and understanding of both the functional groups on and around the chiral centre, the size and shape of the molecule and the likely position of inclusion into the cyclodextrin cavity. To do this it is necessary to understand more about the precise nature of both the inclusion complex and the interactions involved.

Table 8.1 Cyclobond columns

Name	Type	Cavity diameter Å
Cyclobond I	β-cyclodextrin	7.5
Cyclobond II	γ-cyclodextrin	9.5
Cyclobond III	α-cyclodextrin	5.7
Cyclobond I acetylated	β-acetylated	7.5
Cyclobond III acetylated	α-acetylated	5.7

8.3 Inclusion complexes

Inclusion complexes are entities comprising two or more molecules, in which one of the molecules, the 'host', includes partially or completely a 'guest' molecule. This complex is achieved with only physical forces i.e. no covalent bonding takes place (Szejtli, 1982).

Cyclodextrins are typical 'host' molecules and may trap a great variety of molecules having the size of one or two benzene rings, or even larger ones carrying a side chain of comparable size. When a compound less polar than water is added to an aqueous solution of cyclodextrin the energetically less favoured water molecules filling the cavity will be replaced by the new molecule, provided the size and shape is compatible with the host. The extent of the complex formation also depends, however, on the polarity of the guest molecule. Complex formation with molecules significantly larger than the cavity may also be possible in such a way that only certain groups or side chains penetrate the cavity.

Naphthalene is too bulky for α-cyclodextrin and anthracene fits only into β-cyclodextrin. Conversely propionic acid is compatible with α-cyclodextrin but has no satisfactory fitting in the larger cavities. This is a good guide as to the choice of cyclodextrin column to use, depending on the size of the molecule to be separated. It must be remembered, however, that the presence and nature of any side chains and functional groups will affect the ability of the molecule to form an inclusion complex. In substituted benzene certain chemical groups and substituents may promote complex formation. In the case of amino groups it is important whether they are present in the $-NH_2$ or NH_3^+ form. Ionic species are usually unfavoured in complex formation.

The following example demonstrates the role of substituents in complex formation. Relative to 4-nitrophenol, methyl groups in positions 2 and 6 have no significant influence on the stability of the complex but even one methyl group in the 3-position lowers the stability of the complex by about two orders of magnitude. 3, 5-dimethyl-4-nitrophenol fails to give a complex at all.

The stability of the complex is proportional to the hydrophobic character of the substituents and therefore a methyl or ethyl substituent will increase the stability. A methyl group in the ortho position to a carbonyl group has a shielding effect on the hydrophilic carbonyl group, which increases the hydrophobic character of the whole molecule. Hydroxyl groups hinder complex formation but their hydrophilic character decreases in the order ortho > meta > para.

The diameter of the β-cyclodextrin cavity is 7.5 Å and the size of the benzene ring is about 6.8 Å, therefore a substituted benzene ring can penetrate the ring either 'head first' or 'head last' but never 'sideways' (Figure 8.4).

In conclusion however, regardless of how or where inclusion complexation takes place, the precondition for stereoselective separation is that the guest molecule must come into contact with the edge of the cyclodextrin cavity and interact in such a way that the stability of the complexes for each enantiomer is different.

8.4 Examples of chiral separations

The mechanism involved in chiral separations using cyclodextrin may best be explained by looking at illustrative examples. Table 8.2 shows some separ-

Table 8.2 Enantiomeric separations on β-cyclodextrin

Compound	k′	Mobile phase	Column length	Reference
β-Adrenergic blockers				
Propranolol	2.78(2.89)	25:75	50	Armstrong *et al.*, 1986
Metoprolol	3.51(3.62)	32:68	50	Armstrong *et al.*, 1986
Antihistamine				
Chloropheniramine	5.86(6.27)	15:85*	25	Armstrong *et al.*, 1986
Calcium channel blockers				
Verapamil	2.94(3.02)		25	Armstrong *et al.*, 1986
Nisolidipene	4.13(4.25)			Armstrong *et al.*, 1986
Nimodipene	5.09(5.34)			Armstrong *et al.*, 1986
Diuretic				
Chlorthalidone	0.50(0.72)	30:70	25	Armstrong *et al.*, 1986
Anticonvulsants				
Hexobarbital	9.39(10.7)	15.85	10	Armstrong *et al.*, 1986
Mephobarbital	14.8(16.9)	20:80	10	Armstrong *et al.*, 1986
Mephenytoin	0.48(0.64)	40.68	25	Armstrong *et al.*, 1986
Triazoline	5.00(5.75)	40.60	25	Armstrong *et al.*, 1986
Phensuximide	1.97(2.26)	10.90*	25	Armstrong *et al.*, 1986
Anticorticosteroid				
Aminoglutethimide	7.49(7.71)	†	25	Armstrong *et al.*, 1986
Anti-inflammatory				
Ketoprofen	7.67(8.13)	27:73	50	Armstrong *et al.*, 1986
Narcotic analgesic				
Methadone	2.38(2.48)	†	25	Armstrong *et al.*, 1986
CNS stimulant				
Methylphenidate	1.17(1.33)	10:90*	50	Armstrong *et al.*, 1986
Amino acid derivatives				
Dansyl-L(D)Valine	2.2(2.6)	50:50		Beesley, 1985
Dansyl-L(D)Threonine	1.7(2.1)	50:50	10	Beesley, 1985
Dansyl-L(D)Norleucine	1.9(2.4)	50:50	10	Beesley, 1985
Dansyl-L(D)phenylalanine	3.1(3.8)	55:45	10	Beesley, 1985
Dansyl-L(D)leucine	3.0(4.0)	50:50	10	Beesley, 1985
Carboxylic acid				
α-methoxy-α-trifluoro methyl phenylacetic acid	7.5(9.8)	50:50	25	Ward and Armstrong (1986)
Metallocenes				
(−)-*S*-(1-ferrocenylethyl)-thio phenol	3.1(4.3)	90.10	10	Ward and Armstrong (1986)

All mobile phase ratios are methanol:water except
*Acetonitrile:water
†Gradient using acetonitrile/1% TEAA pH4.1 10–20% in 20 min.

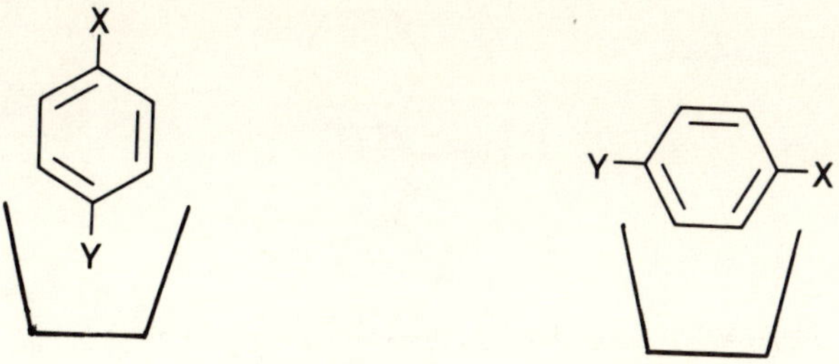

Figure 8.4 Inclusion of a substituted benzene, 'head first' or 'sideways'.

ations achieved using a β-cyclodextrin column and Table 8.3 shows some separations using an α-cyclodextrin column.

Separation of a wide range of compounds has been achieved. The β-cyclodextrin has more applications than the α- at present but this may be due to the fact that the α-cyclodextrin column has been available for a much shorter time. The author has found no separations of enantiomers reported on the γ-cyclodextrin columns.

Table 8.3 Enantiomeric separations on α-cyclodextrin

Compounds	k^a	α	R_s
D, L-tryptophan	2.7	1.20	1.90
α-Methyl D, L-tryptophan	2.6	1.26	1.90
1-Methyl D, L-tryptophan	5.7	1.07	1.10
5-Methyl D, L-tryptophan	4.9	1.07	1.00
6-Methyl D, L-tryptophan	6.4	1.48	4.57
N-Formyl D, L-tryptophan	8.0	1.04	0.70
4-Fluoro D, L-tryptophan	2.4	1.07	0.85
5-Fluoro D, L-tryptophan	2.7	1.08	1.00
6-Fluoro D, L-tryptophan	3.7	1.64	6.18
D, L-phenylalanine	1.1	1.09	0.85
α-Methyl *m*-methoxy D, L-phenylalanine	0.7	1.20	1.10
α-Methyl D, L-phenylalanine	0.9	1.09	1.00
p-Fluoro D, L-phenylalanine	1.5	1.08	1.00
m-Fluoro D, L-phenylalanine	1.0	1.12	1.10
p-Chloro D, L-phenylalanine[c]	5.0	1.05	0.80
N-Benzoyl D, L-phenylalanine[c]	14.2	1.03	0.45
N-Carbamyl D, L-phenylalanine[c]	3.0	1.03	0.70
N-Carbobenzoxy D, L-phenylalanine[d]	5.8	1.02	0.55
D, L-tyrosine	0.1	1.40	0.90
o-Methyl D, L-tyrosine	4.4	1.21	2.70
DNB-Leucine n-butyl thioester[e]	1.3	1.16	0.95

[a]Capacity factor of the first eluted enantiomers.
[b]The separation was achieved using a 25 cm α-CD bonded phase column.
The flow rate was 0.5 mL min^{-1}.
Mobile phase was 1% triethylamine acetate at pH5.1 except
[c]90:10 Buffer:MEOH
[d]74:36 Buffer:MEOH
[e]60:40 Buffer: MEOH.

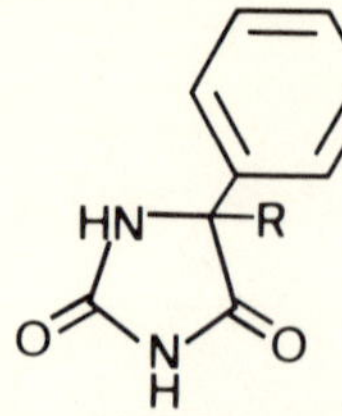

Figure 8.5 The chemical structure of 5-alkyl-5-phenyl hydantoin.

8.4.1 *β-Cyclodextrin column*

Some work on the resolution of phenolic metabolite enantiomers of 5, 5-diphenylhydantoin offers insight in the mechanism of resolution using the β-cyclodextrin column (Maguire, 1987).

The chemical structure of the 5-alkyl-5-phenylhydantoin is described in Figure 8.5. In the series of six 5-substituted phenylhydantoins trends can be observed in the chromatographic properties. With methanol–water eluents the effect of increasing the alkyl substituent is generally to increase the retention time, for substituents up to n-propyl and isopropyl. A comparison of the chromatographic properties of the different substituents is shown in Table 8.4.

The correlation of increasing retention times with increasing lengths of alkyl groups is consistent with a hydrophobic interaction of the 5-alkyl group with the interior cavity of the β-cyclodextrin. The presence of n-butyl and larger substituents may interfere with hydrogen bonding of the hydantoin ring to the secondary hydroxyls of the β-cyclodextrin.

In contrast to the behaviour of the 5-alkyl-5-phenylhydantoins the effect of increasing the 3-alkyl substituent on the chromatographic properties was not noticeable. Figure 8.6 shows the structure of the 3-alkyl substituted phenylhydantoin and Table 8.5 shows the effect of the substituent on the chromatographic properties. For the first four members of the series of n-alkyl substituents there is no obvious trend in the k' or R_s values as hydrophobicity

Table 8.4 Effect of different substituents of 5-alkyl-5-phenyl hydantoin on retention and resolution

R*	k_a	R_s
H	0.67	0.2
CH_3	1.17	0.5
C_2H_5	2.30	2.0
1-C_3H_7	3.30	2.2
2-C_3H_7	4.96	2.5
1-C_4H_9	4.40	1.1

*R = substituent as shown in Figure 8.5.
$k_a = k'$ value of the first eluting enantiomer.

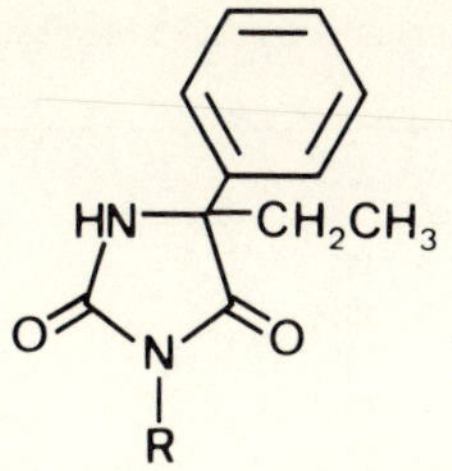

Figure 8.6 The chemical structure of 3-alkyl-5-phenyl-hydantoin.

increases. This suggests an alternative mechanism of enantiomeric discrimination. Maximum resolution is achieved when the substituent is a methyl group and further elongation of the group results in decreasing R_s values. This suggests that disruption of hydrogen bonding by bulky 3-substituents does not allow enantiomeric resolution to occur.

A study of the chromatographic behaviour of a series of substituted 5-(4-hydroxyphenyl)-hydantoins revealed the effect of the hydroxyl substituent on resolution. The structure of these is shown in Figure 8.7 and the chromatographic data in Table 8.6.

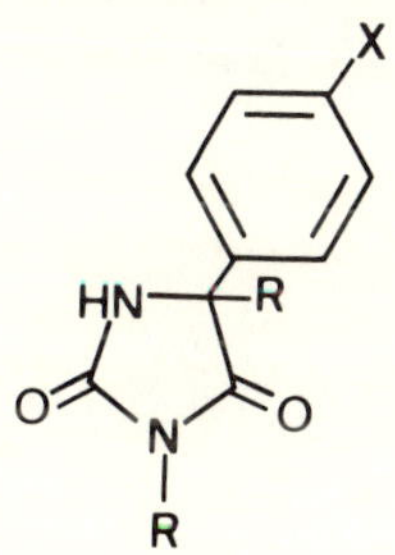

Figure 8.7 The structure of 5-(4-substituted phenyl)-hydantoin.

Table 8.5 Effect of different substituents of 3-alkyl-5-phenyl hydantoin

R^*	k_a	R_s
H	2.30	2.0
CH_3	2.15	4.2
C_2H_5	2.18	2.2
1-C_3H_7	3.13	1.2
2-C_3H_7	2.88	0
1-C_4H_9	5.28	0.4
iso-C_4H_9	6.38	0
C_5H_{11}	14.3	0
$CH_2C_6H_5$	17.8	0

R^* = substituent as shown in Figure 8.6.
$k_a = k'$ values of first eluting enantiomer.

Table 8.6 Effect of different substituents of 5-alkyl-5-(4-substituted phenyl)-hydantoin on retention and resolution

R	*R'*	*X*	k_a	R_s
C_2H_5	CH_3	H	2.15	4.2
C_2H_5	CH_3	OH	2.48	4.7
C_2H_5	H	H	2.30	2.0
C_2H_5	H	OH	2.87	2.3
C_3H_7	H	H	3.30	2.2
C_3H_7	H	OH	5.98	3.6

R and *R'* substituents as shown in Figure 8.6.
$k_a = k'$ value of first eluting enantiomer.

In all the pairs of compounds shown in Table 8.6 the para or 4-hydroxyphenyl derivatives would be considered to be more hydrophilic than the corresponding unsubstituted compounds described earlier. Yet in all cases the k' and R_s values are greater for the 4-substituted compounds. The conclusion is that the phenolic substituent can complex by way of hydrogen bonding with the primary hydroxyls at the edge of the narrower end of the cyclodextrin molecule. This offers the first evidence that hydrogen bonding of *p*-phenols to the primary hydroxyl groups may be important in increasing resolution.

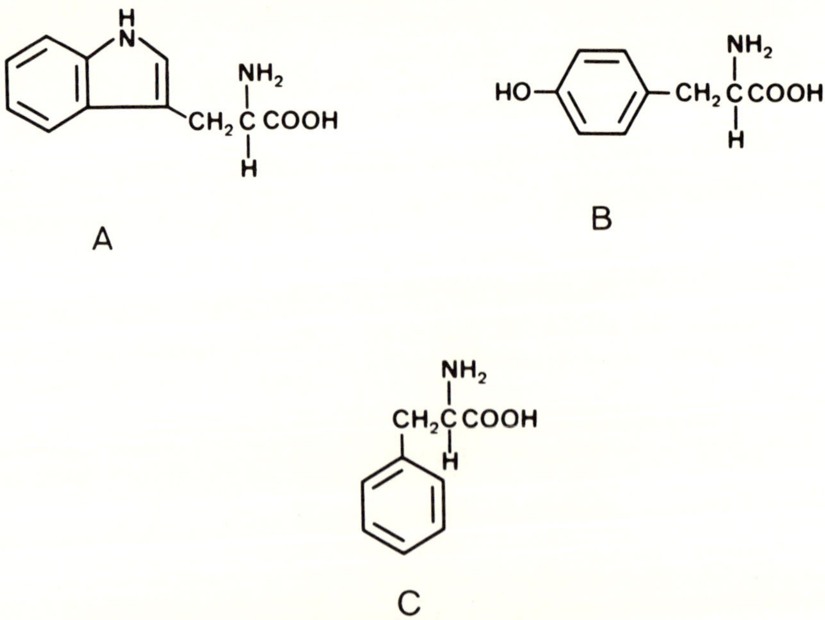

Figure 8.8 The structure of (*A*) tryptophan (*B*) phenylalanine (*C*) tyrosine.

8.4.2 *α-Cyclodextrin column*

The first reported separations on an α-cyclodextrin column have been tryptophan, phenylanaline, tyrosine and their analogues (Armstrong *et al.*, 1987). The structures of these three amino acids is shown in Figure 8.8. A study of the analogues gives some insight into the effect of substituents on resolution.

For tryptophan, substituting methyl or fluoro groups at various locations on the indole ring can affect both resolution and retention. In all but one case, substitution at the 6-position leads to a decrease in enantiomeric resolution. D, L-Phenylalanine is not as well resolved as tryptophan and functionalizing the amine group results in decreased enantioselectivity. Conversely meta substituents tend to enhance selectivity. A meta substituent would tend to skew the orientation of the aromatic ring in the cyclodextrin cavity, forcing the amine and carboxylate groups nearer the mouth of the cavity. Para substituents had little effect unless they extended to the bottom of the cavity and were able to hydrogen bond with the primary hydroxyls as in the *p*-fluoro substituent. Methylation of the *p*-hydroxy group in D, L-tyrosine greatly increases retention and resolution.

Figure 8.9 shows the separation of some of the amino acids and their analogues.

8.4.3 *β-Acetylated cyclodextrin*

Derivatization of β-cyclodextrin to extend the site of interaction allows the separation of compounds that is not possible using β-cyclodextrin. Figure 8.10 shows the separation of D, L-Norgestrel using β-acetylated column (Beesley, 1985). In the case of Norgestrel the chiral centre and its substituents were spacially too far from the mouth of the cyclodextrin cavity to interact with the 2-hydroxyl groups of β-cyclodextrin.

8.4.4 *Aromatic and alkyl groups and their behaviour with cyclodextrin*

It is also instructive to look at some examples of aromatic and alkyl groups in order to show the effect these have with α-, β- and γ-cyclodextrins (Advanced Separation Technologies Inc., 1987).

In Figure 8.11 the retention of polyaromatic hydrocarbons is shown on each of the three cyclodextrin phases. The decreased retention observed for chrysene, phenanthrene and anthracene on γ-cyclodextrin is thought to be the result of less interaction due to the linear nature of these aromatic structures. The more bulky pyrene shows the expected behaviour. Also, due to the bulky nature of the γ-cyclodextrin molecule, less is actually bonded to the silica. This in itself results in reduced retention.

The naphthalene structure enters the γ-cavity horizontally and the β-cavity enters vertically so that the total bonding energy is the same. The slight difference in retention is solely due to the loading differences between the two cyclodextrin types.

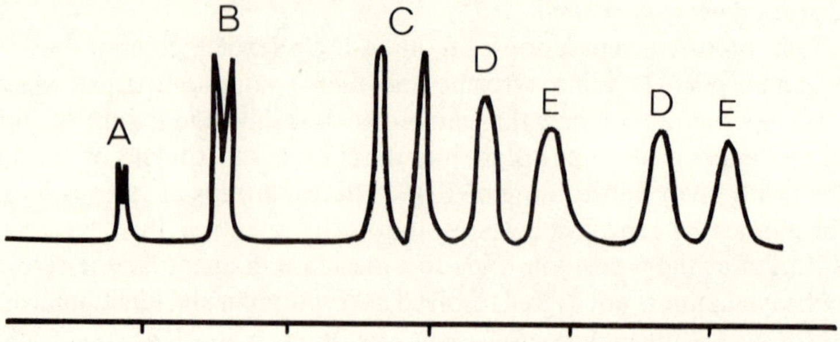

Figure 8.9 Chromatogram showing the separation of (*A*) tyrosine (*B*) phenylalanine (*C*) tryptophan (*D*) 6-fluorotryptophan (*E*) O-methyl tyrosine on a 25 cm α-cyclodextrin column. Conditions as in Table 8.3.

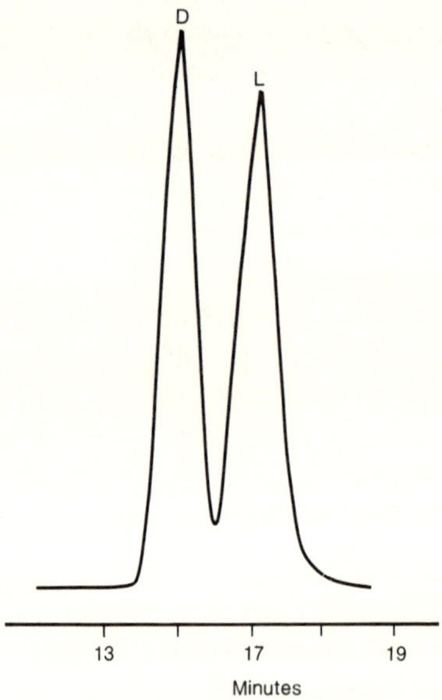

Figure 8.10 Separation of D L Norgestrel.

Not all of the solute molecule has to fit into the cyclodextrin cavity for the separation to occur but at least a portion of it must fit well enough to form a stable complex as is the case with α-cyclodextrin.

Figure 8.12 shows the separation of a series of alkyl nitrosamines on a β-cyclodextrin column and, although this is a non-chiral separation, it shows well the relationship between size, hydrophobicity and retention, and the strength of the inclusion complex formed. The phenyl group has less retention than the isopropyl group because the aromatic group is positioned in the centre of the cavity thereby moving the nitroso group further away from the secondary hydroxyls on the edge of the cavity.

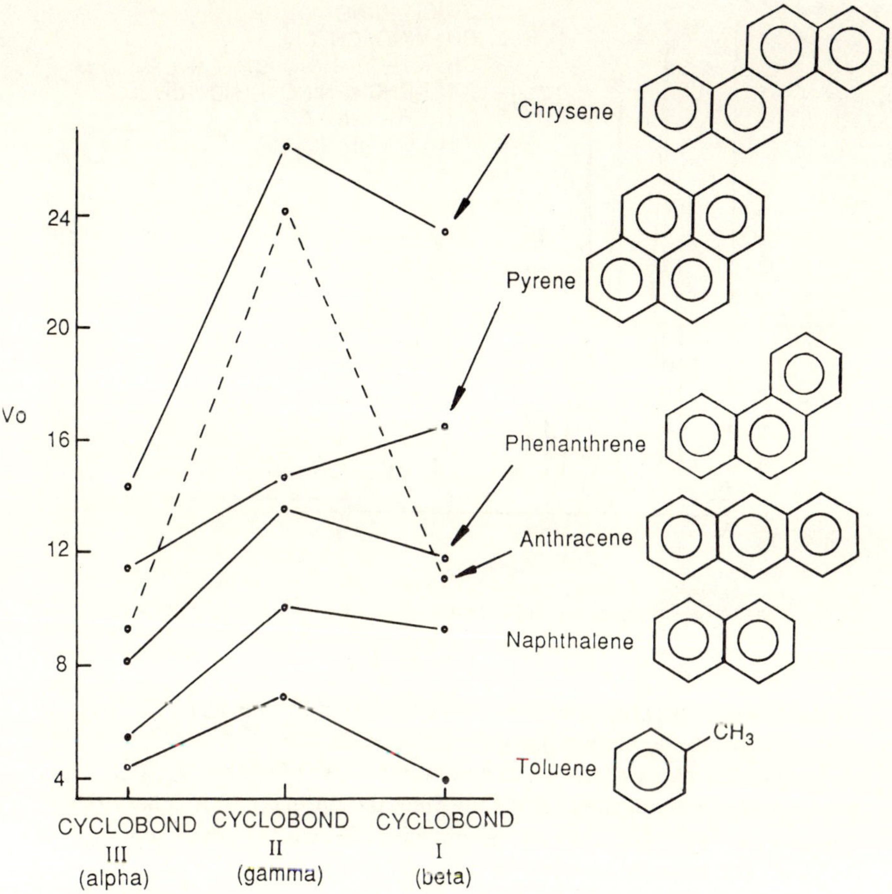

Figure 8.11 Retention of PAHs on Cyclobond columns. From Beesley (1985).

8.5 Choosing conditions for separation

8.5.1 *Mobile phase effects*

Inclusion selectivity is apparently only formed in the presence of water and certain organic modifiers such as dimethylsulphoxide, dimethylformamide, acetonitrile, alcohol and other dipolar solvents. No interactions between aromatic compounds and non-polar solvents such as benzene, chloroform, ether, dioxan, or carbon tetrachloride have been observed. Therefore mobile phases containing water and the above organic modifiers should be used. The binding strengths of substrates to cyclodextrin are reduced in organic: aqueous systems compared with those in water alone. Consequently, just as in reversed-

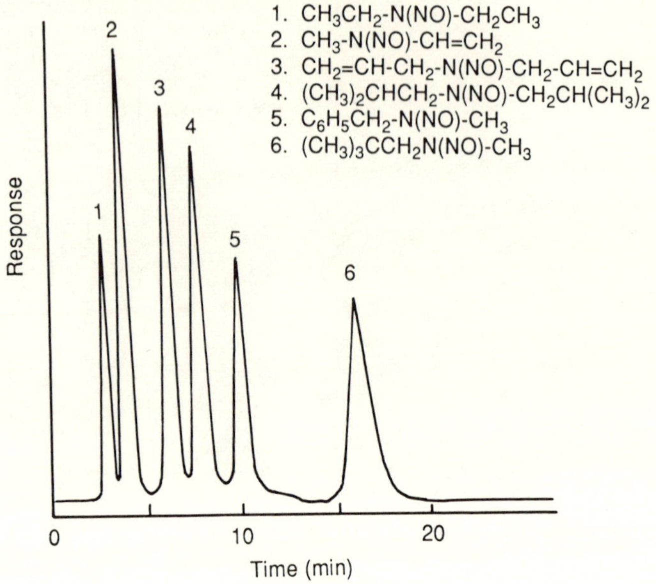

Figure 8.12 The separation of alkyl nitrosamines. From *Cyclobond Handbook.*

phase chromatography, retention can be increased by increasing the proportion of water in the mobile phase.

Acetonitrile has a much greater affinity for the cyclodextrin cavity than methanol and has approximately four times the eluting power. Other organic solvents like DMF and THF can be and have been used to obtain separations, although the mechanism is not fully understood.

It is not possible to predict which modifier will give the best separation but many of the published separations have been achieved using methanol:water and the choice is usually to try methanol first. To decrease retention or increase efficiency or selectivity acetonitrile may be added to the methanol: water mixture or used as a substitute for methanol completely.

Less polar modifiers such as propanol or hexane can be used although the mechanism of separation changes as these solvents occupy the cyclodextrin cavity. In this case, retention occurs due to the adsorption of the solute on the outside of the cavity by interaction with the surface hydroxyls and/or on the solvated surface by way of a partitioning mechanism. It is unlikely that this mechanism would increase the resolution between enantiomers as would inclusion complexing and interaction. However, there may be cases where it would be useful to stop the solute molecule from entering the cavity at a particular site because, for example, it may mean that the chiral centre is too far away for any interaction to take place so that no separation would be achieved anyway. It is the selection of organic modifiers and observation of the elution pattern that occurs that gives an insight into how the molecule is behaving.

Therefore it is recommended that initially a mobile phase containing methanol: water (50:50) is used. The retention time of the solute can be adjusted by changing this ratio. The percent changes should be made in smaller steps than are made with reversed-phase columns as the cyclodextrin column is more responsive to these changes. It is also important to remember that the column must be equilibrated after each change.

8.5.2 *Salt and pH effects*

Because the ionization of the solute molecule is an important factor in inclusion complexing, it is often advantageous to change the pH or add salt mixtures to effect or enhance a separation. By increasing the hydrophobic character of the solute, the retention time increases. In addition, pH control can eliminate or increase the interaction of a functional group with the cyclodextrin or can change its hydrogen-bonding ability. It is necessary to remember that the cyclodextrin is bonded to silica gel and therefore the same pH restraints apply as with reversed-phase chromatography. Also at pHs lower than 3.0 it is possible to cleave the cyclodextrin groups.

With the addition of salts, retention and resolution can be improved markedly due to a narrowing of the peak width. The usual range of salt concentration is from 0.005 to 0.1 M. Buffers that can be used with these columns are trifluoroacetic acid, triethylamine acetate, ammonium acetate and phosphate buffers. The column manufacturers recommend that these buffers should be filtered and passed through a C18 column to remove organic impurities which would collect on the column and impair its performance. Alternatively a guard column packed with C18 material can be used.

As with changes in organic modifier concentration it is important to allow time for the column to equilibrate after a pH change otherwise reproducible results will not be obtained.

8.5.3 *Effect of temperature*

The strength of the inclusion complex is temperature dependent. Stronger binding occurs at lower temperatures. Poor mass transfer and band broadening can be a problem because of the increasing viscosity of the mobile phase and it may be preferable to use acetonitrile and water at lower temperatures rather than methanol:water. If there are significant changes in the room temperature and the method, when developed, is going to be used for routine analysis it may be advisable to use a column oven to ensure reproducible separations and retention times.

8.5.4 *Effect of flow rate*

In normal and reversed-phase chromatography efficiency and flow rate are related. However, flow rate is rarely considered as a means of optimizing a separation, as short analysis times are usually a higher priority. Most analyses are carried out at between 1.0–2.0 mL min^{-1}. With enantiomeric separations it is unlikely that flow rate alone will be the only change required to improve a separation but there can be a marked increase in resolution in going from 1.0 mL min^{-1} to 0.5 mL min^{-1}. For this type of separation generally speed of analysis is sacrificed for baseline resolution.

8.6 Care and use of cyclodextrin columns

Cyclobond columns are cyclodextrins covalently bonded to a silica support and therefore should be operated below pH 7.5. The cyclodextrin groups can be cleaved at pHs lower than 3.0.

Deterioration can also occur when buffer salts are used for a long period and it is recommended that a pre-column be used which will saturate the mobile with silica. This pre-column is placed between the pump and injector and can be packed by hand using 40 micron silica gel as it plays no part in the chromatographic separation.

It is advisable to use a guard column which is placed between the injector and the analytical column and acts as a final filter for the sample. If using biological extracts or samples from fermentation broths a guard column should always be used to prolong the life of the analytical column.

Any salts, acids or bases should always be washed from the column with the mobile phase minus the modifiers. This will help prolong the column lifetime.

These columns can be regenerated by passing several column volumes of water followed by pure ethanol through the column. The columns can also be backflushed when regenerating. If the column is not used for a period of time, it should be stored in methanol.

8.7 Advantages and disadvantages

Most people interested in separating chiral molecules will have a selection of chiral columns from which to choose. There are a number of columns commercially available discussed elsewhere in Chapter 14. In some cases it will be possible to resolve the compound of interest on several different types of stationary phase and in others possibly only on one.

It is incorrect to say that one particular type of phase is better than another in chiral chromatography because the criterion for defining this is not the same as with normal or reversed-phase chromatography. Here, the chromatographer is interested first and foremost in chiral selectivity.

Cyclodextrin columns, however, do offer some advantages in that they are operated in the reverse-phase mode and solvents containing water and organic modifiers can be used. The columns are also rugged and if cared for will last a long time. Should the separation deteriorate the columns can be regenerated. They are also available in preparative sizes, unlike the protein-based columns.

Commercially available columns manufactured before October 1986 were produced with a lower loading of cyclodextrin and therefore will show different selectivity and resolution from those available today. The current series have substantially more cyclodextrin bonded to the silica gel and this increases the retention times and, in most cases, the resolution. It may be possible to reproduce the separation obtained on an early column simply by increasing the amount of organic modifier used.

In some cases the separation may not be as good as that achieved with the column produced before October 1986. To the new user this will be of no consequence but for those involved in chiral chromatography for some time, or for those who are currently using a column purchased before this date, it is important to be aware of these differences.

As a general guide, to decrease the retention times found on the newer columns to those obtained with the older columns the concentration of organic modifier should be increased by 20–25%.

8.8 Use of cyclodextrins in mobile phase

Cyclodextrins have been employed in chiral liquid chromatography primarily as the chiral selector in a chiral stationary phase. However, they can also be used as chiral mobile-phase additives. A number of enantiomeric separations have been reported where α-, β- and γ-cyclodextrins have been used to achieve chiral selectivity in reversed-phase separations. The mechanism of the separation is more complicated than that using bonded cyclodextrin phases as both inclusion complexing and adsorption and partition play a role in the separation. The control of these effects can often be advantageous for achieving selectivity.

The resolution of enantiomers using this technique is much dependent on the concentration of cyclodextrin in the mobile phase (Nobuhara *et al.*, 1983). If the concentration of cyclodextrin is too low, the formation of the inclusion complex is incomplete and the resolution is poor.

The solubility of β-cyclodextrin in a mixture of organic solvent and acetate buffer differs with the choice of organic solvent. β-Cyclodextrin is soluble in ethanol and buffer mixtures, sparingly soluble in methanol mixtures and insoluble in acetonitrile mixtures. Therefore, ethanol is the most widely used organic modifier, in buffered systems. Typical concentrations of cyclodextrins used range from 0.3% w/v to 4.0% w/v.

It has been observed in the resolution of hexobarbital that the order of elution of the enantiomers can change with increasing concentration of cyclodextrin (Zukowski *et al.*, 1986). The elution order of mephenytoin enantiomers remains the same for different concentrations of cyclodextrin and suggests that the separation mechanism for mephenytoin and barbiturates is different.

The enantiomers of D, L-Norgestrel can be completely resolved using γ-cyclodextrin in the mobile phase (Gazdag *et al.*, 1986). In this case the retention of Norgestrel is reduced with increasing γ-cyclodextrin concentration, which indicates strong formation of more polar inclusion complexes. To elucidate the retention mechanism, the adsorption of γ-cyclodextrin on the stationary phase was studied and the results suggested that liquid–liquid partition of the inclusion complex formed in the mobile phase was an acceptable model. It was also found that different organic solvents had no effect on elution order or selectivity.

Resolution was considerably influenced by the column efficiency which was dependent on the organic modifier used. The organic solvents compete with the solutes in the hydrophobic cyclodextrin cavity and maximum resolution was achieved using water–methanol mixtures compared with acetonitrile and tetrahydrofuran.

In the Norgestrel example, the size of the cyclodextrin cavity was found to be important. No inclusion complexes were formed in the presence of α-cyclodextrin. With β-cyclodextrin inclusion complexes were formed but no chiral recognition was observed, which supports previous observations when a bonded β-cyclodextrin column was used. With γ-cyclodextrin, complete resolution of the enantiomers is achieved.

Using the same eluent composition but varying the reversed phase column results in little change in resolution, but produces dramatic differences in retention due to the different carbon loadings of the phases employed. Hypersil ODS (10% carbon) had the least retention and Nucleosil C-18 (14% carbon) the greatest.

In conclusion, cyclodextrins play a useful role in the resolution of enantiomers when used as mobile-phase additives. The same rules apply to them as to the bonded phases, when formation of the inclusion complex and the need for chiral interaction between the guest and host molecule are considered. Examples in the literature show good resolution between enantiomeric pairs and once this has been achieved, the retention can be changed with little effect on resolution by changing the organic modifier or stationary phase used.

References

Advanced Separation Technologies Inc. (1987) *Cyclobond Handbook. A guide to using cyclodextrin bonded phases.* Advanced Separation Technologies Inc., Whippany, NJ.

Armstrong, D.W., T.J. Ward, R.D. Armstrong and T.E. Beesley (1986) Separation of drug stereoisomers by the formation of β-cyclodextrin inclusion complexes. *Science* **232**, 1132.

Armstrong, D.W., X. Yang, S.M. Han and R.A. Menges (1987) Direct liquid chromatographic separation of racemics with an α-cyclodextrin bonded phase. *Anal. Chem.* **59**, 2594–2596.

Beesley, T.E. (1985) Inclusion complexing: a new basis for HPLC selectivity. *Amer. Lab.* May, 78–87.

Gazdag, M., G. Szepesi and L. Huszar (1986) α-, β- and γ-cyclodextrins as mobile phase additives in the high performance liquid chromatographic separation of enantiomeric compounds. 1. Separation of D, L-Norgestrel. *J. Chromatogr.* **351**, 128–135.

Maguire, J.H. (1987) Some structural requirements for resolution of hydantoin enantiomers with a β-cyclodextrin liquid chromatography column. *J. Chromatogr.* **387**, 453–458.

Nobuhara, Y., S. Hirano and Y. Nakanishi (1983) Enantiomeric resolution of 1-[2-(3-hydroxyphenyl)-1-phenylethyl]-4-(3-methyl-2-butenyl) piperazine by reversed phase high performance liquid chromatography using a chiral mobile phase. *J. Chromatogr.* **258**, 276–279.

Szejtli, J. (1982) *Cyclodextrins and Their Inclusion Complexes.* Akademiai Kiado, Budapest.

Ward, T.J. and D.W. Armstrong (1986) Improved cyclodextrin chiral phases: a comparison and review. *J. Liq. Chromatogr.* **9**, 407–423.

Zukowski, J., D. Sybilska and J. Bojarski (1986) Application of α- and β-cyclodextrin and heptakis (2, 6-di-*o*-methyl)-β-cyclodextrin as mobile phase components for the separation of some chiral barbiturates into enantiomers by reversed-phase high performance liquid chromatography. *J. Chromatogr.* **364**, 225–232.

9 Binding to cellulose derivatives

D.M. JOHNS

One of the approaches to solving the problem of chromatographic resolution of chiral molecules is the utilization of naturally occurring, optically active polymers. Like the interaction between an enzyme and its substrate, the polymer can lock on to the different stereochemistries of each enantiomer and provide chiral recognition. Proteins, oligosaccharides and celluloses have had the greatest degree of success, although usually bonded on to silica gel, rather than in their free state. In this chapter, celluloses, their derivatives and analogues will be discussed: oligosaccharides (i.e. cyclodextrins) and proteins (e.g. bovine serum albumin and α_1 acid-glycoprotein) have been covered in earlier chapters.

9.1 Cellulose

Cellulose is an obvious choice as a potential candidate for a chiral stationary phase (CSP), as it is the most abundant organic compound in the biosphere, and is inherently optically active with a highly ordered structure capable of stereoselectivity. It is an unbranched polymer of D-(+)-glucose residues joined by α-1,4 linkages (Figure 9.1). Molecular weights generally range from 250 000 to 1 000 000 or more, showing that each molecule typically contains at least 1500 glucose units. X-ray analysis and electron microscopy indicate that these chains lie side by side in bundles, undoubtedly held together by hydrogen bonds between the numerous neighbouring hydroxyl groups. These bundles become twisted together to form rope-like helical structures, into which the enantiomer of interest partitions and stereospecifically interacts.

In order to be suitable as a CSP, polymers like cellulose and its analogues also need to have the following characteristics:

(i) They need to have sufficient mechanical strength to withstand the operating pressures encountered in HPLC. The degree of rigidity will be strongly dependent on the extent of cross-linking in the structure.
(ii) They need to have a highly ordered structure such that there is only one type of chiral absorbing site, minimizing peak broadening and maximizing resolving power.
(iii) They need to be available in a range of particle sizes suitable for both high-resolution (5 or 10 μm), and also ideally, for preparative (20 or 40 μm) LC work. Particle size distributions will need to be as narrow as

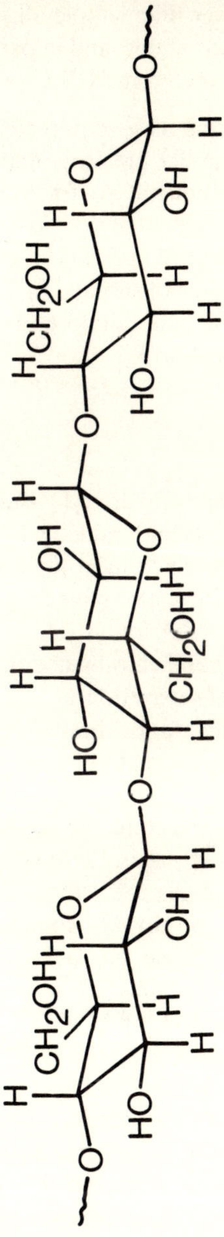

Figure 9.1 Structure of cellulose.

possible for the best efficiency, typically $\pm 1.5\ \mu m$. It is also important to determine the loading capacity at the analytical level in case scale-up should be required at a later stage: some CSPs have a capacity several orders of magnitude lower than simple alkyl-bonded phases.

(iv) They need to be chemically stable, and in particular stable to the organic solvents normally encountered in HPLC.

Cellulose is capable of meeting these requirements, although it is inclined to swell in many solvents and, under pressure, may start to collapse. Without further treatment, it is also difficult to fractionate to sufficiently narrow particle size distribution.

Although the resolving power of natural cellulose is poor, it has been used for many years with some degree of success. In fact, it has been only recently reported (Gübitz *et al.*, 1980) that cellulose is useful in the separation of tryptophan isomers.

9.2 Celluose triacetate

Interestingly, derivatization of the hydroxyl groups of cellulose was found not to destroy its helical structure (Hesse and Hagel, 1973). This gave rise to a new family of cellulose derivatives which meet the above criteria without further treatment and which, fortunately, also have far greater resolving ability than natural cellulose. In 1973, it was found (Hesse and Hagel, 1973) that the triacetate derivative showed good chiral recognition when used directly in its microcrystalline form; partial separation of a cyclic allene hydrocarbon, for example, was achieved (Lindner and Mannschreck, 1980), and resolution of racemic compounds carrying aromatic groups was possible (Mintas *et al.*, 1979).

By cross-linking cellulose triacetate, further advantageous characteristics are achieved (Klimisch and Ambrosius, 1976). Cross-linked cellulose triacetate possesses only a small susceptibility to swelling and has a far lower solubility in organic solvents such as chloroform (1%, compared with 88% for non-cross-linked cellulose triacetate). It also has a finer microcrystalline structure, enabling better fractionation. The pressure drop is, however, limited to 80 bar. It can be obtained in this form as CELAC-40XF from Macherey Nagel, West Germany.

The success of microcrystalline cellulose triacetate (MCT) as a chiral stationary phase was thought to be due to its ordered structure. It would be expected, therefore, that dissolution of MCT in a solvent would lead to a loss of its resolving power. However, when MCT is deposited on silica gel from solution, it not only still shows chiral recognition, but shows different chiral resolution (Okamoto *et al.*, 1984*a*). Trögers base (Figure 9.2), for example, elutes from standard MCT with the (−)-isomer first, whereas, when separated on silica-gel-supported cellulose triacetate, the (+)-isomer elutes first. Similar

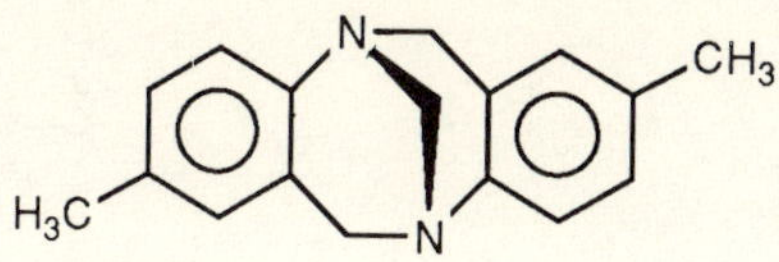

Figure 9.2 Structure of Trögers base.

results were found for other racemates (Shibata *et al.*, 1986). It was suggested (Ichida *et al.*, 1984) that different types of crystalline structure are found in MCT and silica-gel-supported CT, probably due to the difference between chain arrangements or conformations of the glycosidic bond.

9.3 Cellulose derivatives

This work led to an interest in supporting a wide varity of different cellulose derivatives on to silica gel, especially since this technique gives far greater stability and rigidity. The esters, in particular, tribenzoate, trisphenylcarbamate and tricinnamate, as well as the triacetate, have proved to be very useful as CSPs, usually in different application areas. They are readily prepared by the reaction of low-molecular-weight cellulose with the corresponding acyl halide and phenyl isocyanate (Shibata *et al.*, 1986) while the triacetate can be prepared by acidic depolymerization of higher-molecular-weight cellulose triacetate (Okamoto *et al.*, 1984*a*). Okamoto *et al.* (1981) developed a method for coating these derivatives on to macroporous silica gel (Lichrosorb Si 1000) 20–23% by weight which offers good column efficiencies. They also prepared a cellulose tribenzyl ether derivative on silica and have commercially manufactured the complete range as the Chiracel series (Figure 9.3).

9.3.1 *Applications*

The separation of enantiomers on all these phases appears to depend on the 'steric fit' into the cavities of the material and therefore, like the original cellulose, they are very sensitive to small changes in molecular structure at the chiral centre, each derivative showing subtly different characteristics. It is thought (Hesse and Hagel, 1973) that the chiral cavity or ravines of the stationary phase have a high affinity for aromatic groups, and for compounds containing such groups, it is that portion of the molecule that will enter the cavity (Francotte *et al.*, 1985). The affinity of absorbents increases roughly in the order of cellulose trisphenylcarbamate, tribenzyl ether, triacetate, tribenzoate and tricinnamate (Ichida *et al.*, 1984). As the molecular weight of the aromatic moiety increases, k' values also increase, though not generally in a linear fashion over the series. This is presumably due to geometrical changes in the molecule as its size increases.

For separation to occur, the 'fit' of the aromatic moiety in the cavity needs to be reasonably tight but also at least one of the substituents on the chiral centre

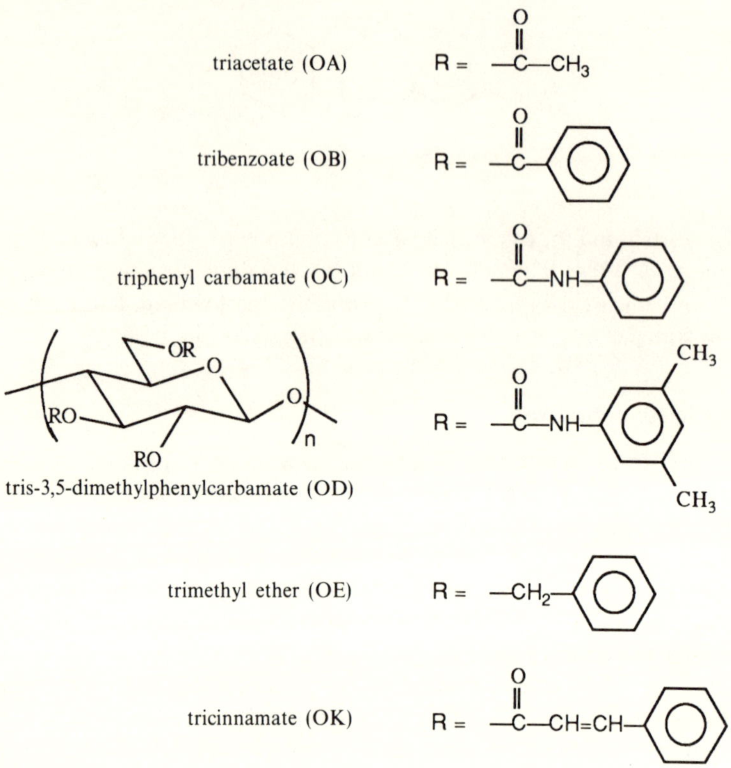

Figure 9.3 Cellulose derivatives and their commercial notation.

needs to be able to interact with the steric environment just outside the cavity. For example in the series shown in Figure 9.4, the phenylcarbamate cellulose derivative exhibited the higher affinity for all the racemates. This was attributed to hydrogen bonding between the urethane groups of phenylcarbamate and the polar hydroxyl and/or amino groups of the substrates (Ichida *et al.*, 1984). In contrast, the smallest retention times were demonstrated by cellulose tribenzylether, where hydrogen bonding will be the least strong.

A phenyl group, however, is not always necessary for resolution, as shown by a homologous series of acyclic enantiomeric amides on cellulose tribenzoate (Wainer and Alembik, 1986). As the steric bulk at the chiral centre increases, the retention time and stereoselectivty increased. Additional interactive forces within the cavity also frequently play an important role. Attractive forces such as π–π and dipole interactions are also used in the formation of the complex between the solute and the cellulose media.

Wainer's group followed up this work with a study of the resolution of a series of enantiomeric aromatic alcohols on the same type of cellulose tribenzoate CSP (Wainer, 1987*b*). Their study led them to propose a chiral

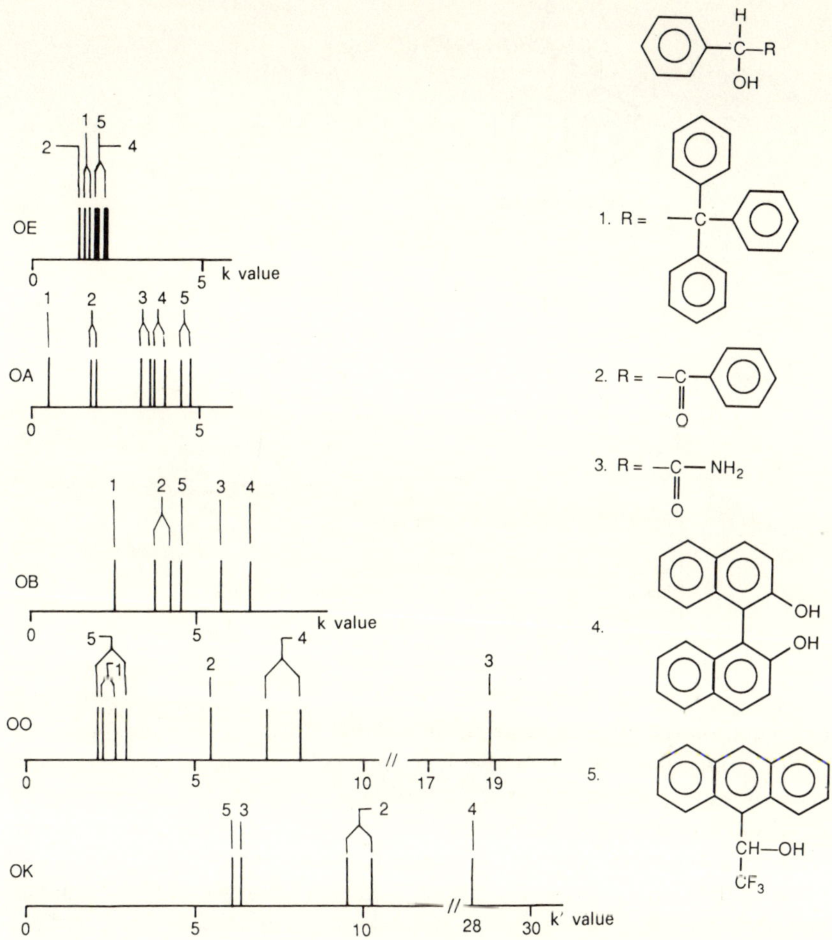

Figure 9.4 Resolution of a series of compounds having hydroxy groups on a series of cellulose derivatives.

Eluent: Hexane-2-propanol (9:1)
Flow rate: 0.2 mL min^{-1}.

From Ichida *et al.* (1984); redrawn and adapted.

recognition mechanism which involved (i) the formation of diastereomeric solute CSP complexes through a hydrogen bonding interaction between the solute's alcoholic hydrogen and an ester carbonyl on the CSP; (ii) the stabilization of this complex through the insertion of the aromatic portion of the solute into a chiral cavity of the CSP; and (iii) chiral discrimination between enantiomeric solutes due to difference in their steric fit in the chiral cavity. Interestingly, this is slightly at variance with Ichida's more general

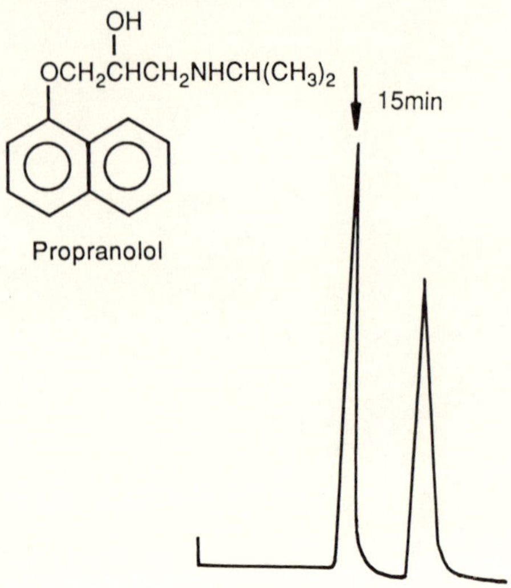

Figure 9.5 Separation of propanolol.

Column: Chiral OD 250 × 4.6 mm
Eluent: Hexane in 2-propanol-diethylamine, 80:20:0.1
Flow rate: 0.5 mL min^{-1}.

From Daicel Chemical Industries Ltd, 1987.

exaplanation of chiral recognition on cellulose CSPs mentioned above.

For compounds with an active hydrogen such as amines and carboxylic acids, however, derivatization of the molecule may often be necessary before injection. Good results are obtained if carboxylic acids are converted into phenyl esters, amines into benzoic acid amides and alcohols into benzoic acid esters. Strong acids or bases are generally unsuitable for separation. Nonetheless, the resolution of a series of β-blocking drugs was achieved without derivatization on the recently developed tris (3, 5-dimethylphenyl carbamate) cellulose (Okamoto *et al.*, 1986*a*). Most of a wide range of β-blockers racemates will separate except, it appears, for those with an aromatic group directly attached to the chiral carbon.

The cellulose derivatives also have good chiral recognition for compounds where the asymmetric centre is a heteroatom such as phosphorus or sulphur. Two agrochemically active thiosulphate compounds, for example, separate particularly well on cellulose triacetate (Figure 9.6), probably by way of mixed mechanism interactions. Among the complete range of cellulose derivatives, the tribenzoate and trisphenylcarbamate exhibit the best selectivity for sulphoxides, whereas the tribenzyl ether showed no signs of separation (Ichida *et al.*, 1984).

Ichida *et al.* (1984) suggested that the following application rules might apply:

(i) Cellulose triacetate (OA) is suitable for many racemates, and especially for compounds where a phosphorus atom is at the asymmetric centre
(ii) Cellulose tribenzoate (OB) shows good chiral recognition for compounds where a carbonyl group (or groups) is close to the asymmetric centre
(iii) Cellulose trisphenylcarbamate (OC) has a strong affinity for polar racemates, and is sensitive to the molecular geometry of the substrate
(iv) Cellulose tribenzyl ether (OE) is effective with protic solvents as mobile phases
(v) Cellulose tricinnamate (OK) is suitable for many aromatic racemates and barbiturates
(vi) Cellulose tris-3, 5-dimethylphenyl carbamate (OD) is especially suitable for β-adrenergic blocking agents.

Utilization of cellulose tribenzoate (Chiralcel OB) under supercritical fluid (SFC) conditions has also been studied (Maccudieri *et al.*, 1987). In general, shorter analysis times and better resolution were observed, when compared with the results obtained with conventional LC solvents, due to higher

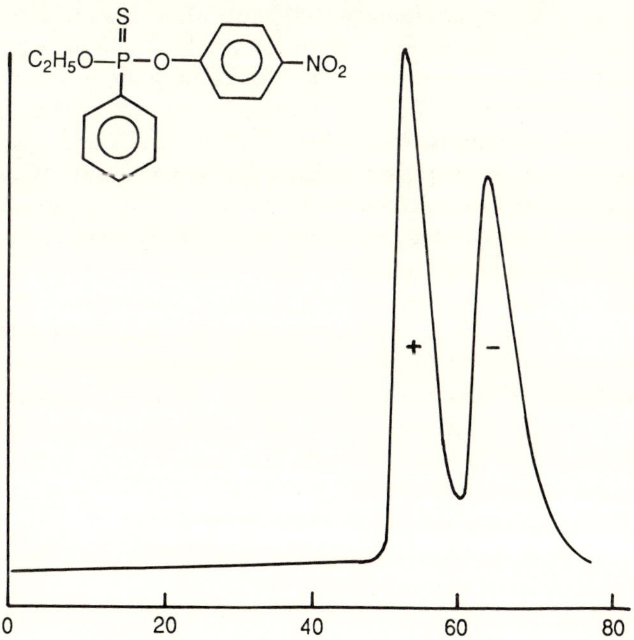

Figure 9.6 Separation of EPN.

Column: Chiralcel OB 250 × 4.6 mm
Eluent: Hexane-2-propanol (9:1)
Flow rate: 0.2 mL min^{-1}.

From Daicel Chemical Industries.

efficiencies. For the resolution of the *p*-methylphenylamide derivative of 2-aminoheptane, for example, it was possible to reduce the retention times by a factor of 10, whilst keeping a constant resolution of 2.0. The constant stereoselectivity indicated that the mechanism of chiral recognition was the same for SFC and LC modes.

The most recent work on cellulose CSPs seems to be directed towards generating an even wider range of such phases. Okamoto and co-workers have developed *trans*- and *cis*-tris(4-phenylazophenylcarbamate), triphenylcarbamate derivative and tribenzoate derivative cellulose CSPs (Okamato *et al.*, 1986*b*, 1986*c*, 1987).

9.3.2 *Practical considerations*

The main influencing factor for chiral recognition in these stationary phases is the choice of mobile phase. For the complete series of cellulose derivatives, mobile phases may be aqueous or non-aqueous, with some notable exceptions, but in general, suitable solvents are limited to hexane, hexane-2-propanol (99 to 50% hexane), and methanol or ethanol or mixtures of these with water. Non-polar mobile phases usually provide greater resolution than methanol/water mixtures, but lower resolution than ethanol, while 2-propanol although acceptable as a mobile phase, gives the lowest resolution. In the separation of *trans*-stilbene oxide on the tribenzyl ether, for example, no separation occurs in 90:10 hexane:2-propanol, but a good separation was achieved with ethanol.

For the cellulose tribenzoate and tribenzyl ether, α-values become larger as the polarity of the mobile phase increases. It was noted also (Wainer *et al.*, 1987*a*) that for a series of alcohol modifiers on the tribenzoate column, an increase in the steric bulk around the hydroxyl moiety tends to result in an increased retention and stereoselectivity for enantiomeric amides. This is due to either the competition for binding to the chiral sites, or to binding of the alcohol to achiral sites near to the cavity, or to a combination of these.

It is important to note with all these phases that chlorinated solvents, like dichloromethane, will remove the derivatized cellulose from its silica support. Similarly, chloroform, acetone, THF, DHF, DMSO, MEK, toluene or acetonitrile (over 1%) cannot be used. A flow rate of 0.5 to 1.5 mL min^{-1} is recommended, although 2 mL min^{-1} can be used for column cleaning. Normal back pressures for all columns are generally 1000 psi at 1.0 to 1.5 mL min^{-1}. As is the case for some other chiral LC systems, resolution can often improve at lower than ambient temperatures.

As these phases are prone to irreversible adsorption, a guard column should be used for biological samples.

9.4 Derivatives of other polysaccharides

A comparison of the characteristics of the phenylcarbamate derivatives of other polysaccharides has also been carried out (Okamoto *et al.*, 1984*b*).

Chiral packing materials were prepared from the phenylcarbamates of amylose, chitosan, xylan, curdlan, dextran and inulin, coating on to silica gel, as before. The chiral discrimination of the stationary phases was highly dependent on the polysaccharides: the xylan phase resolved the same series of compounds as could the cellulose phenylcarbamate. A compound that is typically used as a marker for chiral recognition on Pirkle columns, 1-(9-anthryl)-2, 2, 2,-trisfluoroethanol, was also well resolved on the chitosan and xylan phases. It was concluded that these phases were highly promising as alternatives to cellulose derivatives, although they may prove to be very specific for certain classes of compounds.

References

Francotte, E., Romain, M., Lohmann, D. and Mueller, R. (1985) Chromatographic resolution of racemates on chiral stationary phases. Part 1. Influences of the supramolecular structure of cellulose triacetate. *J. Chromatogr.* **347**, 25.

Gübitz, G., Jellenz, W. and Schönleber, J. (1980) High-performance liquid chromatographic resolution of optical isomers of DL-tryptophan, 5-hydroxy-D,L-tryptophan and DL-dopa on cellulose columns. *J. High Res. Chromatogr. Commun.* **3**, 31.

Hesse, G. and Hagel, R. (1973) A complete separation of a racemic mixture by elution chromatography on cellulose triacetate. *Chromatographia* **6**, 277.

Ichida, A., Shibata, T., Okamoto, I., Yuki, Y., Namikoshi, H. and Toga, Y. (1984) Resolution of enantiomers by HPLC on cellulose derivatives. *Chromatographia* **19**, 280.

Klimisch, H.-J. and Ambrosius, D. (1976) Separation of polycyclic aromatic hydrocarbons by high-pressure liquid chromatography. Comparison of a determination of benzo(*a*)pyrene with separation on columns of cross-linked cellulose acetate and a reversed-phase system. *J. Chromatogr.* **120**, 299.

Lindner, K.R. and Mannschreck, A. (1980) Separation of enantiomers by high-performance liquid chromatography on triacetyl cellulose. *J. Chromatogr.* **193**, 308.

Maccudieri, P., Caude, M., Rosset, R.and Tambute, A. (1987) *Int. Sym. on Chiral Separations*, Guildford, UK (in press).

Mintas, M., Mannschreck, A. and Scheider, M.P. (1979) (1S, 2S)-(+)- and (1R, 2R)-(−)-1, 2-diphenylcyclopropane from their racemic mixture by liquid chromatography on triacetylcellulose. *J. Chem. Soc. Chem. Commun.*, 602.

Okamoto, Y., Kawashima, M., Aburatani, R., Hadata, K., Nishiyama, T. and Mafuda, M. (1986*a*) Optical resolution of β-blockers by HPLC on cellulose triphenylcarbamate derivatives. *Chem. Lett.*, 1237.

Okamoto, Y., Kawashima, M. Yamamoto, K. and Hadata, K. (1984*a*) Useful chiral packing materials for high-performance liquid chromatographic resolution. Cellulose triacetate and tribenzoate coated on macroporous silica gel. *Chem. Lett.*, 739.

Okamoto, Y., Kawashima, M. and Hadata,K. (1984*b*) Useful chiral packing materials for high-performance liquid chromatographic resolution of enantiomers: phenylcarbamates of polysaccharides coated on silica gel. *J. Amer. Chem. Soc.* **106**, 5357.

Okamoto, Y., Okamoto, T., Yuki, H., Murata, S., Noyori, R. and Takaya, H. (1981) Novel packing material for optical resolution: (+)-Poly(triphenyl methyl methacrylate) coated on macroporous silica gel. *J. Amer. Chem. Soc.* **103**, 6971.

Okamoto, Y., H. Sakamoto, K. Hatada and Irie M. (1986*b*) Resolution of enantiomers by HPLC on cellulose *trans*- and *cis*- tris (4-phenylazophenylcarbamate) *J. Chromatogr.* 983.

Okamoto, Y., M. Kawashima and Hatada K. (1986*c*) Chromatographic resolution XI. Controlled chiral recognition of cellulose triphenylcarbamate derivatives supported on silica gel. *J. Chromatogr.* **363**, 173.

Okamoto, Y., R. Aburatni and K. Hatada (1987) Chromatographic chiral resolution XIV. Cellulose tribenzoate derivatives as chiral stationary phases for high-performance liquid chromatography. *J. Chromatogr.* **389**, 95.

Shibata, T., Okamoto, I. and Ishii, K. (1986) Chromatographic optical resolution on polysaccharides and their derivatives. *J. Liq. Chromatogr.* **9**, 313.
Wainer, I.W., Alembik, M.C. and Smith, E. (1987*a*) Resolution of enantiomeric amides on a cellulose tribenzoate chiral stationary phase: mobile phase modifier effects on retention and stereoselectivity. *J. Chromatogr.* **388**, 65.
Wainer, I.W., R.M. Stiffin and T. Shibata (1987*b*) Resolution of enantiomeric aromatic alcohols on a cellulose tribenzoate high-performance liquid chromatography chiral stationary phase. A proposed chiral recognition mechanism. *J. Chromatogr.* **411**, 139.
Wainer, I.W. and Alembik, M.C. (1986) Resolution of enantiomeric amides on cellulose-based chiral stationary phases: Steric and electronic effects. *J. Chromatogr.* **358**, 85.

10 Binding to synthetic polymers

D.M. JOHNS

10.1 Introduction

Given the relative success of natural polymers in chiral LC, it was a logical progression that chiral synthetic polymers would be developed in order to mimic and improve upon them. This expected development has taken place, and synthetic polymer chiral stationary phases (CSPs) are amongst the most recent to be made commercially available. As for natural polymers these synthetic polymers must first be immobilized on silica in some way to be of any practical value in high-speed, high-pressure analytical LC applications.

10.2 Helical polymethacrylates

In following through the concept of developing upon natural polymers, helical polymethacrylates can be thought of as being related to cellulosic materials. Ideally, the helical structure that occurs naturally in celluloses would be utilized to give chiral recognition, at the same time it would be hoped that the limitations that celluloses have even after derivatization and coating on to silica could be overcome.

The preparation of an optically active isotactic polymer, poly(triphenylmethyl methacrylate), PTrMA, by the polymerization of triphenylmethyl methacrylate in the presence of a chiral anion catalyst, was first reported by Okamoto *et al.* (1979). The synthesis yields a high-molecular-weight, highly crystalline polymer (Figure 10.1) which is insoluble in most

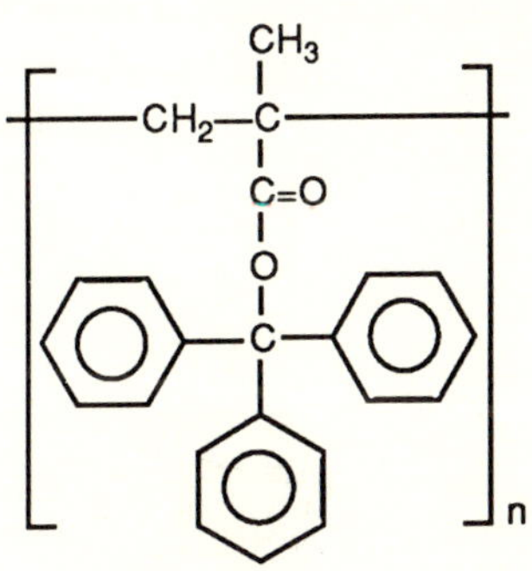

Figure 10.1 Structure of poly(triphenyl methyl methacrylate). From Yuki *et al.* (1980).

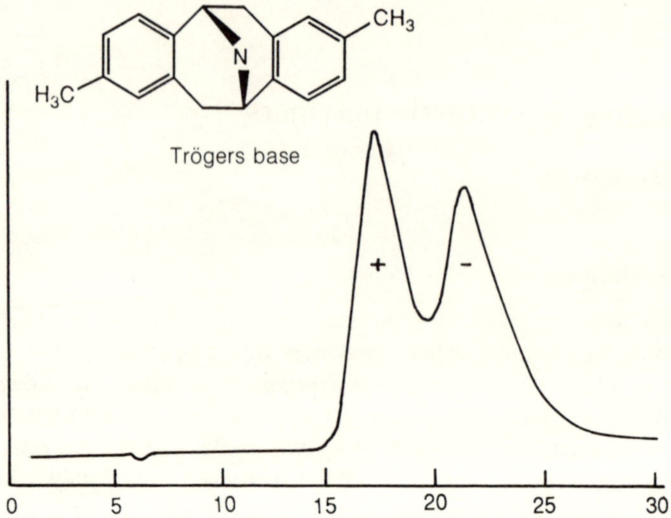

Figure 10.2 Separation of racemic Trögers base by (+)-PTrMA.

Column: 500 × 4.6 mm
Eluent: Cyclohexane: THF (99:1)
Flow rate: 0.2 mL min^{-1}.

Original material from Yuki *et al.* (1980).

organic solvents, and whose chirality arises solely out of its helical structure. Given the enantiomeric separating properties of similar structures, in particular cellulose, the possibility of using this chiral stationary phase was investigated (Yuki *et al.*, 1980) and found to be very useful for the separation of some alcohols, esters, amines and hydrocarbons. Figure 10.2 shows the resolution of a tertiary diamine, Trogers base, achieved using a mobile phase of cyclohexane: THF (99:1). Strong tailing was observed, however, when hexane was used as a solvent. Partial resolution was also achieved for a range of three hydrocarbons, tetramethyl [2.2] paracyclophane, [8]-[8] paracyclophane and '*trans*'-bicyclo [8.8.0] octadeca-1-(10-ene (Figure 10.3).

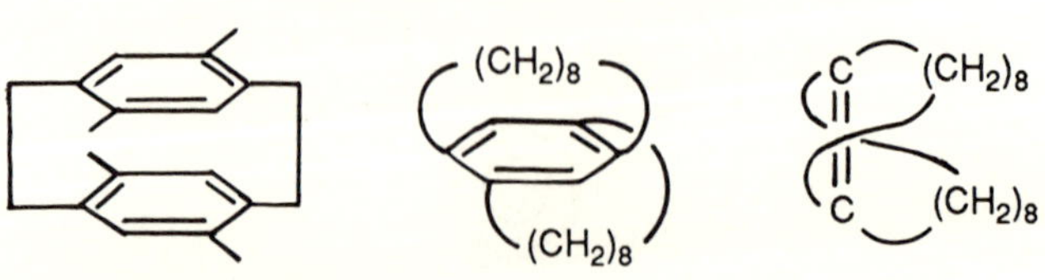

Figure 10.3 Hydrocarbons separated on (+)-PTrMA. From Yuki *et al.* (1980).

In order to accomplish these separations, the polymer was prepared using a (−)-sparteine-butyllithium complex as the catalyst in dry toluene (Yuki *et al.*, 1980). The insoluble polymer, having a degree of polymerization of 220, was ground, and sieved to small particles of about 20–44 μm. The resulting stationary phase, however, was found to swell 2–4 times when hexane was used as mobile phase. Moreover, the material was rather brittle, making column packing difficult.

If the degree of polymerization is 40 or less, the polymer becomes soluble in many common solvents but, importantly, is conformationally stable enough to maintain its helical structure in solution at room temperature. To overcome the difficulties experienced for PTrMA, this low-molecular-weight, soluble PTrMA was adsorbed on to 10-μm silica gel, using THF as solvent. By altering the choice of catalyst for the polymerization step to (+)-6-benzylsparteine-butyllithium complex, a better yield of the soluble (+)-PTrMA was achieved. Additionally, the silica coating procedure was achieved by first treating the 1000-nm pore size silica with a large excess of dichlorodiphenylsilane and triethylamine (Okamoto *et al.*, 1984).

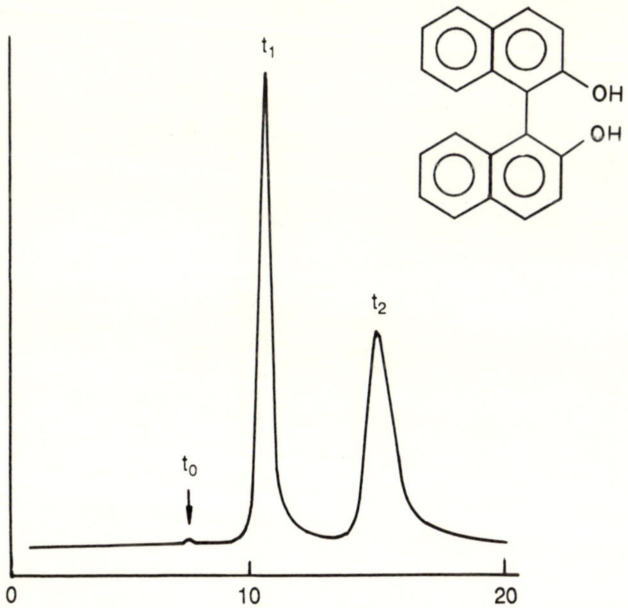

Figure 10.4 Separation of 2, 2′-dihydroxy-1, 1′-binaphyl on (+)-PTrMA-coated silica.

Column: 250 × 4.6 mm
Eluent: Methanol
Flow rate: 0.5 mL min^{-1}
Temperature: 20°C.

From Okamoto *et al.* (1987).

One of the novel effects of coating (+)-PTrMA on to silica is that the new packing showed different chiral recognition when compared with the high-molecular-weight (+)-PTrMA. High-pressure HPLC also became a more viable technique, as the material possessed a higher resistance to compression. A series of racemic 2, 2′-disubstituted 1, 1′-binaphthyls was completely resolved by a column packed with this material, using methanol as eluent (e.g. Figure 10.4). The results were compared with those for ground insoluble (+)-PTrMA, and in general showed more efficient and more rapid separations.

10.2.1 *Applications of poly(triphenylmethyl methacrylate)*

Commercially, (+)-PTrMA coated on to silica gel is available as Chiralpak OT(+) (Daicel Chemical Industries Ltd). In addition, a second phase, Chiralpak OP(+), is also available, where the phenyl rings are exchanged for pyrazine rings. This material, poly(2-pyridyldiphenylmethyl methacrylate), is also coated on to 1000 nm pore size silica.

These phases show an exceptional selectivity for many enantiomers. Separation of compounds possessing a rigid non-planar structure (often with a C_2 symmetry axis) is usually successful. A series of *trans*-disubstituted cyclic

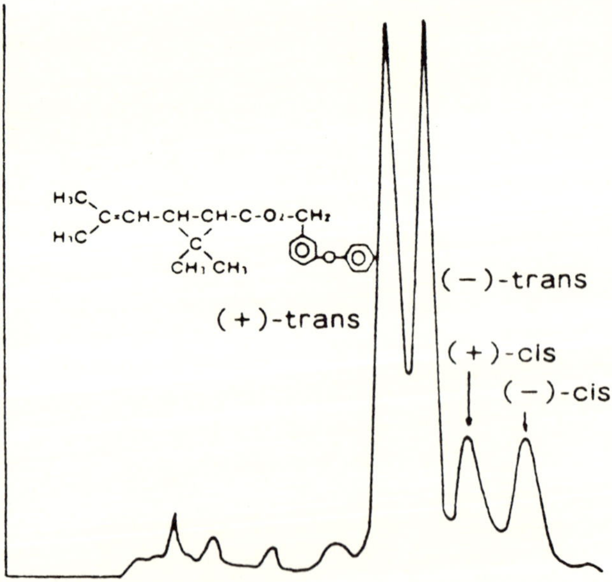

Figure 10.5 Separation of the pyrethroid, phenothrin.

Column: Chiralpak OT(+), 250 × 4.6 mm
Eluent: Methanol
Flow rate: 0.5 mL min^{-1}.

Original material from Okamoto (1987).

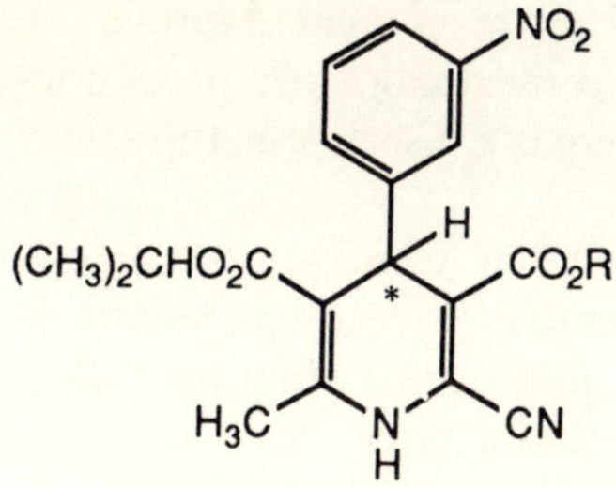

Figure 10.6 Structure of Nilvadipine (R = CH_3). Original material from Tokuma *et al.* (1987).

molecules with six-, four- or three-membered rings for example, were resolved with separation factors ranging from a minimum of 1.18 up to 6.31 (Okamoto *et al.*, 1984).

Phenothrin, 3-phenoxybenzyl chrysanthemate, is an important insecticidal pyrethroid and the four stereoisomers have been shown to separate well on Chiralpak OT(+) (Figure 10.5). One of a new series of dihydropyridine drugs, useful as calcium antagonists, has also been reported (Tokuma *et al.*, 1987) to separate well on Chiralpak OT(+). Separation of (+)Nilvadipine (Figure 10.6) was achieved using a mobile phase of methanol:water (95.5), giving greater than baseline resolution. The separation of 4, 6-dideoxyl-3, 5-O-isopropylidene-2-O-(methoxymethyl)-DL-glucitol, an important precursor to naturally occurring griseusin A, had been found to be poor by diastereomeric derivatization techniques, but was readily achieved as its phenyl ester on Chiralpak OT(+) (Takeuchi *et al.*, 1983).

Racemic compounds containing a phosphorus or sulphur atom at the chiral centre have also been successfully separated on Chiralpak OT(+). Insecticides such as O-ethyl O-(4-nitrophenyl)phenyl phosphonothionate (EPN) and 2-methoxy-4H-1, 3, 2-benzo-dioxaphosphorin 2-sulphate (Salithion) will resolve on Chiralpak OT(+) in aqueous methanol (Okamoto *et al.*, 1984) although they will also separate on the cellulose derivatives, Chiracel OB and OC. In the case of (+)-PTrMA, resolution was found to increase by the addition of water to the eluent, probably because of a stronger hydrophobic interaction between the polymer and enantiomers in a more polar medium.

Another useful separation achieved on Chiralpak OT(+) (3) is the resolution of the tris(acetyl acetonates) of cobalt (III) ($\alpha = 1.32$) and of chromium (III)($\alpha = 1.28$), indicating another area of application in organometallic chemistry.

10.2.2 *Practical considerations*

The recommended mobile phase for the polymethacrylate CSPs is hexane (also used as storage solvent), hexane/2-propanol and methanol. For Chiralpak OP(+) only, water may be added to methanol eluents, up to a maximum of 20%: increasing the water content improves resolution and increases retention time. In both cases, ethanol or 2-propanol give lower α-values than methanol, and are therefore used primarily when required for sample solubility.

For Chiralpak OT(+), the temperature should be below 15°C (down to 0°C) when using methanol, as separations tend to be better and column life longer than at room temperature, while for OP(+) room temperature is recommended. Optimum flow rates are 0.5 to 1.5 mL min^{-1}.

As for the Chiralcel series, solvents to be avoided are methylene chloride, chloroform, acetone, THF, DHF, DMSO, MEK, toluene and acetonitrile (over 1%), as these cause irreversible damage. For Chiralpak OT(+), water, or any aqueous mobile phase, should not be used. A maximum of 0.05% water in anhydrous methanol is advised, so care needs to be taken in the choice of supplier.

Usually, better separation of enantiomers is achieved with polar eluents, such as methanol, rather than with a non-polar eluent (2), suggesting that non-polar or hydrophobic interactions between the triphenylmethyl group of (+)-PTrMA, and a non-polar group of the solute is important for achieving chiral recognition.

In general, samples with strong acids and bases should be avoided when using these columns. Derivatisation in these cases is necessary, and good results are obtained if carboxylic acids are converted into phenyl esters, amines into benzoic acid amides and alcohols into benzoic acid esters.

10.3 Polyacrylamides and Polymethacrylamides

The helical polymethacrylates mentioned above are the synthetic polymer-based CSPs in most widespread use. However, polyacrylamide-based CSPs are also available commercially, albeit on a limited basis (in mainland Europe only, at the time of writing). ChiraSpher (E. Merck, Darmstadt, West Germany) is produced on the surface of silica gels by the radical polymerization of N-acryloyl-(S)-phenylalanine ethyl ester (Kinkel, 1987). This phase arose from earlier work by Blaschke and co-workers (Blaschke, 1986). They prepared not only the polyacrylamide with (S)-phenyl-alanine partial structures, but also polymethacrylamides with (S)-1-phenylethyl and (S)-1-cyclohexylethyl moieties. The chirality of these polymers obviously arises from their individual units rather than any macro-structural features. They were originally prepared by carrying out the polymerization of the monomeric acrylamides in solution in a suspension of LiChrosorb® Diol which had been esterified with methacrylic acid. Later, it was found, surprisingly, that polymerization of the monomers in the presence of LiChrosorb® Diol or silica gel itself resulted in anchoring of the polyamides (*c*. 10% by weight). The exact mechanism of this anchoring has yet to be fully clarified.

The range of racemates that have been resolved on these polyamide CSPs is very similar to the types of compound that have been resolved on the helical polymethacrylate CSPs. Also the same types of mobile phase have been used.

10.4 Other synthetic polymer CSPs

There is an almost infinite number of possibilities for preparing synthetic polymer CSPs from chiral monomeric units, and much progress could well be made in this area in the near future. This is especially so since (i) such polymers are also being developed for use in GC and SFC where the use of polymers is more common and has more obvious advantages, and (ii) many advances have recently been made in 'coating' silica with polymers (Schomburg, 1988).

Amongst other such CSPs which have already been prepared for use in LC are the chiral polyamide-coated silica phases from Saigo's group (Saigo, 1986). The polyamides were synthesized from (−)-1, 2-diphenylethylenediamine and diacid chlorides by low-temperature solution polycondensation. The CSPs were used for the direct LC resolution of enantiomers such as *trans*-1, 2-cyclobutanedicarboxanilide, Trögers base, 2,2′-dihydroxy-1, 1′-binaphthyl, mandelamide, benzyl mandelate and benzoin. Again, mobile phases used were in the main *n*-hexane/2-propanol mixtures.

A polymer containing chiral groups has been prepared by the copolymerization of styrene with the new chiral monomer, N-(*p*-methylstyryl)-(1R,2S)-ephedrine (Villedon-Denaide, 1986). However, to date, its application as the basis of a CSP for use in LC would seem to be hypothetical only.

References

Blaschke, G., Bröker W. and Fraenkel W. (1986) Enantiomeric resolution by HPLC on silica-gel-bound, optically active polyamides. *Angew. Chem. Int. Ed.* **25**, 830.

Kinkel, J.N., Fraenkel W. and Blaschke G. (1987) The separation of enantiomers on polymer coated chiral silica gels. *Kontacte* **1**, 3.

Okamoto, Y., Honda, S., Hatada, K., Okamoto, I., Toga Y. and Kodayashi S. (1984). Chromatographic resolution of racemic compounds containing phosphorus or sulphur atom as chiral centre. *Bull. Chem. Soc. Japan* **57**, 1681.

Okamoto, Y., Honda S. and Yuki H. (1983) Complete chromatographic resolution of tris(acetylacetonato)cobalt(III) and chromium(III) on an optically active poly(triphenylmethyl methacrylate) column. *Chem. Lett.*, 1221.

Okamoto, Y., Honda, S., Hatada, K. and Yuki H. (1985) Chromatographic resolution Part 9: High performance liquid chromatographic resolution of enantiomers on optically active poly(triphenyl methyl) methacrylate. *J. Chromatogr.* **350**, 127.

Okamoto, Y., Suzuki, K., Ohta, K., Hatada, K. and Yuki, H. (1979) Optically active poly(triphenylmethyl methacrylate) with one-handed helical conformation. *J. Amer. Chem. Soc.* **101**, 4763.

Saigo, K., Chen, Y., Kubota, N., Tachibana, K., Yonesawa N. and Hasegawa M. (1986) New chiral stationary phases for the high-performance liquid chromatographic resolution of enantiomers. *Chem. Lett.* 515.

Schomburg, G. (1988) Stationary phases in high-performance liquid chromatography. Chemical modification by polymer coating. *LC-GC* **6**, 37.

Takeuchi, Y., Furumura, M. and Yoshii, E. (1983) Synthesis and optical resolution of 4, 6-dideoxy-3, 5-*O*-Isopropylidene-2-*O*-(methoxymethyl-DL-glucitol. *Chem. Pharm. Bull.* **31**, 3967.

Tokuma, Y., Fujiwara, T. and Noguahi, H. (1987) Determination of (+) and (−)-nilvadipine in human plasma using chiral stationary-phase liquid chromatography and gas chromatography-mass spectroscopy, and a preliminary pharmacokinetic study in humans. *J. Pharm. Sci.* **76**, 310.

Villedon-Denaide, F., Lecavalier, P. and Fréchet, J.M.J (1986) Reactive monomers and polymers containing chiral groups. Synthesis and copolymerisation of N-*p*-methylstyryl-(1R, 1S)-ephedrine. *Polymer Bull.* **15**, 491.

Yuki, H., Okamoto, Y. and Okamoto, I. (1980) Resolution of racemic compounds by optically active poly(triphenylmethyl methacrylate). *J. Amer. Chem. Soc.* **102**, 6356.

11 Ion-pairing

G. SZEPESI

11.1 Introduction

In comparing the use of chiral counter-ions in HPLC eluents for the separation of enantiomeric compounds to other approaches of direct chiral resolution, there are some general features of the ion-pairing technique which should be considered when selecting a method. These can be summarized as follows:

(i) Direct chiral resolution is possible if the chiral counter-ion can form stable diastereomeric complexes with the enantiomeric solutes.
(ii) The stationary phase can easily be selected. Commercial, achiral columns may be used.
(iii) Mobile-phase composition can be varied widely, thus providing excellent possibilities for the optimization of the separation system.
(iv) Elution order of enantiomeric compounds is a function of chiral reagent configuration, and can be altered by using the opposite configuration of the same reagent.
(v) Selection of an appropriate reagent is most difficult due to the very strict requirements regarding optical purity, solubility and detector background demands. These will be discussed in this chapter.
(vi) Compared with other techniques the mobile phase is more complex. This limits its preparative application.

11.2 Retention principles

Organic acids and bases are able to interact with ion-pairing reagents in the mobile phase by forming less polar non-dissociated 'ion pairs'. In reversed-phase systems this technique is very popular for the separation of ionic compounds. Apart from the many existing advantages of reversed-phase ion-pair chromatography regarding the opportunities it provides for achieving significant improvement in separation selectivity, it does not seem to be suitable for chiral resolution. The only reversed-phase ion-pair resolutions of enantiomeric pairs which have been reported involved the use of a zwitterionic ion-pairing agent to resolve the racemate of a zwitterionic solute, e.g. D, L-tryptophan (Knox and Jurand, 1982). This unfavourable situation can be changed when the ion-pair formation using chiral counter-ions dissolved in the eluent is performed in normal-phase systems. As basic literature sources of information, the books of Souter (1985) and of Lindner and Pettersson (1985) are recommended. The structural differences of the diastereomeric ion pairs (complexes) formed are substantial enough to result in their different distribution in the stationary and mobile phases. If a mixture of (R)- and (S)-isomers of an organic base, (R)-B and (S)-B, is separated in a normal phase system containing a chiral counter-ion dissolved in the less polar mobile phase $[(R)\text{-HA}]_m$, diastereomeric ion pairs are formed according to the following equation:

$$[(R)\text{-B}_m]\cdot[R\text{-HA}_m] = K_{RR}[(R)\text{-A:}(R)\text{HB}]_s \qquad (11.1)$$

and

$$[(S)\text{-B}_m]\cdot[R\text{-HA}_m] = K_{SR}[(S)\text{-B:}(R)\text{-HB}]_s \qquad (11.2)$$

where subscripts 's' and 'm' refer to the stationary and mobile phases respectively; the equilibrium constants (K_{RR} and K_{SR}) include ion-pair formation and distribution of the ion pairs formed.

The stereoselectivity of the separation can be expressed by the following formula:

$$\alpha_{S/R} = \frac{k'_{S\text{-B}}}{k'_{R\text{-B}}} = \frac{K_{SR}}{K_{RR}}$$

where $k'_{S\text{-B}}$ and $k'_{R\text{-B}}$ are the capacity factors of the corresponding ion pairs.

As has recently been studied in the literature (Pettersson and Schill, 1981; Szepesi *et al.*, 1982*a*, *b*) the retention model is more complex than that described by simple ion-pair formation.

Pettersson and Schill (1981) described an interaction at three points in the vicinity of the chiral carbon atom which might be necessary and responsible for stereoselectivity. Using (+)-10-camphorsulphonic acid as chiral counter-ion, and some amino alcohols of a β-blocker series (e.g. propanolol, metoprolol), the following essential interactions were identified:

(i) Ionic interaction (electrostatic attraction) between the solute and chiral reagent
(ii) Hydrophobic interactions between the different ring systems
(iii) Hydrogen bonding between the hydroxyl group of the solute chain and keto-group of the reagent.

According to the findings of Szepesi *et al.* (1982*b*), relatively strong acid–base character and suitable lipophilic nature are the most important features required for chiral counter-ions. Both properties together are desirable for diastereomeric complex formation. In addition to the ionic interaction (primary interaction), the hydrophobic interaction between the lipophilic portions of the reagent and solute molecules (secondary interaction) leads to a stable complex structure. Although the possibility for hydrogen bonding can increase the strength of intermolecular interaction between the functional groups attached to the reagent and/or solute, to result in a significant increase in complex stability, stable complexes can be formed with solutes which do not contain a hydrogen-bonding functional group or in which this group is too distant from the place of ionic interaction. In those cases, other selective or non-selective interaction (dipole stacking, electron-pair interactions, dispersion) must replace hydrogen bonding to provide a rigid structure for the diastereomeric complexes if enantioselectivity is to be achieved. As the latter interactions are highly dependent on physicochemical properties of solvents used for eluent preparation, a blend of solvents can be prepared which facilitates the best interactions (tertiary interaction).

11.3 Desired properties of chiral counter-ions

Selection of the appropriate chiral ion-pairing reagent has a great importance, since the success of enantiomeric separation is highly dependent on the features of the counter-ion chosen for the experiments. The most important properties of a suitable chiral reagent can be summarized as follows.

11.3.1 *Acid–base character of counter-ions*

As diastereomeric complex formation by ionic interaction between the enantiomeric solute and counter-ion is necessary (primary interaction), relatively strong acid–base character of the reagent is important.

11.3.2 *Kind and nature of counter-ions*

The size of the lipophilic portion of the reagent contributes to the possibility of the hydrophobic interaction which is required for the diastereomeric complex formed from the reagent and enantiomeric solute to be stable and rigid. Small, polar molecules require the use of a counter-ion with a large, non-polar surface, and vice versa. The structure of the counter-ion not only has a great effect on retention, but also influences stereoselectivity. According to the investigations of Pettersson *et al.* (1981, 1983), improved stereoselectivity can be expected with the introduction of bulky and rigid groups in the vicinity of the chiral centre in the counter-ion.

11.3.3 *Presence of functional groups suitable for tertiary interaction*

The presence of a polar functional group (such as a carbonyl or hydroxyl group) in the vicinity of the chiral centre seems to be necessary for chiral recognition. In the absence of such functional groups, or when the other groups can interfere with the tertiary interaction (total loss in stereoselectivity was reported by Pettersson and Schill (1981) using (+)-3-bromo-10-camphorsulphonic acid), only weak diastereomeric complex formation insufficient for enantiomeric separation can be obtained.

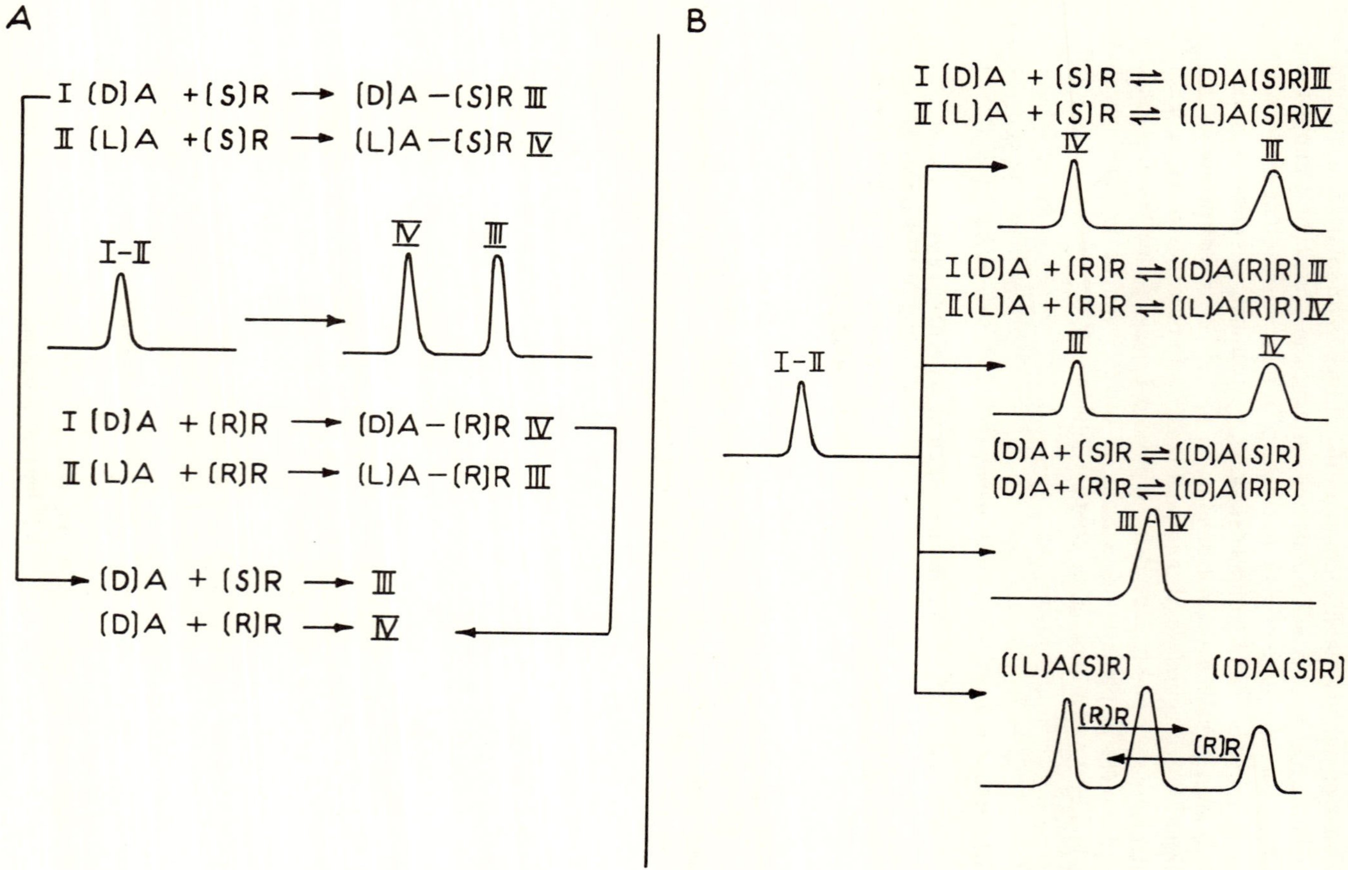

Figure 11.1 Influence of enantiomeric purity of reagent on the separation in case of chiral derivatization (*A*) and in diastereomeric complex formation (*B*).

11.3.4 *Enantiomeric purity of chiral counter-ions*

Separation methods based on pre-derivatization of enantiomeric solutes with an optically active reagent which transforms these solutes to their diastereomeric derivatives have several limitations, and might even be risky to the extent that they can lead to incorrect, false results for the enantiomeric composition. Possible racemization during the derivatization reaction as well as the enantiomeric impurity of the reagent may affect the quantitative determination, especially when enantiomeric impurity in low concentrations is being determined. This is shown in Figure 11.1. As can be seen in Figure 11.1*A*, the enantiomeric impurity of the reagent (S)-R forms a product (IV) with the solute (D)-A possessing the same chromatographic properties as the product formed by the reagent (R)-R with the enantiomeric impurity of the solute (L)-A. The formed products are enantiomers and cannot be separated from each other due to their identical retentions. The knowledge of the amount of enantiomeric impurity in the reagent is important because it will determine the lowest detectable amount of enantiomeric impurity content of the solute.

In contrast, Figure 11.1*B*, using a chiral counter-ion in the eluent, only the stereoselectivity of the separation ($\alpha_{S/R}$) is affected, being decreased by increasing impurity content. The influence of enantiomeric purity of the counter-ion on the value of stereoselectivity ($\alpha_{S/R}$) is demonstrated in Table 11.1 through an example taken from the literature (Pettersson *et al.*, 1987). Pettersson *et al.* (1987) have defined the term of conditional separation factor ($\alpha^x_{S/R}$). It can be calculated according to the following equation:

$$\alpha^x_{S/R} = \frac{\alpha x P + (100 - P)}{\alpha x(100 - P) + P} \tag{11.3}$$

where P is the fraction for one enantiomer compared to total counter-ion

Table 11.1 Influence of enantiomeric purity of chiral counter-ion on stereoselectivity (Pettersson *et al.*, 1987). Conditions: column, LiChrosorb DIOL; eluent, 2.5 mM mixtures of benzoxycarbonylglycyl-L- and D-proline and 0.25 M triethylamine in dichloromethane saturated with 80 ppm water.

Percentage of pure form of counter-ion (%)	$\alpha^x_{S/R}$ Calculated	Found
0	—	1.38
5	1.34	1.33
10	1.29	1.28
15	1.23	1.26
50	1.00	1.01
100	1.38**	1.38**

**$\alpha^x_{R/S}$

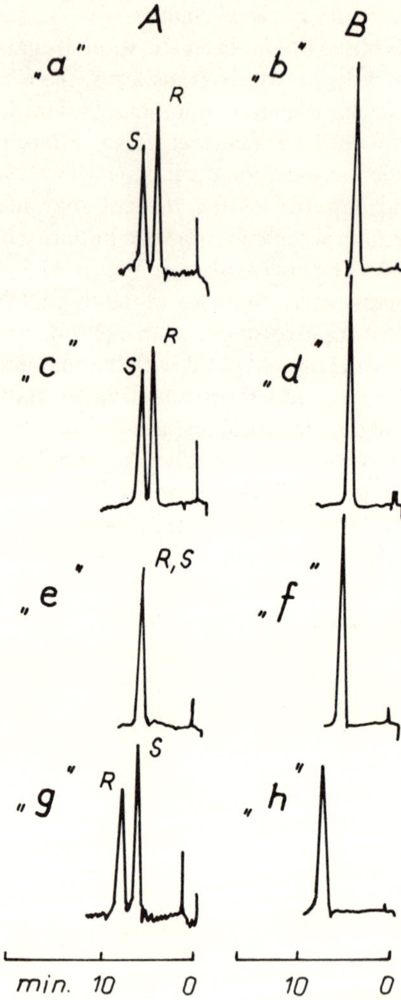

Figure 11.2 Influence of enantiomeric purity of counter-ion on stereoselectivity (Pettersson *et al.*, 1987). Conditions as in Table 11.1.

a, *b*:	100% L-ZGP;	0% D-ZGP
c, *d*:	85% L-ZGP;	15% D-ZGP
e, *f*:	50% L-ZGP	50% D-ZGP
g, *h*:	0% L-ZGP	100% D-ZGP

A: racemate; *B*: pure *R*-enantiomer

ZGP: benzoxycarbonylglycyl-D- and L-proline

concentration in percentage; α is the separation factor when the reagent is optically pure; and $\alpha^{x}_{S/R}$ is the conditional separation factor.

Stereoselectivity as a function of enantiomeric purity of reagent is shown on chromatograms in Figure 11.2 via the example of (R)- and (S)-propranolol using (D)- and (L)-forms of benzoxycarbonylglycylproline as chiral counter-ions.

In accordance with the theory of diastereomeric complex formation, good separation can be achieved when the chiral reagent is contaminated with 15% antipode (single peak on chromatogram *d*, when only the R-isomer is injected on to the column), while by chemical derivatization two peaks with a peak height ratio of 85 to 15 should be obtained. This means that the optical purity of chiral counter-ions is not so important as in the case of chiral derivatization, usability of the reagent being limited only by the separation factor (stereoselectivity) achieved by the application of optically pure chiral reagent. If it is sufficiently large, even higher impurity contents of the reagent (10–20%) can be tolerated for the separation.

11.3.5 *Other important properties of chiral counter-ions*

Three important features of the chiral reagent have been mentioned: its acid–base character, lipophilic nature and optical purity. The applicability of chiral reagents is limited by two additional factors. The first limiting factor is low UV-absorbance in the spectrum range desired for the detection. This restriction excludes the use of several reagents generally applied in synthetic chemistry. The second factor is solubility of the reagent in the mobile phase. As was mentioned, the enantiomeric separations are carried out in normal phase systems. Low solubility of the reagent in low-polarity mobile phases can exclude its application.

Table 11.2 Chiral counter-ions dissolved in the eluent for enantiomeric separations

	Chromatographic conditions			
Chiral counter-ion	Stationary phase	Mobile phase	Enantiomeric solute	References
(+)-10-Camphor-sulphonic acid	LiChrosorb DIOL	Dichloromethane + different alcohols	β-Blockers alprenolol, metoprolol)	Pettersson *et al.* (1981, 1982, 1986)
	Cyano-propyl silica	Mixture of apolar, polar and medium-polar solvents + DEA	Ergot peptide and eburnane alkaloids	Szepesi *et al.* (1982*a*, b, 1988)
	Silica	Mixture of polar, medium-polar and apolar solvents + TEA	Amino steroids	Ladanyi *et al.* (1986)
Different amines	LiChrosorb DIOL	Dichloromethane + 1-pentanol	Carboxylic and sulfonic acids	Pettersson and No (1983)
Benzoxycar-bonylglycyl--L- and D- proline	LiChrosorb DIOL	Dichloromethane (water) + TEA	β-Blockers	Pettersson *et al.* (1986, 1987)

11.4 Availability of chiral counter-ions

Some of the reagents are available from general suppliers, others can be synthesized as described in the appropriate references. These are listed in Table 11.2, indicating the kind of chiral reagents, along with some information about the separation systems and enantiomeric solutes.

Based on the requirements discussed earlier, it would seem that a search for new types of chiral counter-ions would be desirable, especially if the problems of (semi) preparative resolution of the diastereomeric complexes and recovery of the parent enantiomers are to be addressed.

11.5 Separation variables

The initial selection for trial separation is made by judgement. If the choice turns out to be a poor one, great efforts must be directed to improve the stereoselectivity of the separation. Achieving sufficient separation for the enantiomeric solutes is also a function of the number of separation variables, because if one variation fails to yield acceptable separation, a variety of other variables can be considered. Separation variables which can be manipulated to improve the selectivity and efficiency of chiral separations using chiral counter-ions dissolved in the mobile phase are summarized as follows.

11.5.1 *Influence of mobile-phase composition*

Although several publications described the use of a single-component eluent (dichloromethane) modified with a small amount of a polar constituent (alcohols, water, acetonitrile), the findings of Szepesi *et al.* (1982*b*) indicate that a mobile phase consisting of three solvents possessing different polarities (a hydrophobic, a polar and a medium-polarity solvent) offers excellent possibilities for phase system optimization. The data obtained by variation of hydrophobic, polar and medium-polarity solvents (Szepesi *et al.*, 1982*b*) in the separation of different eburnane alkaloids is shown in Table 11.3.

From the experimental data collected in Table 11.3, the following general conclusions can be drawn:

(i) Evidently, the nature of the medium-polarity solvent is a key factor in achieving the separation of enantiomers. High priority should be given to the selection of the most appropriate medium-polarity solvent. According to the data collected in Table 11.3, when dioxan and tetrahydrofuran are used as medium-polarity solvents, there is sufficient separation for apovincaminic acid ethylester enantiomers, even when no hydrogen bonding group is attached to the molecules. In accordance with the data of Pettersson *et al.* (1981, 1982, 1986, 1987), chlorinated hydrocarbons (chloroform, dichloromethane etc.) can only be used when a polar function such as hydroxy-or phenoxy groups are in the proximity of the chiral centre present on the molecules.

Table 11.3 Influence of apolar, polar and medium-polar components in the eluent on the selectivity and efficiency of enantiomeric separations (Szepesi *et al.*, 1982*b*). Conditions: column, cyano-propyl silica; eluent, mixture of apolar (*X*), polar (*Y*) and medium-polar (*Z*) solvents containing 2 mM (+)-10-camphor sulphonic acid and 1 mM DEA.

	cis-Apovincaminic acid ethylester				*cis*-Vincamine			
	$k'(+)$	$k'(-)$	α	H (μm)	$k'(+)$	$k'(-)$	α	H (μm)
A. Apolar component (*X*: dioxane: 1-butanol = 75:20:5).								
Hexane	8.70	10.2	1.18	80				
Isooctane	5.49	6.59	1.20	83				
Heptane	5.37	6.36	1.18	83				
Cyclohexane	4.95	5.69	1.15	87				
B. Medium-polar component (hexane: *Z*: isopropanol = 70:25:5).								
Chloroform	3.67	3.67	1.00	—	5.13	5.60	1.09	79
Dioxan	10.0	11.6	1.15	95	12.5	14.6	1.16	88
Dichloroethane	6.03	6.03	1.00	—	9.52	9.97	1.05	172
Dichloromethane	3.74	3.74	1.00	—	5.90	6.40	1.09	170
Acetonitrile	4.75	4.75	1.00	—	8.57	8.57	1.00	—
Tetrahydrofuran	3.53	3.81	1.08	120	4.57	4.83	1.05	81
C. Polar component (hexane: dioxane: *Y* = 57.5:37.5:5)								
Methanol	1.55	1.55	1.00	—	2.12	2.30	1.09	130
Ethanol	2.48	2.82	1.14	85	2.02	2.46	1.14	115
1-Propanol	3.36	3.91	1.16	79	4.39	5.08	1.16	115
2-Propanol	3.91	4.54	1.16	100	4.88	5.69	1.19	160
1-Butanol	3.82	4.50	1.18	80	4.96	5.53	1.12	81
2-Butanol	4.85	5.63	1.16	119	6.63	7.41	1.12	100
tert-Butanol	10.0	12.1	1.16	246	10.7	11.6	1.18	100
Methoxyethanol	2.03	2.49	1.23	127	4.41	4.94	1.12	89
Ethoxyethanol	2.85	3.21	1.13	76	3.66	4.03	1.10	95

(ii) Apparently, the nature of the hydrophobic solvent does not have a significant influence on the stereoselectivity. However, the retention of the compounds is dependent on the nature of the hydrophobic solvent.

(iii) Convenient control of the retention can be achieved by changing the ratio of hydrophobic and medium-polarity solvents without significant loss in selectivity. This is shown in Figure 11.3. A linear correlation has been obtained between the capacity factors and the ratio of hexane and dioxane in the eluent.

(iv) Polar constituents mainly affect the efficiency of the column and have less of an effect on the stereoselectivity. The correct choice of polar constituent is also an important factor in the optimization of the separation system. The use of special polar solvents (e.g. ethoxyethanol) can play a role in improving stereoselectivity.

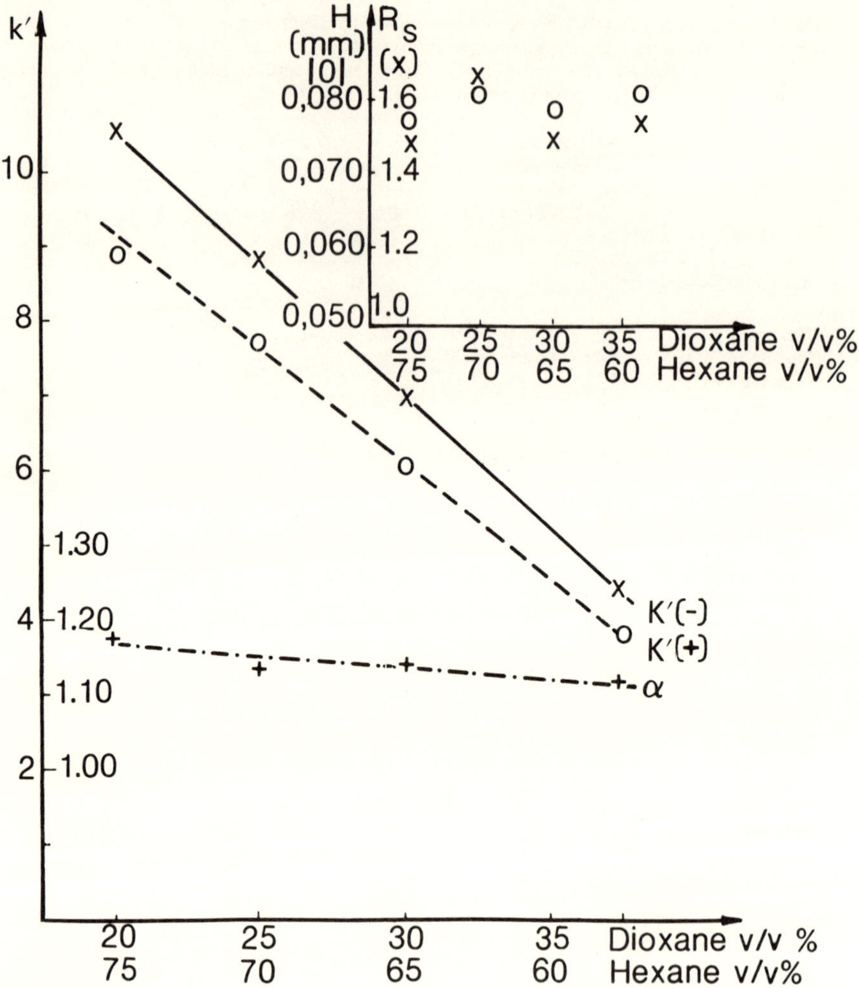

Figure 11.3 Dependence of capacity factors, selectivity, column efficiency and resolution measured for apovincaminic acid ethyl ester on the ratio of apolar and medium-polar solvents in the eluent (Szepesi *et al.* 1982*b*). Conditions: column, Nucleosil 10CN; eluent, (hexane-dioxane)-1-butanol (95:5) containing 2 mM CSA and 1 mM DEA.

With respect to the mobile phase composition, there is another point worthy of consideration, namely, the effect of water content of the eluent on the separation. According to the experience of Pettersson *et al.* (1981), in no case could resolution be obtained if water was present, even in a low concentration, in the mobile phase. Interference of water with hydrogen bonding between the hydroxyl group of the solute and the keto-group of camphorsulphonic acid was given as the explanation for its deleterious effect. However, it was found that when using the mobile-phase composition introduced by Szepesi *et al.*

(1982*b*) on a cyanopropyl silica stationary phase, water below 1% concentration did not affect the separation. This seemed to be a significant improvement on other systems. Similar results were reported by Ladanyi *et al.* (1986) on silica.

11.5.2 *Stationary-phase selection*

Applications of silica (Ladanyi *et al.*, 1986) silica- DIOL (Pettersson *et al.*, 1981–87) and cyanopropyl-silica (Pettersson *et al.*, 1986, 1986, 1987; Szepesi *et al.*, 1982*a*, *b*, 1988) have been reported as stationary phases in normal phase systems. With respect to the sensitivity to water reported by Pettersson *et al.* (1981), long equilibration times were necessary in order to achieve stable operating conditions. This experience has not been reported in other cases, possibly due to the application of mobile phase additives (diethyl- and triethylamine). In the original report of Szepesi *et al.* (1982*a*, *b*), diethylamine (DEA) was used as the mobile phase additive, and the influence of the concentration of DEA on retention behaviour was investigated in detail (see section 11.5.3). From the experimental data it can generally be concluded that the additive forms an ion pair with the reagent which can readily adsorb to the silanol groups on the stationary phase surface, with the result that peak shapes and separation efficiency are significantly improved.

Chiral ion-pairing agents in normal phase eluents may also be used in conjunction with chiral stationary phases. It was shown recently (Pettersson *et al.*, 1988) that the presence of quinine in the mobile phase (0.35 mM acetic acid in dichloromethane) significantly increased the stereoselectivity for the enantiomers of naproxen when using acetylquinidine-silica as the chiral stationary phase. On the other hand, using quinidine in the mobile phase destroyed the stereoselectivity.

11.5.3 *Influence of concentration of chiral counter-ion and of mobile phase additive on retention*

The dependence of the capacity ratio of an enantiomeric solute on the concentration of a chiral counter-ion is very complex. It can be influenced by several simultaneously existing equilibria. In Figure 11.4, the dependence of the capacity ratios of some eburnane alkaloids on the camphorsulphonic acid (CSA) concentration in the presence of a constant amount of DEA (Figure 11.4*A*) as well as on DEA concentration using constant amount of CSA (Figure 11.4*B*) are shown (Szepesi *et al.*, 1982*a*).

Data shown in Figure 11.4 reveal that CSA, in accordance with its strong acidic character, can form complexes with eburnane alkaloids in the presence of excess DEA. This suggests that the separation mechanism involves a complex exchange. A 2:1 ratio of CSA and DEA used in as low a concentration as possible (2×10^{-3} M CSA and 10^{-3} M DEA) seemed sufficient for diastereomeric complex formation to take place.

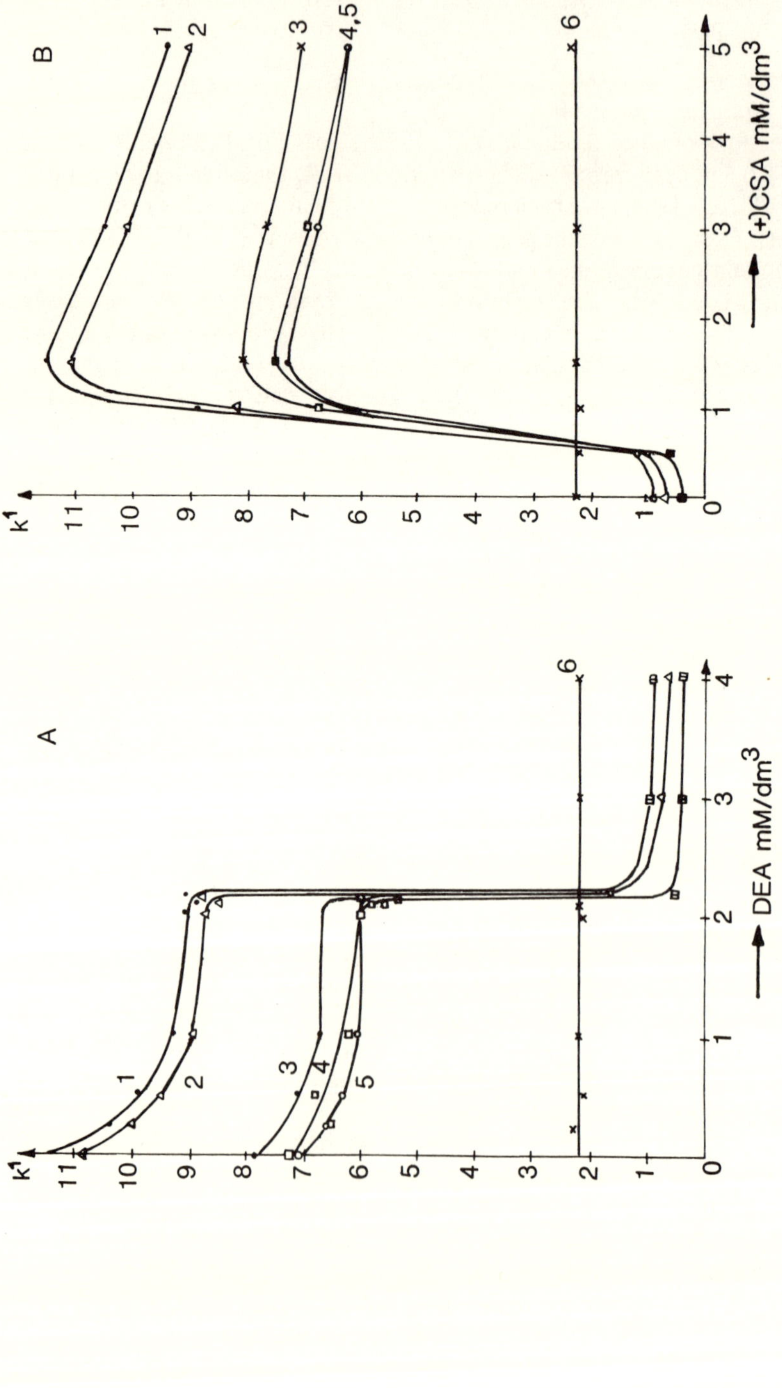

Figure 11.4 Influence of CSA (*A*) and DEA (*B*). Concentrations on the capacity ratios of some eburnane alkaloids using constant concentration of DEA (*A*) and CSA (*B*) (Szepesi *et al.*, 1982*b*). Conditions: eluent, hexane-isopropanol (8:2) containing 1 mM DEA and 2mM CSA, respectively.

11.5.4 *Influence of optical antipode selection of chiral counter-ion on selectivity of separation*

When enantiomers are investigated, it is sometimes possible to improve the selectivity of the separation by changing the optical antipode of chiral counter-ion. It was demonstrated (Figure 11.2), that the elution order of R- and S-propranolol can be regulated by the relative proportion of the enantiomeric forms of the counter-ion in the mobile phase (Pettersson *et al.*, 1987). Similar results were obtained by Szepesi *et al.* (1982*b*).

This possibility of altering the retention behaviour should have great importance, especially when the optical purity of the solute is investigated. The enantiomeric impurity can be eluted before or after the principal component, depending on the antipode of the counter-ion. Moreover, the observed shift of the retentions for the replacement of one enantiomeric form of the reagent by the other one can provide information about the nature of the impurity peak in low concentration, thus enabling its certain assignation as an enantiomeric impurity (elution order will be reversed), or that of another impurity, such as a structurally related compound (elution order remains the same). This information is valuable, especially when purity tests of unknown samples are performed.

11.5.5 *Influence of column temperature and flow rate*

A detailed study of the influence of column efficiency on the flow rate and column temperature has been performed by Pettersson *et al.* (1987). They found that there was a significant increase in the plate number when the linear flow velocity was decreased from 4 to 0.7 mm s^{-1}. It was also found that the mass transfer characteristics could be improved by increasing the column temperature. This resulted in decreased retention and plate height. However, in contrast with this improvement, the resolution was diminished due to the significant decrease in stereoselectivity at elevated temperature.

11.6 Problems of diastereomeric complex formation using chiral counter-ions: advantages and limitations

Table 11.4 gives an overview of the applicabilities, advantages and limitations of this technique.

Considering the application areas where enantiomeric separations are desired (in-process control of asymmetrical synthesis; control of optical purity of intermediates and drug substances; determination of different enantiomeric forms of administrated racemic drugs in biological media; preparative-scale separation of optically pure enantiomers), it would seem that the methods based on diastereomeric complex formation using chiral counter-ions in the eluent could be advantageously applied to two fields.

(i) Optical purity control of intermediates and active ingredients (in addition to, or replacement of the optical rotation measurement), when the

Table 11.4 Most important characteristics of direct resolution using chiral counter-ions in the eluent

I. *Parameters for reagent selection*	
Type	Anionic for basic solute, cationic for acidic solute
Acidic-basic character	Relatively strong
Lipophilic nature	Suitable lipophilic portion is necessary
Optical purity	Depending on the achieved resolution, not very important
UV absorbance	Negligible UV-absorbance in the applicable detection range
Chemical structure	Bulky and rigid groups in the vicinity of chiral carbon are important. Hydrogen-bonding group is superior to form stable complexes
Solubility	Must be soluble in the mobile phase, possibly by the aid of additives
Availability	Limited
II. *Separation variables*	
A. Parameters for optimization.	
Primary parameters (possessing large effect on solute retention)	Secondary parameters (possessing large effect on stereoselectivity)
Reagent concentration	Nature of reagent
Nature of polar solvent solvent	Nature of medium-polar solvent
Ratio of medium-polar and apolar solvents	Configuration (antipode) of reagent
Concentration of polar and apolar solvents	
B. Other separation variables	
Kind of stationary phase	
Kind of mobile phase additives	
Nature of apolar solvents	
Temperature	
Flow rate	
III. *Selectivity*	
Type of compounds (ionic)	
Apolar	Not applicable
Medium-polar	Good
Polar	Good
Size of compounds	
Small	Not selective
Medium-large	Not selective
Large	Not applicable
Chemical structure of compounds	
Functional group	
Presence	Less selective
Place	Less selective
Type and size	Significant
Place of chiral centre	Highly dependent
Size of Lipophilic portion	Dependent
Steric effects	Affected

Table 11.4 (*Contd.*)

IV. *Applicability*	
Sample preparation	Easy, but sample must be soluble in the eluent
Column efficiency	Problematic
Phase system sensitivity	For water and temperature
Reproducibility	Good
Type of compounds	Only ionic samples
Size of compounds	
Small	Problematic
Medium-large	Good
Large	Not applicable
Chemical structure of compounds	High dependent on the position of chiral atom, ionizable functional group and other parameters connected with the structure of the molecule
Analytical tasks:	
In-process control of asymmetrical synthesis	Highly applicable
Control of optical purity	Attractive even in low concentration range
Determination of administered racemic drugs in biological media	Limited applicability due to the solubility, detectability and other problems relating to the interference of unknown materials
(Semi)preparative applications	Limited applicability due to the solubility and column loadability problems and to re-transformation step required to yield pure enantiomers

enantiomeric impurity is present in a low concentration range. Figure 11.5 shows the separation of the enantiomeric impurity in (+)-*cis*-apovincaminic acid ethylester (Vinpocetine). As can be seen, the enantiomeric impurity can be detected at levels as low as 0.1% by the method developed (Szepesi *et al.*, 1982*b*).

(ii) In-process control of asymmetrical synthesis. The separation of samples taken from different stages of the synthesis is demonstrated in Figure 11.6.

11.7 Combination of ion-pair chromatography and inclusion complex formation

As mentioned previously, ion-pair chromatography itself in reversed-phase systems is not sufficient for enantiomeric separations, since the undesired dissociation and solvation effects of water present in the eluent detrimentally affect the intimate contact between the hydrophobic surfaces of solute and reagent molecules, so that the formation of diastereomeric complexes in the eluent is prevented.

An interesting new possibility for enantiomeric separations using chiral additives in the eluent has been published by Szepesi and Gazdag (1988),

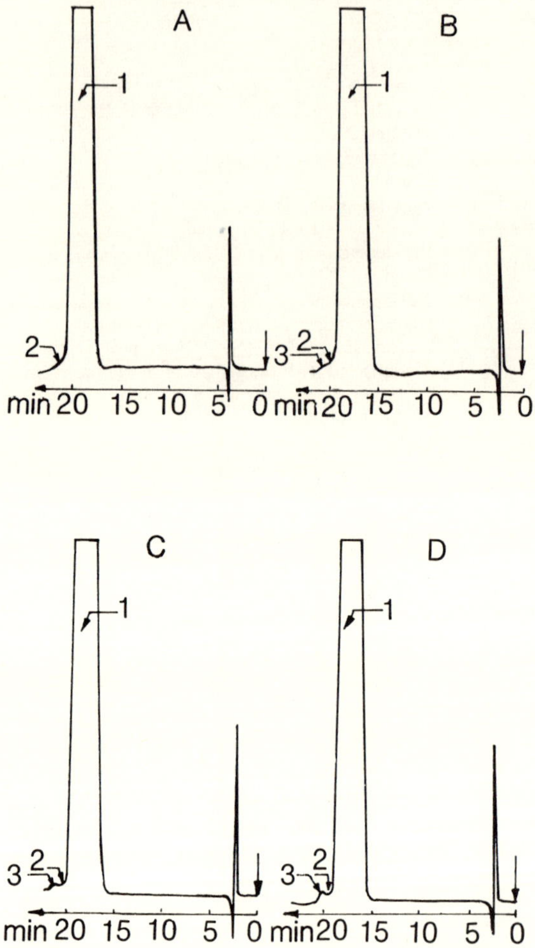

Figure 11.5 Separation of (−)-*cis*-apovincaminic acid ethyl ester impurity in vinpocetine (Szepesi *et al.*, 1982*b*). Conditions: eluent, hexane-dioxane-1-butanol (70:25:5) containing 2mM (+)-CSA and 1 mM DEA. Chromatograms: (*A*) vinpocetine sample; (*B*)-(*D*) vinpocetine sample spiked with 0.1% (*B*), 0.2% (*C*) and 0.5% (*D*) of (−)-*cis*-isomer. Compounds: 1, vinpocetine; 2, (+)-*cis*-apovincamine; 3, (−)-*cis*-apovincaminic acid ethyl ester.

combining ion-pair chromatography with cyclodextrin inclusion complex formation. The most important conclusions can be summarised as follows:

(i) The influences of ion-pair chromatography and inclusion complex formation on solute retention are opposite. In reversed-phase systems ion-pair formation increases solute retention, whereas inclusion complexation decreases solute retention.

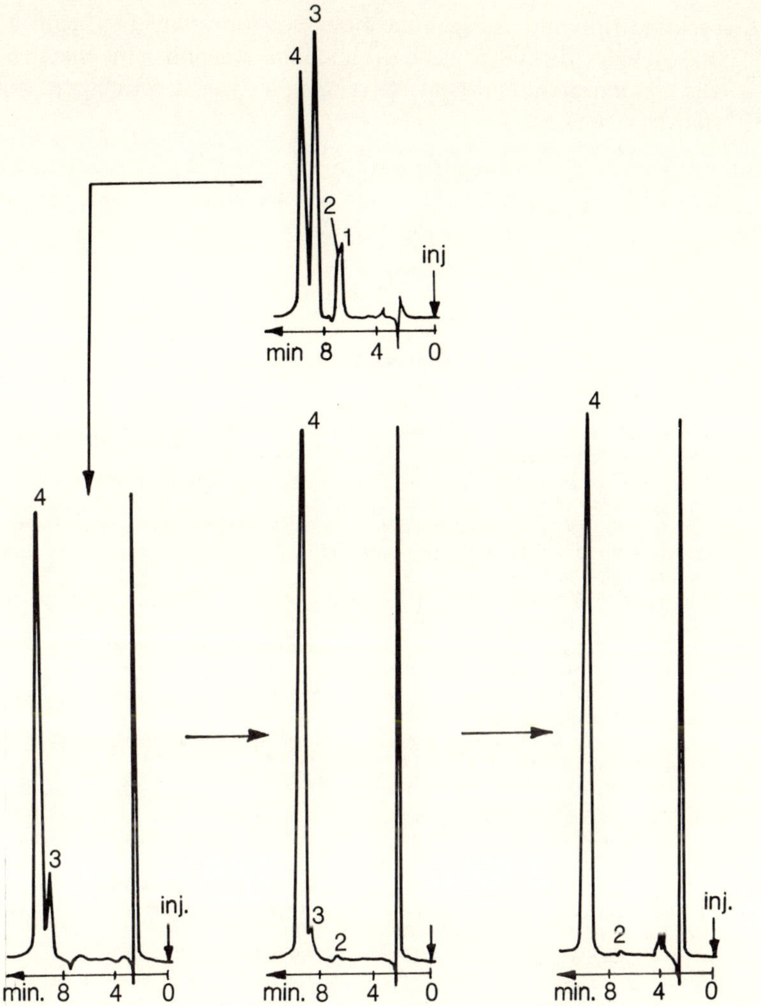

Figure 11.6 Separation of optical isomers of indolo-chinazoline derivatives. Conditions: column, Nucleosil 10 CN; eluent, hexane-tetrahydrofurane-1-butanol (10:89:1) containing 2 mM CSA and 1 mM DEA. Compounds: 1, (+)-*trans*-isomer; 2, (−)-*trans*-isomer; 3, (+)-*cis*-isomer; 4, (−)-*cis*-isomer.

(ii) The presence of ion-pair reagent in eluents containing cyclodextrins results in practically no change in retention characteristics and in selectivity for those compounds having no ionizable functional groups suitable for ion-pair formation.

(iii) The number of separation variables is significantly extended to improve the selectivity and efficiency of enantiomeric separations, such as: nature and concentration of ion-pair reagent; nature and concentration of cyclodextrins; ratio between the molar concentration of ion-pair reagent and cyclodextrin in the eluent; pH and ionic strength of the eluent; nature and concentration of organic solvents in the eluent; selection of suitable stationary phases.

The main advantage as reported by the authors is concluded to be the possibility of application of this combined technique to the simultaneous investigation of ionic and non-ionic solutes.

References

Knox, J.H. and Jurand, J. (1982) Separation of optical isomers by zwitterion-pair chromatography. *J. Chromatogr.* **234**, 222–224.

Ladanyi, L., Sztruhar, I., Vedres, A. and G. Vereczkey-Donath, G. (1986) High-performance liquid chromatography of 8-azagonane-12-one derivatives and their oximes. Separation of optical isomers. *J. Chromatogr.* **353**, 27.

Lindner, W. and C. Pettersson (1985) An introduction. In *Liquid Chromatography in Pharmaceutical Development*, ed. Wainer, I.W., Aster, Springfield, OR.

Pettersson, C. and Gioeli, C. (1988) Improved resolution of enantiomers of naproxen by the simultaneous use of a chiral stationary phase and a chiral additive in the mobile phase. *J. Chromatogr.* **435**, 225.

Pettersson, C. and Josefsson, M. (1986) Chiral separation of aminoalcohols by ion-pair chromatography. *Chromatographia* **21**, 321.

Pettersson, C. and No, K. (1983) Chiral resolution of carboxylic and sulphonic acids by ion-pair chromatography. *J. Chromatogr.* **282**, 671.

Pettersson, C. and Schill, G. (1981) Separations of enantiomeric aimines by ion-pair chromatography. *J. Chromatogr.* **204**, 179.

Pettersson, C. and Schill, G. (1982) Chiral resolution of aminoalcohols by ion-pair chromatography. *Chromatographia* **16**, 192.

Pettersson, C. and Schill, G. (1986) Separation of enantiomers in ion-pair chromatographic systems. *J. Liq. Chromatogr.* **9**, 269.

Pettersson, C., Karlsson, A. and Gioeli, C. (1987) Influence of enantiomeric purity of a chiral selector on stereoselectivity. *J. Chromatogr.* **407**, 217.

Szepesi, G., Gazdag, M. and Ivancsics, R. (1982*a*) Normal-phase dynamic (solvent-generated) molecular complexation chromatography using anionic ion exchanger. I. Characterization of the separation system. *J. Chromatogr.* **241**, 153.

Szepesi, G., Gazdag, M. and Ivancsics, R. (1982*b*) Normal-phase dynamic (solvent-generated) molecular complexation chromatography using anionic ion exchangers. II. Separation of optical isomers. *J. Chromatogr.* **244**, 33.

Szepesi, G. and Gazdag, M. (1988) Problems of enantiomeric separations and its application in the pharmaceutical analysis. *J. Pharm. and Biomed. Anal.* (in press).

12 Other direct chiral resolution methods

W.J. LOUGH

12.1 Introduction

As far as commercially-available chiral stationary phases (CSPs) are concerned there are, at the present time, no 'other direct resolution methods' than those already mentioned in Part 3 of this book. Even for less common direct methods, such as those which have only been used in a research environment, the categorization used in the rest of Part 3 is virtually all-encompassing. However there are some other direct chiral resolution methods about which it is worth knowing. They may for instance provide the basis for some chiral resolution method which may become commercially available and popularized in the future, or else they may provide further useful insights into the whole process of chiral recognition. Most of these other methods are in fact only minor variations on the categories discussed already, but there are also some more novel methods of achieving enantiomeric resolution.

12.2 Minor variations on more common methods

Inclusion complexing is found not only with cyclodextrins but also with several other types of 'host' molecule. The terms 'host' and 'guest' to describe respectively the molecular complexation agent and the molecule being complexed were coined by Cram who, with co-workers, employed inclusion complexing for chiral chromatography as long ago as 1974 (Sousa, 1974). Chiral crown ethers were used in the liquid–liquid partition chromatography of hexafluorophosphate salts of racemic amino acids. Presumably this type of system was not commercialized because of the complexity of the host and the poor solubility of the solutes in the apolar eluents used. More recently, similar work has been carried out using reversed-phase eluents. Shinbo *et al.* (1987) used a dilute aqueous solution of perchloric acid as the eluent for the enantiomeric resolution of underivatized racemic amino acids on reversed-phase packings dynamically coated with the chiral crown ether shown in Figure 12.1. At the time of writing, a commercial CSP based on a chiral crown ether was due for imminent release (Daicel Chemical Industries Ltd, Japan).

Potentially most other types of inclusion complexing agents can be used in a chiral variation for the purposes of enantiomeric resolution by liquid chromatography. Significant advances have recently been made in the design and synthesis of chiral inclusion complexing agents. This research has

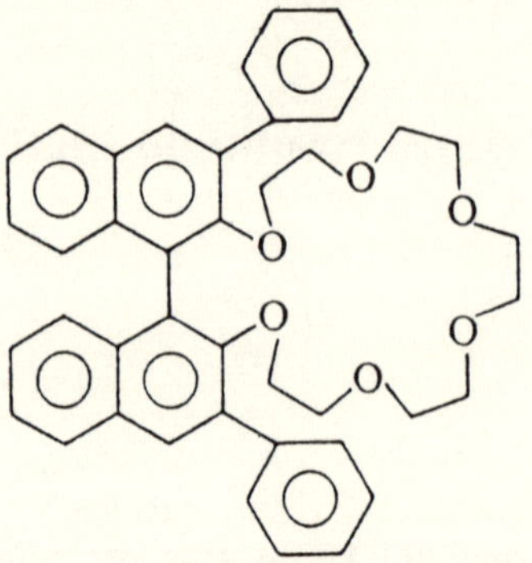

Figure 12.1 Chiral crown ether used for the resolution of racemic amino acids (From Shinbo, 1987).

primarily been aimed at producing biomimetic systems with (relatively) small molecules, but it is probably only a question of time before these agents find chromatographic application. Examples of this work include the application of molecular mechanics by Diederich and co-workers to design chiral hosts specifically to distinguish between the enantiomers of aryl-propionic acids such as naproxen (Dharanipragada, 1988), and the synthesis of small molecule, concave, chiral 'molecular clefts' from Tröger's base analogues on a multigram scale in four chemical steps by Wilcox *et al.* (1987).

The use of helicenes in chiral LC is a case where there is some element of a steric fit into a chiral groove involved in the enantiomeric recognition. However, there is also a very strong dependence on π–π charge transfer interactions. As in the case of crown ethers, the use of helicenes in chiral LC is fairly old (Kim, 1982), and there has been a recent revival of interest. The use of a hexahelicene-immobilized CSP by Matlin and co-workers (1988) was viable because a much more efficient synthesis which could be carried out on a large scale had been developed for (+)-hexahelicene-7-acetic acid (Figure 12.2).

The possible minor variations on protein CSPs are legion. However although many similar biomacromolecules have been immobilized on silica or other materials, it is only proteins and the enzyme α-chymotrypsin that have

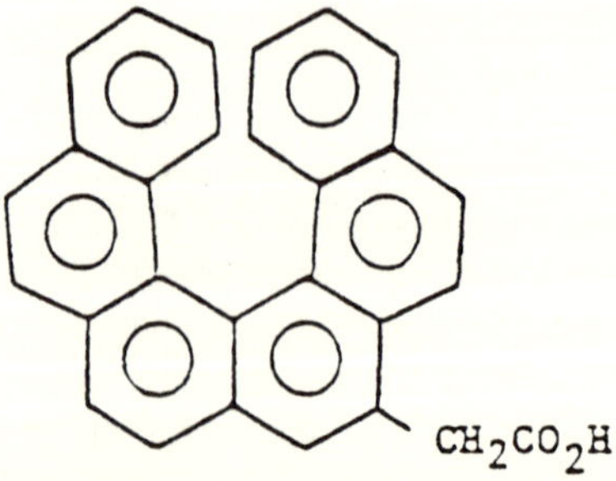

Figure 12.2 (+)-hexahelicene-7-acetic acid.

found use in chiral LC. Perhaps a very significant variation is the use of serum albumin fragments by Erlandsson (1987). It is likely that more robust CSPs could be made from protein fragments than from whole proteins. Also, if the active fragment is small enough, there is then the possibility of synthetically preparing large quantities of the fragment and even closely related analogues thereof in order to attempt to produce improved CSPs.

There are also many possible minor variations on cellulose CSPs, some of which have already been mentioned (see Chapter 9). However, potato starch (Konrad, 1986), chitosan-coated silica gel (Carunchio, 1987) or amylose derivatives offer few advantages over the CSPs using cellulose or cellulose derivatives. CSPs have been prepared from monosaccharides as well as from polysaccharides. König and co-workers (Schulze, 1986) covalently bonded 1-isothiocyanato-D-glucopyranosides with different substituents to aminopropyl silica. The enantioselectivitics obtained have more to do with the substituents used than the monosaccharide ring, and these phases are really only another type of synthetic, multiple-interaction CSP.

Other small-molecule natural products immobilized on silica are also another type of synthetic, multiple-interaction CSP, although the multiple interactions often all arise from the natural product itself and not from synthetic modifications. As well as the quinine and quininium CSPs mentioned elsewhere in this book, other examples are the coating of silica gel with riboflavin by Gil-Av and co-workers (Kim, 1980) to accomplish the enantiomeric resolution of helicenes, and the impregnation of silica with berberine by Bhushan and Ali (1988) to resolve amino acids (albeit by TLC). There must be plenty of scope for producing similar such CSPs in the future, especially with the advances that are being made in methodologies for immobilizing organic compounds on silica. The most common methods for immobilization are (i) the formation of an amide bond by reaction with aminopropyl silica, (ii) preparation and subsequent reaction of a silyl derivative with silica, and (iii) attack of a nucleophile on glycidoxypropyl-silica. Recently, Japanese workers (Iwaki, 1987) have used disuccinimido carbonate to activate the amino group in aminopropyl silica and thus bring about more facile amide bond formation. Pesek and Swedberg (1986) have introduced an allyl bonded stationary phase which could act as an intermediate in the synthesis of novel CSPs.

12.3 More unusual direct chiral resolution methods

Inorganic or organometallic compounds have been used only occasionally in chiral LC (excluding of course ligand exchange chromatography). In the few examples that there are, these compounds have only been used to resolve enantiomers of similar inorganic/organometallic compounds. For example, Yamagishi and Ohnishi (1983) modified a montmorillonite clay by adsorbing on to it an enantiomer of the optically active chelate, [tris(1, 10-phenanthroline) ruthenium(II)]$^{2+}$. The resultant material was found to resolve the enantiomers of a range of neutral, positive and negative metal chelates. On the same column, organic molecules such as proline, pyroglutamic acid, thioproline and γ-valerolactone were only partially resolved. Similarly, antimony d-tartrate ion was applied only to the enantiomeric resolution of *cis*-$[Co(O)_2(N)_4]^+$ type of complexes (Nakazawa, 1981). The mechanism proposed for this separation was directional ion-association. As mentioned earlier, Davankov (Kurganov, 1987) used optically active cobalt-tris-ethylenediamine cations in the enantiomeric resolution of cyclopentadienyl-rhodium-norbornadiene complexes. In this situation the mechanism proposed was outer-sphere ligand exchange. Outer-sphere ligand exchange is the mechanism used by chiral lanthanide shift reagents in the determination of enantiomers by NMR. However, it was found by Matlin and co-workers (unpublished work) that an immobilized chiral lanthanide shift reagent was not successful as a CSP. Furthermore a similar immobilized chiral copper(II)-β-diketonate complex (Figure 10.3) was also unsuccessful in the straight-phase mode as a CSP even although it had been highly successful as a catalyst in asymmetric synthesis (Matlin, 1984). This may be regarded as a cautionary tale. Catalysts for asymmetric synthesis may well provide a valuable source of future CSPs but they will not automatically work well. The energy differences between transient diastereomeric complexes required to give chiral discrimination may be different in the two different situations. Also, in catalysis a complex of a transition state will often be involved in the rate-determining step, whereas in a CSP essentially only the ground state is involved.

One type of direct chiral resolution method which may definitely be described as more unusual is the separation of enantiomers in a 'totally achiral system' (Cundy, 1983). Indeed, in the first reported example of such an occurrence, Cundy and Crooks referred to it as an 'unexpected phenomenon'. However, since this work on ^{14}C-labelled nicotine there have been several more observations of this phenomenon (Charles, 1984; Tsai, 1985; Dobashi, 1987). In the most recent example (Dobashi *et al.*, 1987) enantiomeric mixtures of N-acetylvaline tert-butyl ester were chromatographed on silica with a non-aqueous mobile phase. The injection of enantiomerically enriched mixtures of the chiral solutes resulted in two split zones differing in enantiomeric excess being eluted from the column (Figure 10.4). ^1H-NMR was used to confirm that

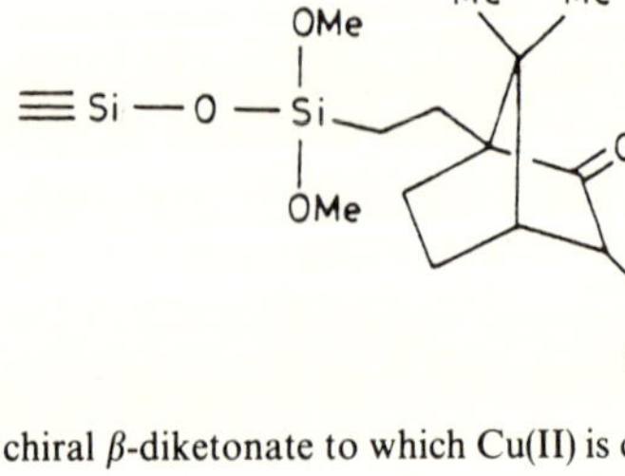

Figure 12.3 Immobilized chiral β-diketonate to which Cu(II) is complexed (From Matlin, 1984).

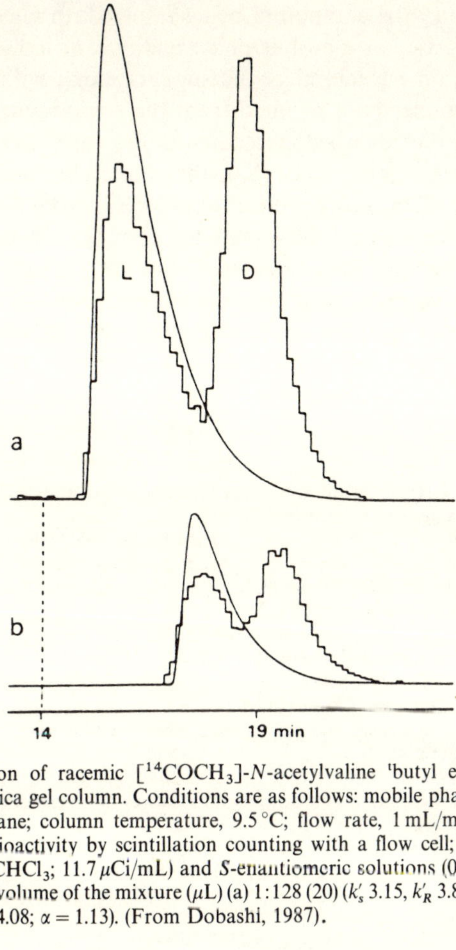

Figure 12.4 Separation of racemic [$^{14}COCH_3$]-*N*-acetylvaline ᵗbutyl ester diluted with *S*-enantiomeric on a silica gel column. Conditions are as follows: mobile phase solvent, 7.5% (v/v) 2-propanol in *n*-hexane; column temperature, 9.5 °C; flow rate, 1 mL/min; detection, UV at 230 nm and ^{14}C radioactivity by scintillation counting with a flow cell; mixing ratios of the racemic (2.4 mM in $CHCl_3$; 11.7 μCi/mL) and *S*-enantiomeric solutions (0.095 M in $CHCl_3$) by weight and injection volume of the mixture (μL) (a) 1:128 (20) (k'_s 3.15, k'_R 3.89; $\alpha = 1.23$), (b) 1:128 (2.5) ($k'_s = 3.62$, $k'_R = 4.08$; $\alpha = 1.13$). (From Dobashi, 1987).

this was caused by the enantiomers undergoing diastereomeric solute–solute interactions to different extents. This explanation also applies to the other examples. Although the separations are said to take place in an achiral system, this is strictly speaking not correct. For separation to take place, there must already be an enantiomeric excess of one of the enantiomers. The enantiomer in excess acts as the chiral selector. If the enantiomers are present in equal proportions, a single peak is observed, since the chromatography taking place is effectively equivalent to chromatography on a racemic version of a dynamically generated CSP.

Another direct chiral resolution method which qualifies as unusual is inductive adsorption. This new approach has been reported only recently, being used by Gassend *et al.* (1987) to resolve chiral amines. It is related to the autoseparations in achiral systems described above in that the chiral selector is similar in nature to the compound being resolved. In Gassend's example, a chiral amine was used as a mobile phase additive, or inductive agent, which was coadsorbed on an achiral acidic support along with the amine being resolved. Enantioselectivity resulted from the steric hindrance between the carbon chains of the adsorbed molecules. In this case it was measured in the study of batch equilibria rather than the whole chromatographic process. Clearly, in this inductive adsorption approach it would be possible for an excess of one of the solute enantiomers to act as the inductive agent. In fact, Gassend's experimental data suggested that the inductive effect increased when the inductive agent had very similar structure to the solute amine. However, attempts to produce 'optical activity self-amplification' were unsuccessful.

References

Bhushan, R. and Ali, I. (1988) Resolution of enantiomeric amino acids on berberine impregnated silica plates. *Fresenius Z. Anal. Chem.* **329**, 793.

Carunchio, V., Girelli, A.M., Messina, A. and Sinibaldi, M. (1987) Chitosan-coated silica gel as a new support in high-performance liquid chromatography. *Chromatographia* **23**, 731.

Charles, R. and Gil-Av, E. (1984) Self-amplification of optical activity by chromatography on an achiral adsorbent. *J. Chromatogr.* **298**, 516.

Cundy, K.C. and Crooks, P.A. (1983) Unexpected phenomenon in the high-performance liquid chromatographic analysis of ^{14}C-labelled nicotine: separation of enantiomers in a totally achiral system. *J. Chromatogr.* **281**, 17.

Dharanipragada, R., Ferguson, S.B. and Diederich, F. (1988) A novel optically active host: design, computer graphics, synthesis, and diastereomer complex formation in aqueous solution. *J. Amer. Chem. Soc.* **110**, 1679.

Dobashi, A., Motoyama, Y., Kinoshita, K., Hara, S. and Fukasaka, N. (1987) Self-induced chiral recognition in the association of enantiomeric mixtures on silica gel chromatography. *Anal. Chem.* **59**, 2209.

Erlandsson, P. and Nilsson, S. (1987) Albumin fragments used as chiral stationary phases. *11th Int. Symp. on Column Liquid Chromatography*, Amsterdam, The Netherlands.

Gassend, R., Duprat, F. and Gau, G. (1987) Inductive adsorption as a new approach to amine enantiomer resolution. *J. Chromatogr.* **404**, 87.

Iwaki, K., Yoshida, S., Nimura, N., Kinoshita, T., Takeda, K. and Ogura, H. (1987) Preparation of chiral stationary phase via activated carbamate intermediate for liquid chromatographic optical resolution. *Chromatographia* **23**, 727.

Kim, Y.H., Tishbee, A. and Gil-Av, E. (1980) Chiral recognition by small biological molecules. Resolution of helicenes on silica gel coated with riboflavin. *J. Amer. Chem. Soc.* **102**, 5915.

Kim, Y.H., Tishbee, A. and Gil-Av, E. (1982) Chiral differentiation by the P-(+)-hexahelicene-7, 7′-dicarboxylic acid disodium salt. Resolution of N-2, 4-dinitrophenyl-α-amino-acid esters by high performance liquid chromatography. *J. Chem. Soc. Chem. Commun.*, 1336.

Konrad, G. and Musso, H. (1986) Chromatographic resolution of enantiomers with natural polymers. *Liebigs Ann. Chem.*, 1956.

Kurganov, A.A., Facklam, C. and Davankov, V.A. (1987) Enantiomeric resolution of cyclopentadienyl-rhodium-norbornadiene complexes using outer-sphere ligand-exchange HPLC. *11th Int. Symp. on Column Liquid Chromatography*, Amsterdam, The Netherlands.

Matlin, S.A., Lough, W.J., Chan, L., Abram, D.M.H. and Zhou, Z. (1984) Asymmetric induction in cyclopropanation with homogeneous and immobilised chiral metal β-diketonate catalysts. *J. Chem. Soc. Chem. Commun.*, 1038.

Matlin, S.A., Stacey, V.E. and Lough, W.J. (1988) Hexahelicene chiral stationary phase. Part I. Phase synthesis and use in HPLC resolution of enantiomers. *J. Chromatogr.* **450**, 157.

Nakazawa, H., Sakaguchi, U. and Yoneda, H. (1981) Chromatographic study of optical resolution. VII. Directional ion-association model for the optical resolution of *cis*-$(Co(O)_2(N)_4)^+$ type of complexes by the antimony d-tartrate ion. *J. Chromatogr.* **213**, 323.

Pesek, J.J. and Swedberg, S.A. (1986) Allyl-bonded stationary phase as possible intermediate in the synthesis of novel high-performance liquid chromatographic phases. *J. Chromatogr.* **361**, 83.

Schulze, J. and Konig, W.A. (1986) Enantiomer separation by high-performance liquid chromatography on silica gel with covalently bound monosaccharides. *J. Chromatogr.* **355**, 165.

Shinbo, T., Yamaguchi, T., Nishimura, K. and Sugiura, M. (1987) Chromatographic separation of racemic amino acids by use of chiral crown ether-coated reversed-phase packings. *J. Chromatogr.* **405**, 145.

Sousa, L.R., Hoffman, D.H., Kaplan, L. and Cram, D.J. (1974) Total optical resolution of amino esters by designed host-guest relationships in molecular complexation. *J. Amer. Chem. Soc.* **96**, 7100.

Tsai, W.-L., Hermann, K., Hug, E., Rohde, B. and Dreiding, A.S. (1985) Enantiomer differentiation induced by an enantiomeric excess during chromatography with achiral phases. *Helv. Chim. Acta* **68**, 2238.

Wilcox, C.S., Greer, L.M. and Lynch, V. (1987) Synthesis of chiral molecular clefts. New armatures for biomimetic systems. *J. Amer. Chem. Soc.* **109**, 1865.

Yamagishi, A. and Ohnishi, R. (1983) Resolution of inorganic and organic compounds on a Λ-$Ru(phen)_3^{2+}$ montmorillonite column. *VII Int. Symp. on Column Liquid Chromatography*, Baden-Baden, West Germany.

Part 4 Strategy for development of LC enantiomeric determination methods

13 Consideration of other techniques

W.J. LOUGH and S.A. MATLIN

13.1 Introduction

The first step to take in developing an assay in chiral liquid chromatography is self-evident; the question must be asked, 'Is LC the best technique to use?' This is especially so since there is a surprising variety of techniques available for the determination of enantiomers, and the number of options is ever-increasing.

Amongst these options is isotope labelling. This may be carried out either by an isotope dilution method or by a radiotracer method (Raban, 1967). In the isotope dilution method, the labelled racemate is added to a partially resolved enantiomer mixture of unknown purity. This method is used infrequently because of poor precision, related to the precision of the isotope analysis and two rotation measurements. In the radiotracer method, a pure labelled enantiomer is added to an enantiomer mixture of unknown purity. This method is useful for samples which have a high enantiomeric purity, but suffers the obvious disadvantage that it depends on the availability of a pure labelled enantiomer.

Differential scanning calorimetry (dsc) may also be applied to the determination of enantiomers (Jacques, 1981). This can be done either directly or indirectly. The direct method involves the reconstitution of the phase diagram using all the information available from dsc traces of the melting of the pure racemate and of a partially resolved mixture of unknown enantiomeric purity. This method is imprecise, and is generally only applied to samples of intermediate enantiomeric purity in order to monitor the progress of a resolution. The indirect method involves the study of the melting scan by a general procedure for nearly pure crystalline compounds, and gives very precise results for samples which are almost racemic or are of high enantiomeric purity. Neither method can be applied to liquid samples, and both may suffer complications in situations where polymorphism or thermal instability arise.

The determination of enantiomers by kinetic techniques may be carried out if each enantiomer of a pair reacts at a different rate with a chiral reagent (Wilen, 1971). More and more such reactions which have very high stereospecificity are now being reported in the literature, but in practice this method of determining enantiomers has so far been restricted to enzymic reactions, almost always on α-amino acids. Usually analysis involves monitoring an oxidation or decarboxylation reaction by measuring absorbed oxygen or evolved carbon dioxide respectively. The method is best applied to mixtures of high enantiomeric purity where the minor enantiomer reacts.

The techniques mentioned so far have their uses, and are good illustrations of the diversity of methods which can be used for the determination of enantiomers. However, they are not options which are often used in preference to chiral LC. Of the alternatives to chiral LC, measurement of optical rotation, NMR spectroscopy and gas chromatography are the most frequently used and easily accessible. This chapter will therefore deal mainly with these techniques.

13.2 Measurement of optical rotation

The actual measurement of optical rotation, the rotation of the plane of linearly polarized light by polarimeters, is still frequently used as a means of assessing optical purity. Until the 1960s it was the most frequently used probe of chirality (Drake, 1986). It is a very rapid and simple method, so that when pure bulk compounds of known specific rotation are being studied and a very low limit of detection of the unwanted enantiomer is not necessary (e.g. in monitoring samples generated from a fractional crystallization), there is no need to look for anything more sophisticated. In some instances, the technique can also be quite sensitive, for example when the specific rotation is very high, as for helicenes, where specific rotations are in the order of 3600°, and the sample is known to be chemically very pure.

In some circumstances, for example for regulatory purposes, it may be necessary to confirm that a sample is a racemate. This can be done very simply by demonstrating zero rotation. A major disadvantage in doing this is that a large quantity of sample (up to 1 g) may be required. Also, sensitivity would be a problem for compounds with a low specific rotation so that it would be difficult to differentiate between racemates and enantiomeric mixtures which were close to 50:50.

Much has been done recently to enhance the sensitivity of polarimeters. Simple polarimeters have been made for over a century, with sensitivities of the order of 0.1° to 0.01°. Following developments made since the mid-1950s, sensitivities of commercially-available instruments are now typically around 0.001°, with response times of a few seconds. Even this performance can be improved upon considerably by using intense laser light sources. However, the advent of laser-based systems (Yeung, 1980; Goodall, 1987) has not brought about more widespread use of direct polarimetry but has opened up the possibility of using polarimeters as detectors for LC. This development has been accessible to the practising chromatographer for a very short time only, and accordingly is covered in Part 5.

The measurement of circular dichroism spectra is seldom used to determine optical purity. However, it has been used for the direct measurement of optically pure compounds in the presence of a different class of optically pure compounds. Purdie and Swallows (1987) found that discriminations among either penicillin or cephalosporin analogues were not possible, but the distinction between the two groups was easy.

Like the measurement of optical rotation, the measurement of circular dichroism can be used as a means of detection in LC. Detection by this means gives the prospect of lower limits of detection but with the disadvantage that it is not so simple to determine which enantiomer is which.

13.3 Nuclear magnetic resonance spectroscopy

Like the measurement of optical rotation, nuclear magnetic resonance spectroscopy (NMR) is another technique which is best applied to pure bulk compounds. However it is not so limited in its applicability. Modern high-field, Fourier-transform instruments allow compounds of moderate chemical purity to be studied with limits of detection for the minor enantiomer in the order of 0.1%.

Optical purity can be measured by carrying out NMR spectroscopy on a diastereomeric derivative, but more commonly NMR is carried out on the underivatized sample in the presence of a chiral shift reagent. Chiral lanthanide shift reagents such as [tris (3-trifluoroacetyl)-d-camphorato] europium (III), shown in Figure 13.1, have been in use for many years. They exert their influence by outer sphere complexation with functional groups with potential for electron donation, and are therefore applicable to a wide range of compounds. Their main disadvantage is that with increasing concentration they lead to band-broadening in the NMR spectrum, but this is very much offset by the very large induced shifts and very large differences in induced shifts for enantiomer pairs. With the advent of high-field, Fourier-transform instruments, such large shifts are not necessary to discriminate between enantiomers, and lanthanide chiral shift reagents are largely being superseded by other chiral shift reagents or chiral solvating reagents. Nonetheless, they still find use. Recently Martin *et al.* (1987) found 'dramatic' differences in lanthanide-induced shifts for diastereotopic CH_2 protons in the novel 1, 4-benzodiazepine, Ketazolam. Also, a chiral lanthanoid shift reagent for aqueous solution, propylenediaminetetra-acetatoeuropium (III) has been applied to α-amino acids (Kabuto, 1987). A highly consistent correlation was observed between the absolute configuration of α-amino acids and the shift of their α-proton signals.

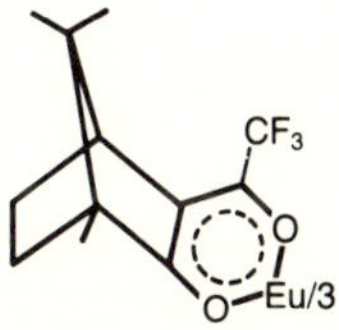

Figure 13.1 Structure of [tris(3-trifluoroacetyl)-D-comphorato] europium(III), a chiral shift reagent for NMR.

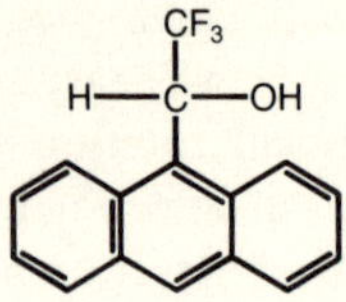

Figure 13.2 Structure of 2, 2′, 2″-trifluoro, 1-(9-anthryl) ethanol, a chiral solvating agent for NMR.

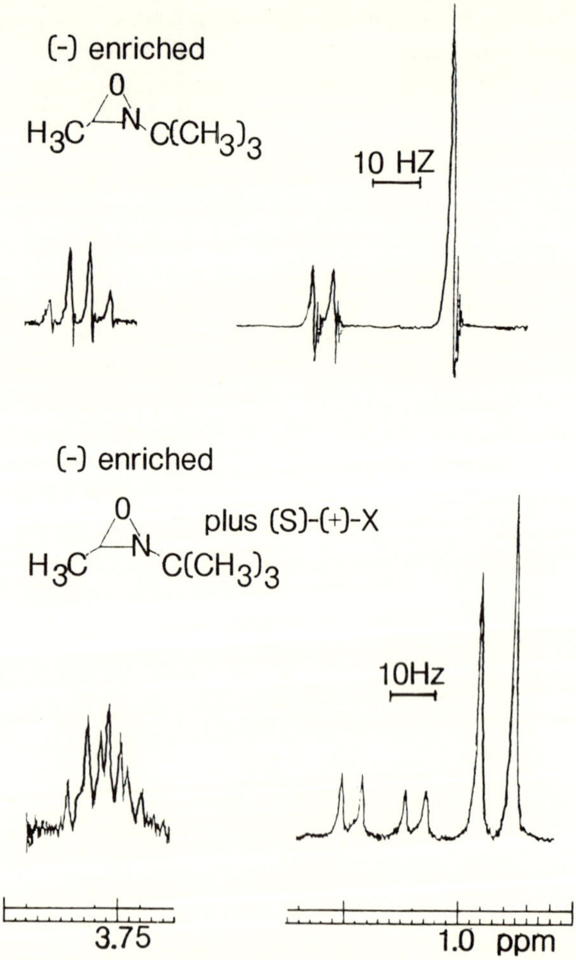

Figure 13.3 100 MHz NMR spectrum of a (−)-enriched oxoaziridine in the absence (top) and in the presence of (S)-(+)-2, 2′, 2″-trifluoro, 1-(9-anthryl) ethanol (X). (From Pirkle, 1977a).

Of the newer chiral reagents used in NMR, those which interest with the solute via π–π charge-transfer are amongst the most popular. This is especially so of 2,2′, 2″-trifluoro, 1-(9-anthryl) ethanol, shown in Figure 13.2, which, following its success in NMR applications (Pirkle, 1977*a*, *b*) was used by Pirkle (1979) in the first of his range of synthetic multiple-interaction chiral stationary phases for LC. Figure 13.3 shows an excellent example of the chiral discrimination that can be obtained with this reagent in NMR. The chemical shift differences achieved for the methyl protons allowed these signals to be used for the very facile quantitation of the enantiomeric ratio of the oxaziridine. In his discussion of this work, Pirkle suggested structures for the diastereomeric solvates formed. No charge transfer interactions were inolved. However in many cases 2,2′,2″-trifluoro, 1-(9-anthryl) ethanol will act via a charge transfer interaction with any electron-deficient aryl groups present in the chiral compounds being studied. In these cases, the chemical shift of the protons in the anthryl- shift reagent is often very close to the chemical shift of the aryl protons in the solute molecule. In the absence of suitable protons at higher field which can be monitored, it can be extremely difficult to distinguish between shift reagent and solute protons using a low-field ($\leqslant$ 100 MHz) NMR instrument with which multiplets are more likely to overlap. As a result, these reagents are usually used with high-field instruments (with remarkable success).

The power of modern high-field instruments is such that many chiral reagents which induce only small shifts may be used, even to the extent that NMR is now being used as a tool to study the interactions which take place in chiral LC systems (Armstrong, 1988). Amongst the other chiral reagents recently used in NMR are O-acetyl mandelic acid for the enantiomeric assay of amines and β-amino alcohols (Parker, 1987), α-methoxy-α-(trifluoromethyl)phenylacetic acid for tertiary amines (Villani, Jr., 1987), and L-phenylalanine for DL-valine, DL-leucine, and DL-isoleucine (Shiraiwa, 1984).

In general the use of chiral shift reagents in high-field NMR is rapid, versatile and requires minimal method development. In situations where the sample is reasonably pure and milligram or greater quantities are available, it would therefore be preferred to chiral LC. The use of low-field NMR ($\leqslant$ 100 MHz) in the determination of enantiomers is much less attractive. There is of course, one more very important practical point worth making. NMR instruments are very expensive, and many workers may not have any NMR instrument available to them, far less a high-field instrument.

13.4 Gas chromatography

The determination of enantiomers by gas chromatography (GC) is very popular, and is very much a competitive alternative to chiral LC. Chiral GC was earlier to be established in common use than chiral LC, sufficient work having been generated to warrant a review as long ago as 1975 (Gil-Av, 1975). The progress that has been made since then has been well documented, particularly by König (König, 1982, 1987). However, despite the amount of work done in this area it has distinct disadvantages in comparison to the resolution of enantiomers by LC:

(i) Since the analyte must be volatile, derivatization is very frequently required to enhance volatility. Another consequence of this is that the range of chiral derivatizing agents that may be used in GC is more limited than in LC.
(ii) In general, preparative-scale separations are not feasible.
(iii) High column temperatures can lead to some racemization of the stationary phase.
(iv) Similarly, high column temperatures can lead to racemization of the analyte.
(v) As a result of the high column temperatures, there is a decrease in stability differences between the adsorbed diastereomeric complexes, with the result that selectivity is generally poorer than in chiral LC.
(vi) There is not such a wide range of commercially-available chiral GC columns as there is of chiral LC columns.

Having listed these many advantages, it is only fair to add that GC offers much easier access to high sensitivity and, with capillary columns, high efficiency. With respect to the former advantage, GC has great potential in the analysis of chiral drugs in biological fluids. With respect to both advantages, it has great potential in the determination of trace enantiomers in the presence of a large excess of the antipode.

Although many chiral GC phases have been described in the literature, they are often suitable only for very volatile compounds and, as mentioned above, very few are commercially available. Chirasil-Val (Chrompak, Netherlands) is easily the most commonly used commercial chiral stationary phase for GC.

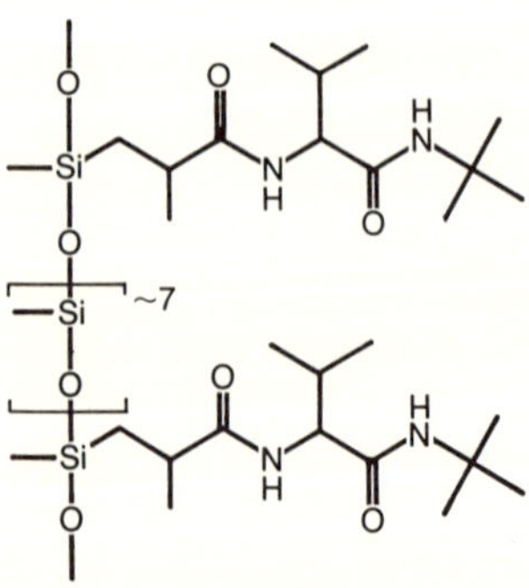

Figure 13.4 Structure of Chirasil-Val (Chrompak, Netherlands), chiral stationary phase for capillary GC columns.

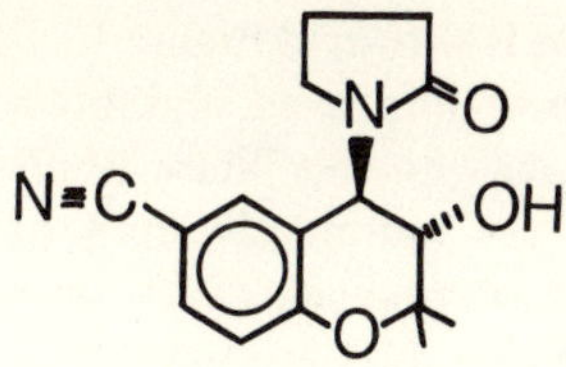

Figure 13.5 Chromakalim, a novel anti-hypertensive agent (Beecham Pharmaceuticals).

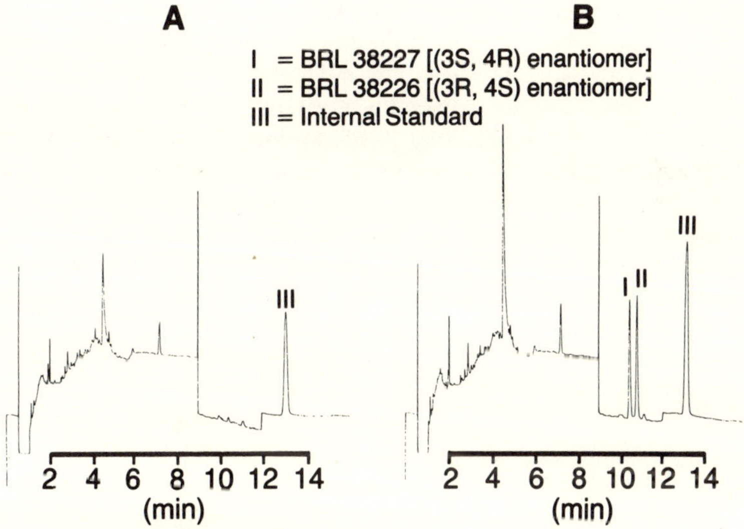

A = Control diluted rat blood containing 50 ng/ml of internal standard.

B = Control diluted rat blood spiked with 10 ng/ml of cromakalim and 50 ng/ml of internal standard.

HP 5880 GC equipped with N/P detector

Splitless injection
Valve time: 60 sec

Chirasil-Val-D fused silica capillary column
25m x 0.22mm x 0.06μm
Temperature programmed from 110 ➤ 195°C at 25°C/min

Detector temp.: 300°C
Injector temp. : 250°C

Helium head pressure: 30 p.s.i.

Figure 13.6 Resolution of enantiomers of chromakalim by GC on Chirasil-Val-D fused silica capillary column.

The structure of this phase is shown in Figure 13.4. Since its first introduction, its thermal stability has been improved, and it is now available in the D-form as well as the L-form. Even with this phase, applications have until recently been largely limited to the (admittedly important) field of enantiomeric resolutions of derivatized amino-acids. These applications are too numerous to list here. Good recent examples are the work of Bayer's group (Köppenhoeffer, 1987*a*) in resolving enantiomers and diastereomers of N-TFA-dipeptide methyl esters, and of Wagner *et al.* (1987) in resolving derivatives of various α-substituted ornithine and lysine analogues of biological interest. Amongst the publications that are now emerging on non-amino acid applications are those by Köppenhoeffer *et al.* (Köppenhoeffer, 1987*b*) on halo-carboxylic acid amides and on hydroxy and carbonyl compounds (Köppenhoeffer, 1985). Perhaps the best example is by Davies and co-workers (Gill, 1987; Davies, 1988), who not only obtained the resolution of the enantiomers of a silyl derivative of a novel anti-hypertensive agent (Figure 13.5), but also demonstrated that this separation could be used to monitor both enantiomers at very low levels (nanogram on-column quantities of total drug) in extracts from samples of drug in biological fluids (Figure 13.6). In the chromatograms shown, N/P detection was used. Mass spectrometric detection was also to be used, allowing a positive identification of the chromakalim peaks or enhanced sensitivity via mass selective detection with single-ion monitoring.

13.5 Other chromatographic techniques

Strictly speaking, chiral thin layer chromatography (TLC) should be covered by the heading 'chiral liquid chromatography'. However, the 'liquid chromatography' (LC) in the title of this book is, as is common practice, taken to imply modern column liquid chromatography.

In principle, all the chiral LC systems already described could be set up in a TLC mode. In practice, the use of TLC has not offered sufficient advantages for this to take place to any great extent. Indirect chiral TLC separations by diastereomer formation are commonplace. However, Chiralplate (Macherey-Nagel, West Germany) is the only type of chiral TLC plate commercially available. This type of TLC plate consists of an immobilized amino acid (L-hydroxy-proline), and resolves enantiomers via a ligand-exchange mechanism.

The other chromatographic techniques mentioned here are not yet viable alternatives to chiral LC for routine applications.

Super-critical fluid chromatography (SFC) has been used for chiral separations both in the packed column mode (Hara, 1986) with mobile phase modifiers and the capillary mode (Röder, 1987). So far there have been no reports of pronounced differences in selectivity between chiral SFC and the equivalent chiral LC system. As SFC finds more widespread use, its advantages of very rapid re-equilibration and compatability with f.i.d., MS and other detectors may result in its being used occasionally instead of chiral LC.

Following various uses of liquid–liquid partition and liquid–liquid partition chromatography for chiral resolution, Oya and Snyder (1986) used droplet counter-current chromatography for the separation of enantiomers of bicyclo(2, 2, 1)hept-5-ene-2-carboxylic acids.

Gozel *et al.* (1987) resolved amino acid enantiomers by high-voltage capillary zone electrophoresis. The separation was based on the diastereomeric interaction between D, L-amino acids and a chiral Cu(II)-aspartame complex present in the support electrolyte.

13.6 Radioimmunoassays

Like GC, radioimmunoassays are suitable in applications which demand high sensitivity. As such, they have been used to directly determine enantiomer ratios of chiral drugs in biological fluids. Examples in the literature include warfarin (Cook, 1979), ephedrine (Midha, 1983), and pentobarbital (Cook, 1987). However, the synthesis of a suitable immunogen seldom a simple matter, and it is difficult to avoid some cross-reaction with the opposite enantiomer and metabolites or other closely structurally-related compounds.

References

Armstrong, D.W. (1988) *XII Int. Symp. on Column Liquid Chromatography*, Washington DC, USA.

Cook, C.E., Ballentine, N.H., Seltzman, T.B. and Tallent, C.R. (1979) Warfarin enantiomer disposition: determination by stereoselective radioimmunoassay. *J. Pharmacol. Exp. Ther.* **210**, 391.

Cook, C.E., Seltzman, T.B., Tallent, C.R., Lorenzo, B. and Drayer, D.E. (1987) Pharmaco-kinetics of pentobarbital enantiomers as determined by enantioselective radio-immunoassay after administration of racemate to humans and rabbits. *J. Pharmacol. Exp. Ther.* **241**, 779.

Davies, B. (1988) Development of a chiral capillary GC method for the quantitation of the enantiomers of chromakalim in biological fluids. *Proc. VII Int. Bioanalytical Forum*, Guildford, Surrey, UK, 1987 (in press).

Drake, A.F. (1986) Optical activity and the spectroscopy of chiral molecules. *Eur. Spectrosc. News* **69**, 10.

Gil-Av, E. (1975) Present status of enantiomeric analysis by gas chromatography. *J. Mol. Evol.* **6**, 131.

Gill, T., Allen, G. and Davies, B. (1987) Stereospecific plasma and urine assay for BRL 34915, a potent, new anti-hypertensive agent. *Brit. J. Clin. Pharmacol.* **24**, 273.

Goodall, D.M. and Lloyd, D.K. (1988) An optical rotation detector for HPLC. *Proc. 1st. Int. Symp. on Chiral Separations*, Guildford, Surrey, UK, 1987 (in press).

Gozel, P., Gassman, E., Michelson, H. and Zare, R.N. (1987) Electrokinetic resolution of amino acid enantiomers with copper(II)-aspartame support electrolyte. *Anal. Chem.* **59**, 44.

Hara, S., Dobashi, A., Kinoshita, K., Hondo, T., Saito, M. and Senda, M. (1986) Carbon dioxide supercritical fluid chromatography on a chiral diamide stationary phase for the resolution of D- and L-amino acid derivatives. *J. Chromatogr.* **371**, 153.

Jacques, J., Collect, A. and Wilen, S. Eds. (1981) Binary mixtures of enantiomers. *Enantiomers, Racemates and Resolutions*, John Wiley, New York, Chapter 2.

Kabuto, K. and Sasaki, Y. (1987) Highly consistent correlation between absolute configuration of α-amino acids and their shift induced by the N.M.R. chiral shift reagent propyleneaminetetraacetatoeuropium(III) in aqueous solution. *J. Chem. Soc., Chem. Commun.* 670.

König, W.A. (1982) Separation of enantiomers by capillary gas chromatography with chiral stationary phases. *J. High Resol. Chromatogr. & Chromatogr. Commun.* **5**, 588.

König, W.A. (1987) *The Practice of Enantiomer Separation by Capillary Gas Chromatography.* Hüthig, Heidelberg.

Koppenhoeffer, B., Allmendinger, H. and Nicholson, G. (1985) Direct enantiomer resolution of hydroxy and carbonyl compounds by gas chromatography on Chirasil-Val. *Angew. Chem. Int. Ed. Engl.* **24**, 48.

Koppenhoeffer, B., Allmendinger, H. and Bayer, E. (1987*a*) Enantiomer and diastereomer resolution of dipeptides by gas chromatography on Chirasil-Val. *J. High Resol. Chromatogr. Commun.* **10**, 324.

Koppenhoeffer, B., Koch, E.M., Nicholson, G.J. and Bayer, E. (1987*b*) Chiral recognition of carboxylic acid amides studied by gas chromatography on Chirasil-Val. *J. Chromatogr.* **406**, 157.

Martin, R., Myers, C. and Rothchild, R. (1987) ^{1}H NMR spectral simplification with achiral and chiral lanthanide shift reagents. Ketazolam. Method for direct optical purity determination *Spectrosc. Letts.* **20**, 645.

Midha, K.K., Hubbard, J.W. Copper, J.K. and Mackonka, C. (1983) Stereospecific radioimmunoassays for l-ephedrine and d-ephedrine in human plasma. *J. Pharm. Sci.* **72**, 736.

Oya, S. and J.K. Snyder (1986) Chiral resolution of a carboxylic acid using droplet counter-current chromatography. *J. Chromatogr.* **370**, 333.

Parker, D. and Taylor, R.J. (1987) Direct ^{1}H NMR assay of the enantiomeric composition of amines and β-amino alcohols using O-acetyl mandelic acid as a chiral solvating agent. *Tetrahedron* **43**, 5451.

Pirkle, W.H. and Rinaldi, P.L. (1977*a*) Nuclear magnetic resonance determination of enantiomeric compositions of oxaziridines using chiral solvating agents. *J. Org. Chem.* **42**, 3217.

Pirkle, W.H. and Boeder, C.W. (1977*b*) Estimation of allene optical purities by nuclear magnetic resonance. *J. Org. Chem.* **42**, 3697.

Purdie, N. and Swallows, K.A. (1987) Direct determination of β-lactam antibiotics by circular dichroism. *Anal. Chem.* **59**, 1349.

Raban, M. and Mislow, K. (1967) Modern methods for the determination of optical purity. *Topics in Stereochemistry*, Vol. 2. N.L. Allinger and E.L. Eliel, eds. Wiley-Interscience, New York, 199.

Röder, W., Ruffing, F.-J. Schomburg, G. and Pirkle, W.H. (1987) Chiral SFC-separations using open tubular fused silica columns. Comparison of enantiomeric selectivity in SFC and LC using the same stationary phase of the Pirkle type. *J. High Resol. Chromatogr. Commun. & Chromatogr. Commun.* **10**, 665.

Shiraiwa, T., Ikawa, A., Sakaguchi, K. and Kurokawa, H. (1984) Optical resolution of DL-Valine, DL-leucine and DL-isoleucine by formation of adduct with L-phenylalanine. *Bull. Chem. Soc. Jpn* **57**, 2234.

Villani, Jr., F.J., Costanzo, M.J., Inners, R.R., Mutter, M.S. and McClure, D.E. (1986) Determination of enantiomeric purity of tertiary amines by ^{1}H NMR of α-methoxy-α-(trifluoromethyl)phenylacetic acid complexes. *J. Org. Chem.*, **51**, 3715.

Wagner, J., Gaget, C., Heintzelmann, B. and Wolf, E. (1987) Resolution of the enantiomers of various α-substituted ornithine and lysine analogues by high-performance liquid chromatography with chiral eluent and by gas chromatography on Chirasil-Val. *Anal. Biochem.* **164**, 102.

Wilen, S.H. (1971) Resolving agents and resolutions in organic chemistry. *Topics in Stereochemistry*, Vol. 6, N.L. Allinger and E.L. Eliel, eds., Wiley-Interscience, New York, 107.

Yeung, E.S., Steenhoek, L.E., Woodruff, S.D. and Kuo, J.C. (1980) Detector based on optical activity for high performance liquid chromatographic detection of trace organics. *Anal. Chem.* **52**, 1399.

14 Choice of chiral LC systems

W.J. LOUGH and I.W. WAINER

14.1 Reference to known examples

If, having assessed the relative merits of all the techniques available for a particular chiral resolution, it is decided that LC is likely to be the most suitable, it would then be most prudent to take a careful look at the literature. Taking a logical approach to what might work and what might not work is highly commendable, but is no substitute for checking what has been shown to work and what has been shown not to work. A literature search should be carried out for LC enantioseparations, not only of the compound in question but also of other closely related structures. A computer search would be ideal, but for those with no access to such facilities reference to the index of this book would be a more than useful substitute.

Referring to the literature is a starting point. Only in a limited number of cases (for example for very urgent work when there is only one method found in the literature and there is no time available for supplementary method development) will it provide an immediate solution to a problem. If no methods are found, then further assessment of available LC methods is obviously necessary. With the many recent advances that have been made in chiral LC it will often be the case that several possible methods are found. The question is usually no longer just one of feasibility but of which is the most suitable method for the particular application problem in hand. This also applies to the situation where only one method is found in the literature. Even in this case it may be necessary to carry out a further assessment, since the problem is seldom simply one of obtaining a separation of enantiomers. Other requirements, such as achieving high sensitivity or a short analysis time, may be equally important and turn out to be the key factors in dictating the eventual choice of chiral LC method.

14.2 Relative merits of LC enantioseparation methods

All the information required for the further assessment of available chiral LC methods has already been described in detail in Parts 2 and 3 of this book. The purpose of this chapter is to provide an at-a-glance guide to the most important points and to give some guidance to the most logical sequence in which method development should be carried out. This will be done with the simple problem in mind of just obtaining an enantiomer separation. Sugges-

Table 14.1 Diastereomer formation *v.* direct methods

	Diastereomer formation	Direct methods
Advantages	In most cases a separation is obtainable	Rapid, simple
	Minimal method development required	Pure enantiomer elutes from the end of the column and is easily recovered
	Achiral columns used are much cheaper	Enantiomers have the same UV absorption
	Choice of column is often not critical	Much less danger of racemization during analytical procedure
	Can be used to enhance sensitivity	Purity of the chiral selector is not critical
	Cheap method for preparative work if diastereomer formation is reversible	
Disadvantages	Can be time-consuming	Still cases where it is difficult to obtain any kind of separation
	Often difficult to recover the pure enantiomer	Lengthy method development and optimization sometimes required
	Possibility of diastereomers not having the same UV absorption	Expensive columns or high mobile-phase additive consumption
	Possibility of racemization, decomposition or kinetic fractionation during reaction	Separation often possible on one type of chiral LC system only No universal chiral LC system with complete breadth of spectrum
	Purity of chiral selector can be important, especially when trying to detect a trace enantiomer	Preparative work is very expensive

tions are then made on how more complex enantiomer separation problems could be approached.

14.2.1 *Direct v. indirect methods*

The first stage of assessing the various chiral LC methods would be to consider whether a direct chiral resolution method or an indirect method involving derivatization would be the most appropriate approach to the particular problem. The various strengths and weaknesses of methods involving derivatization are discussed in detail in Part 2. The main points to consider when making the comparison between direct and indirect methods are shown in Table 14.1.

If a separation of enantiomers could be achieved equally well by a direct and an indirect method, then the direct method would almost always be the method of choice, provided there were no other significant factors to be taken into consideration.

14.2.2 *Chiral stationary phases v. mobile phase additives*

The direct LC resolution of enantiomers may be carried out by a chiral stationary phase (CSP) consisting of a chiral selector immobilized on silica (or some other suitable support) or by the use of a chiral mobile phase additive in an otherwise achiral LC system. The main points to consider when making a choice between these two methods are shown in Table 14.2. It will most often be found that a bonded CSP is preferable to the use of a chiral mobile phase additive.

Chiral mobile phase additives are often used when there is no suitable CSP commercially available. There are, however, other situations when they might be preferred. For instance, if an enantiomeric resolution can be achieved by a straight-phase ion-pair mechanism using a cheap pairing ion such as d-10-camphorsulphonic acid or l-quinine on a silica or LiChrosorb-DIOL column, then (despite the inherent disadvantages of using ion-pair systems for preparative work), scale-up of this method could be a very acceptable alternative to carrying out preparative LC on a large column packed with a very expensive CSP. Chiral mobile-phase additives might also be used in situations where it was convenient to use the concentration of additive to manipulate retention times.

In making comparisons between bonded phases and equivalent mobile phase additives, one factor which need not be considered is differences in mechanisms of enantioselectively leading to differences in obtainable resolution. Except in isolated cases, the mechanisms would be the same.

14.2.3 *Choice of chiral stationary phase*

The choice of CSP may at first seem bewildering, since in recent years there has been a veritable proliferation of new CSPs. However, the choice is made much simpler by the fact that except in some rare circumstance the choice would be

Table 14.2 Chiral mobile phase additives *v.* chiral stationary phases (CSPs)

	Chiral mobile phase additives	Chiral stationary phases
Advantages	More variables to manipulate	Shorter mobile phase equilibration times
	Chiral selector constantly regenerated	Less wasteful of chiral selector
	Access to a wider range of chiral selectors	Separations more reproducible
Disadvantages	Longer equilibration times	Fewer variables to manipulate
	High consumption of chiral selector (unless a miniaturized LC system is used)	No regeneration of chiral selector, therefore shorter column life times
	Conditions more difficult to control, therefore chromatography less reproducible and assays less robust	More limited range of commercially-available chiral selectors

restricted to commercially available CSPs. Also, the commercially available CSPs may be classified by their mode of action (Wainer, 1987), with the types of racemic compounds which can be resolved on each CSP class being well documented. The classification adopted in this book to classify direct LC chiral resolution methods (i.e. CSPs and chiral mobile phase additives; each chapter of Part 3 represents a classification) is very similar to that of Wainer. This is not surprising, as they are both based on the retention mechanism for chiral recognition.

All the information necessary to make a choice between commercially available CSPs is contained in Part 3 of this book. The key features of each class of CSP to bear in mind when making a choice on which to use are listed in Table 14.3.

Table 14.3 does not contain the two factors which in the practical situation are often the most important: availability and price. Availability speaks for itself. In most cases, the practising chromatographer would try to develop a method on a column already in his laboratory containing a CSP which might give enantioselectivity before purchasing a new column containing a CSP which he thinks might give better enantioselectivity. Fortunately, however, a wide range of commercial CSPs are readily available in most countries today and therefore availability should no longer be a major stumbling block. The same cannot be said for price. Despite the progress made in the field of chiral

Table 14.3 Key points on classes of commercially-available chiral stationary phases (CSPs)

Ligand exchange	Operation in reversed phase Requires metal ions in the mobile phase Used for amino acids, amino alcohols and α-hydroxy acids
Synthetic, multiple interaction	Usually operated in straight phase conditions High efficiencies Robust if covalent bonding used rather than ionic bonding Many of the most successful utilize charge transfer interactions Mainly used for non-ionic compounds with aryl and/or carboxamide groups
Protein affinity	Operation in reversed phase Relatively poor efficiencies Often very high enantioselectivity achieved Relatively poor stability Broader spectrum than other classes
Cyclodextrins	Robust CSPs for reversed phase Enantioselectivity via steric fit plus other ancillary interactions Reasonable efficiency For a series of analogous solutes, a slight difference in structure may produce a large difference in enantioselectivity obtained
Cellulose-based phases	Mobile phase often methanol or isopropanol/*n*-hexane Can be damaged by use of solvents such as methylene chloride Steric fit in a hydrophobic cleft is important in the chiral recognition mechanism Generally used for hydrophobic compounds; often aryl and/or esters
Synthetic polymers	Similar in properties and applications to cellulose-based CSPs

liquid chromatography, CSPs remain relatively expensive. To give some indication of availability and price, some of the major suppliers of CSP-containing LC columns in the UK, Europe and the USA are listed in an Appendix to this book along with their inventory and prices. Absolute prices can change rapidly, but those quoted will give some impression of the relative prices of the various CSP classes. Price is most often a deterrent in the case of cellulose-based CSPs and synthetic polymer CSPs. Although some CSPs are easy to synthesize, the use of 'home-made' CSPs is not a viable alternative. Batch-to-batch reproducibility is difficult to control in 'home-made' CSPs, and, more importantly, when developing a routine method it is often essential to use a CSP which will be readily available to not only other users but also to regulatory authorities.

14.3 Method development

Having assessed all the information available it is possible to proceed with the method most likely to solve the problem. As mentioned already, for the simple case where the only objective is to obtain a separation of enantiomers the preference will usually be for a CSP, assuming there is one that looks as if it might be successful. As mentioned earlier, in many cases there is more than one objective so that the method most likely to solve the problem would not necessarily be the simplest method which looked likely to give good resolution. The choice would rest on practical considerations of price and bias towards chiral LC systems ready at hand, but more likely on the other objectives discussed later in this chapter. Having chosen the most promising method, it should then be possible to begin the process of optimization and then validation.

However, resolving enantiomers is not an easy matter, and even with the range of CSPs now available it is very possible that no enantioselectivity might be observed. It would then be necessary to look for alternative approaches. If the obvious approaches had failed, then it is only logical to suppose that it might take some time to obtain a successful solution to the problem via the less obvious approaches. At this stage, therefore, it is often prudent to settle for an interim solution: a method which is not ideal but can be used during the period of time it takes to arrive at a more satisfactory method. This may involve, for instance, using a time-consuming derivatization procedure, or, if the chiral compound is an intermediate in a synthetic process, monitoring optical purity at an earlier or later stage in the synthetic process.

In the future a useful interim, or even final, solution might be to use an achiral column to measure the peak for the unresolved pair of enantiomers with both a polarimetric detector and a UV detector, the ratio of the enantiomers present being related to the ratio of the signals from the two detectors (Reitsma, 1986). This possibility is discussed in Part 5. Unfortunately, the sensitivity of polarimetric detectors likely to be available in the near future is such that it will be some time before this method becomes common practice. For sugars, quantities of the order of 5 μg can be detected (Turner, 1988).

If it is possible to buy some time by adopting a stop-gap method, it should then be possible to persevere with work on some of the remaining hopeful options. At this stage it would be necessary to take a closer look at what interactions could be employed to achieve enantioselectivity.

As recent literature suggests, theoretical studies can be very useful. Boehm *et al.* (1988) formulated a statistical thermodynamic theory of chiral solute retention and separation and applied it to a model of a cyclodextrin CSP. Lipkowitz *et al.* (1987) used molecular orbital and empirical force field methods to study the conformational properties of covalent Pirkle phases. Similar methods were used by Norinder and Sundholm (1987) to predict enantiomer retention order on Pirkle phases. Taylor (1987) has applied computerized molecular graphics to investigate two new families of synthetic, multiple-interaction CSPs developed in his own laboratories. By studying the relationship between the column packing's surface coverage, column efficiency and enantioselectivity, optimum performance of the CSPs was obtained.

The above examples serve to illustrate just how much can be done if given the right tools available. However, at the present time such sophistication is beyond the reach of most LC laboratories working on the development of LC enantiomer separation methods. One tool which is more readily available and could be put good use much more often in the study of interactions which might lead to enantioselectivity is NMR. A splendid example of this is the work of Salvadori *et al.* (1987). ^{1}H-NMR was used to study the structures of the adducts between quinine and 2′-(2-propoxy)-1, 1′-binaphthyl-2-ol in order to assist in the provision of an experimental basis to a chiral recognition mechanism by which novel cinchona alkaloid CSPs operated.

The simplest and most often used method for studying chiral LC systems to determine whether or not enantioselectivity can be achieved is to consider the three-point interaction rule (Dalgleish, 1952). This rule specifies the minimum requirement for chiral recognition, when trying to match chiral LC systems to the compound, the enantiomers of which it is wished to resolve. The logic behind this rule is illustrated in Chapter 3. Its validity is sometimes questioned, but these questions usually arise from confusion as to how the rule should be interpreted. For instance, it should be noted that not all the interactions need be attractive interactions. Also, it should be noted that the third crucial interaction can arise through contact of a diastereomer or diastereomeric association complex with an achiral surface. This is illustrated in Figure 14.1. Chiral separation through a single substituent association has been claimed (Feibush, 1988), but perhaps this was an example of situations where there is one very strong interaction so that the other two may be fairly weak or not immediately apparent, and enantioselectivity may still be achieved. However, such theoretical niceties should not be allowed to detract

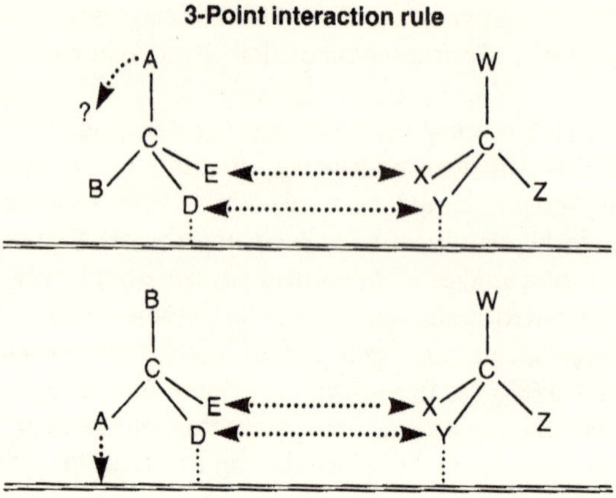

Figure 14.1 Three-point interaction rule with third interaction with achiral stationary phase.

from the fact that in practice the three-point interaction rule can prove very useful. The main difficulty which might arise is in its application to chiral LC systems which employ inclusion complexing. In these cases, the concept of a point interaction might be too much of a simplification to be useful.

By adopting the systematic approach described above, with the range of chiral LC methods that are currently available, it should be possible for almost all enantiomer separation problems to reach a stage where a satisfactory solution to the problem would emerge. This solution must then be optimized and validated.

Optimization is described in Chapter 15. If, even after optimization, it is felt that a further increase in resolution would be desirable, it may be worthwhile to study the effect of temperature. Lowering column temperature invariably results in enhanced resolution of enantiomers. This has been successfully employed, for instance by Mazzo and co-workers (Mazzo, 1986) in the LC resolution of the enantiomers of the anti-hypertensive compound, (+)−[(6, 7-dichloro-2, 3-dihydro-2-methyl-1-oxo-2-phenyl-1 H-inden-5-yl)oxy]acetic acid, on a Pirkle covalent L-phenylglycine CSP with a predominantly organic buffered ternary reversed-phase mobile phase. From a practical point of view, it is worth noting that when working at low temperatures with predominantly aqueous eluents there is always a danger of a very high back pressure suddenly arising as a result of precipitation of buffer salts.

Under most circumstances, certainly if a routine method is being developed, it would be preferable not to have to work at sub-ambient temperatures. In the future though, sub-ambient temperature chromatography might not present so much of an experimental complication, since column controllers especially designed for this type of work are now being produced (Flowgen Instruments Ltd., Sittingbourne, Kent, UK).

Like other chromatographic methods, chiral LC methods must be validated for features such as specificity, robustness, linearity, reproducibility and sensitivity. What distinguishes the validation of an enantiomeric determination method from validation of other LC methods is the need to confirm that it is enantiomers that are being separated. In the past, this might have presented a problem if a sample of each enantiomer or enantiomerically enriched samples was not available, as may well happen in the early stages of development of a novel pharmaceutical. However, this should be less of a problem in the future, as it will be possible to use polarimetric detectors to confirm that the two peaks being resolved are enantiomers. As mentioned earlier, this type of detector is not very sensitive, but this should not present a major drawback in their use for validation purposes.

14.4 More complex enantiomer separation problems

As was suggested earlier, the problem is not always the simplest case of obtaining enantiomer resolution without any other constraints. There are a variety of other factors which often have to be taken into consideration. Many such more complex enantiomer separation problems are illustrated in Part 3. Here, some suggestions are made as to what might be the best general approach to these situations.

14.4.1 *Preparative enantiomer separations*

When scale up is a consideration, the main priority is to achieve very high enantioselectivity. It is also important to work with a mobile phase from which the purified enantiomers can easily be isolated.

High enantioselectivity must be sought in the same way as in analytical scale work. As cited earlier, Pirkle demonstrated how to achieve high enantioselectivity (Pirkle, 1986) in spectacular fashion. However, the generation of bis-analytes seldom offers a realistic practical solution for preparative work.

The constraint on the types of mobile phase which are suitable has resulted in much of the preparative work on LC enantioseparations being carried out on synthetic multiple-interaction CSPs using straight-phase eluents. Herbert and co-workers (Herbert, 1988) have carried out an in-depth study of preparative LC on a dimethoxypropylsilyl-*N*-ethylenaminopropyl-(L-) valylphenylurea CSP, the results of which are generally applicable to straight-phase preparative LC on synthetic multiple-interaction CSPs. One significant finding was that the capacity of their CSPs was relatively low, thus leading to a reduced throughput compared to (say) an ODS material bonded on to the same silica.

Although commercial manufacturers produce cheaper CSPs, based on silicas of larger particle size, specifically for preparative work, this is still a very expensive option. Fortunately, many of the earlier commercial CSPs can be prepared very easily and therefore the use of home-made CSPs for preparative work has its attractions.

Chiral derivatization followed by LC on an achiral stationary phase is a much cheaper option. For instance, alcohols can be converted to diastereomeric carbamates, which can be mildly hydrolysed following resolution without racemization to yield the optically pure alcohols. However, for many other types of diastereomeric derivatives it is not so easy to recover the optically pure enantiomers. Hence there is frequently the need to resort to the expensive CSP option.

14.4.2 *Sensitivity*

When obtaining high sensitivity is a major priority, a GC solution to the problem is worth investigating since most GC detectors are more sensitive than the standard UV detector used in LC. As far as LC is concerned, this is an area where derivatization comes into its own. The need for chiral derivatizing agents for LC which allow for the enhancement of sensitivity by the introduction of a fluorophore or a strongly UV-absorbing chromophore has been recognized by Goto (1986) and Einarsson (1987), who introduced the reagents shown in Figure 14.2*a* and 14.2*b* respectively.

It is perhaps easier to enhance sensitivity by using more readily accessible achiral derivatizing agents with subsequent LC on a CSP. This is most commonly done by using a derivatizing agent containing a naphthyl group and carrying out LC of the derivative on a Pirkle π-acid CSP. There is no reason why this could not be extended to achiral derivatizing agents containing anthryl or other fluorescent aryl groups.

14.4.3 *Limited time for method development*

This situation where a chiral LC method is required urgently in a very short space of time occurs only too frequently. It is then often necessary to aim not for the optimum method but for a method which, irrespective of other drawbacks, is virtually certain to work.

Again derivatization comes into its own in this type of situation. If a diastereomer is prepared, a resolution of the diastereomers on an achiral phase can usually be achieved relatively quickly compared with the time it might take to obtain a direct resolution of enantiomers. Perhaps an even safer option is the achiral derivatization with a reagent containing a naphthyl group followed by LC on a Pirkle π-acid CSP mentioned above. The wealth of data in the literature suggests that if the naphthyl-group is within two or three

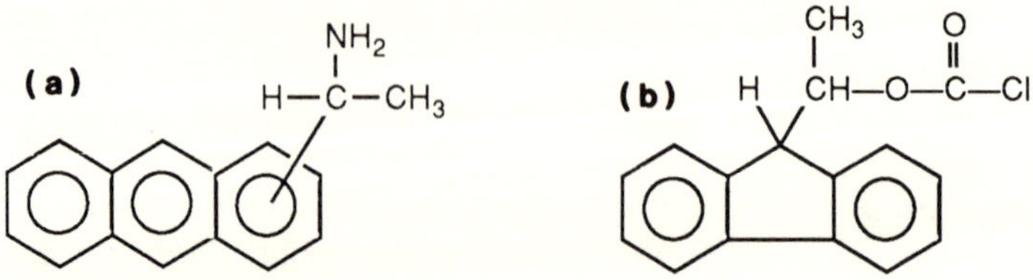

Figure 14.2 Chiral derivatizing agents suitable for work at high sensitivity. (*a*) Goto, 1986; (*b*) Einarsson, 1987.

atoms of the chiral centre (especially if the chiral centre is part of a rigid structure), enantiomeric resolution will be achieved.

Amongst CSPs the α_1-acid glycoprotein CSP (commercialized by LKB/Pharmacia as EnantioPac) has the broadest spectrum and therefore would often be a good first choice in looking for a quick solution to an LC enantiomeric resolution problem. Wainer and Schill (Schill, 1986) demonstrated that it could be successfully used for a very wide range of basic compounds of pharmaceutical interest. They also cited some examples of neutral compounds and by the use of mobile-phase modifiers extended its applicability to acidic compounds. This phase does however have practical disadvantages (relatively poor stability, broad peak shapes, low flow rates, long run times) with respect to routine use. It therefore is often prudent to use an α_1-acid glycoprotein method until a better method can be found.

14.4.4 *Determination of trace enantiomers in the presence of excess of the antipode*

The problem of determining a trace enantiomer in the presence of a large excess of its antipode has been illustrated in Part 3. The chiral selector being used in the chiral LC system should be chosen so that the trace enantiomer elutes first. Figure 14.3 shows the typical benefit in detectability that might be obtained. It is self-evident that this option is not available when working with proteins, cyclodextrins, celluloses and helical polymers as chiral selectors. It is less evident but equally true that when working with commercially-available ligand-exchange or synthetic, multiple-interaction CSPs, that the equivalent CSP employing the antipode of the chiral selector is not always also commercially available.

14.4.5 *Other typical enantiomer separation problems*

It is impossible to give an exhaustive list of other types of problem that might arise. Often application problems have idiosyncrasies which render them unique and demand a novel approach. Nonetheless, this chapter is concluded with a few more example situations with suggestions on how best they might be approached.

If an automated system was required for the throughput of large numbers of

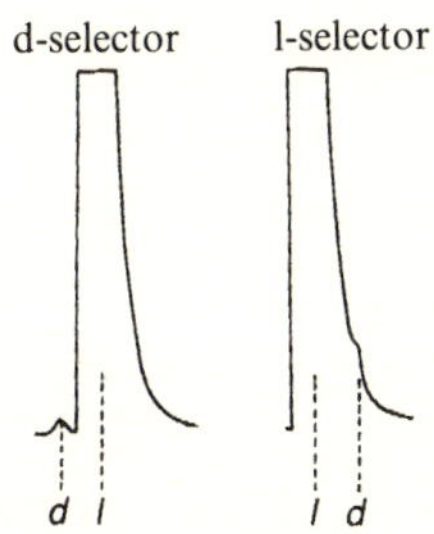

Figure 14.3 Accurate quantitation of a trace enantiomer in the presence of a large excess of its antipode.

aqueous samples, for instance, a cyclodextrin or ligand-exchange CSP might be investigated. Protein CSPs are also suitable for reversed-phase work with aqueous samples but, especially in the case of α_1-glycoprotein phases, might not have the necessary robustness unless handled very carefully.

If a short analysis time was required, high enantioselectivity would be necessary so that baseline resolution could be obtained at low k' values. Straight-phase work on synthetic, multiple-interaction CSPs might be useful in this situation. These CSPs can sustain high flow rates without too much loss of efficiency and resolution, and without being irreversibly damaged.

Finally, if it was required to separate a large number of enantiomeric pairs in one chromatographic run, it might be necessary to resort to the coupling of an achiral and a chiral column in series.

References

Boehm, R.E., Martire, D.E. and Armstrong, D.W. (1988) Theoretical considerations concerning the separation of enantiomeric solutes by liquid chromatography. *Anal. Chem.* **60**, 522.

Dalgleish, C.E. (1952) The optical resolution of aromatic acids on paper chromatograms. *J. Chem. Soc.* 3940–3942.

Einarsson, S., Josephsson, B., Möller, P. and Sanchez, D. (1987) Separation of amino acid enantiomers and chiral amines using precolumn derivatization with (+)-1-(9-fluorenyl)ethyl chloroformate and reversed-phase chromatography. *Anal. Chem.* **59**, 1191.

Feibush, B. (1988) Chiral separation through a single substituent. *J. Chromatogr.* **436**, 517.

Goto, J., Ito, M., Katsuki, S., Saito, N. and Nambara, T. (1986) Sensitive derivatizing agents for optical resolution of carboxylic acids by high performance liquid chromatography with fluorescence detection. *J. Liq. Chromatogr.* **9**, 683.

Herbert, N.R., Rao, K.R.N. and Towill, R.C. (1988) Preparative scale chiral separations using new preparative materials. *2nd Int. Symp. on Preparative and Up-scale Chromatography*, Baden-Baden, West Germany.

Lipkowitz, K.B., Demeter, D.A., Parish, C.A., Landwer, J.M. and Darden, T. (1987) Column. 3. Theoretical studies of a chiral stationary phase used in column chromatography. *J. Comput. Chem.* **8**, 753.

Mazzo, D.J., Lindemann, C.J. and Brenner, G.S. (1986) Subambient temperature high performance liquid chromatographic determination of the enantiomers of (±)-((6,7-dichloro-2,3-dihydro-2-methyl-1-oxo-2-phenyl-1H-inden-5-yl)oxy)acetic acid. *Anal. Chem.* **58**, 636.

Norinder, U. and Sundholm, E.G. (1987) The use of computer aided chemistry to predict chiral separation in liquid chromatography. *J. Liq. Chromotogr.* **10**, 2825.

Pirkle, W. and Pochapsky, T.C. (1986) Generation of extreme selectivity in chiral recognition. *J. Chromatogr.* **369**, 175.

Reitsma, B.H. and Yeung, E.S. (1986) High performance liquid chromatographic determination of enantiomeric ratios of amino acids without chiral separation. *J. Chromatogr.* **362**, 353.

Salvadori, P., Rosini, C., Pini, D., Bertucci, C., Altemura, A., Uccello-Barretta, G. and Raffaelli, A. (1987) A novel application of cinchona alkaloids as chiral auxiliaries: preparation and use of a new family of chiral stationary phases for the chromatographic resolution of racemates. *Tetrahedron* **43**, 4969.

Schill, G., Wainer, I.W. and Barkan, S.A. (1986) Chiral separations of cationic and anionic drugs on an α_1-acid glycoprotein-bonded stationary phase (EnantioPac^R) II. Influence of mobile phase additives and pH on chiral resolution and retention. *J. Chromatogr.* **365**, 73.

Taylor, D.R. and Bridger, G. (1988) Applications of computerised molecular graphics to enantioseparation in HPLC. *Proc. Int. Symp. on Chiral Separations*, Guildford, Surrey, UK, 1987.

Turner, B. (1988) Chiramonitor^{TM} Model 705/25. Preliminary Information. ACS, Macclesfield, Cheshire, UK.

Wainer, I.W. (1987) Proposal for the classification of high performance liquid chromatographic chiral stationary phases: how to choose the right column. *Trends in Anal. Chem.* **6**, 125.

15 Optimization

T.A.G. NOCTOR, A.F. FELL AND B. KAYE

15.1 Introduction

As suggested in the previous chapter, the process of method development does not end when enantiomeric resolution is obtained. In almost every case, a further fine-tuning of experimental variables will be necessary, or at least advisable. This process of optimization should be entered into under the following circumstances, for example:

(i) Baseline resolution is not obtained. Due to the degree of difficulty involved in separating enantiomers, this is a frequent occurrence. To obtain a separation which is suitable as the basis of a quantitative assay, optimization must be carried out until the conditions which give the best possible resolution are found. Ideally, this best possible resolution should be at least baseline.

(ii) Baseline resolution is just obtained. Some of the chiral LC systems are sensitive to very minor changes in experimental variables. Optimization should therefore be carried out to improve resolution even when baseline resolution is just obtained. In this way, the assay being developed will be as robust as possible, and will not depend on say, very high efficiency or extremely accurate preparation of the mobile phase for its success.

(iii) Good baseline resolution is obtained with lengthy retention times. Obviously in most such cases it is desirable to improve the method being developed by optimizing to obtain baseline resolution in the minimum possible analysis time.

Previous chapters have discussed the effects of changing experimental variables in the different kinds of chiral LC systems. Some variables have little effect on enantioselectivity, whereas others have a very pronounced effect. In Table 15.1, for each type of chiral LC system the variables which have the most effect on enantioselectivity, and therefore are the most useful to manipulate in optimization procedures, are summarized.

The power of optimization procedures is best shown by illustrative examples. Optimization can be very effective for LC on EnantioPac, the commercially available α_1-acid glycoprotein (α_1-AGP), which has found widespread use, because of its reversed-phase mode of operation and applicability to a variety of classes of basic compounds of therapeutic interest (Hermansson, 1984).

Table 15.1 Mobile-phase variables with the most pronounced effect on enantioselectivity in chiral LC systems

Ligand exchange	Nature of metal ion M^{z+}, $[M^{z+}]$, ionic strength, pH, nature of ligand[a], ligand concentration[a,c]
Synthetic multiple-interaction phases[b]	Nature and % of organic component(s)
Protein affinity	Ionic strength, pH, nature of protein[a], protein concentration[a,c]
Cyclodextrin inclusion	PH, ionic strength and nature of buffer, nature of cyclodextrin[a], cyclodextrin concentration[a,c]
Cellulose immobilized phases[b]	Nature and % of organic component(s)
Synthetic polymer phases[b]	Nature and % of organic component(s)
Ion-pairing	Nature and % of organic components, nature and concentration of ion-pairing agent[c]

[a]When chiral selector is in mobile phase.
[b]Optimization not as effective for these systems.
[c]Intermediate effect on enantioselectivity, but much more effect on retention, which usually must be taken into consideration when selecting the chromatographic response factor which is to be optimized.

Resolution on the α_1-AGP chiral stationary phase (CSP) is critically controlled by eluent composition; small alterations in the concentration of organic modifier or pH can lead to quite large chromatographic changes. It is therefore important to carefully optimize the resolution of enantiomers on this system. It is possible that the antipodes of a compound may be described as unresolvable on AGP, when in fact they may have only been examined under conditions remote from the optimal. Once any degree of resolution is known to be tenable it is necessary that the conditions be optimized, as efficiently, as possible in terms of analysis time. For this purpose a number of approaches, sequential and simultaneous, are available (Berridge, 1985). Sequential variation of the individual parameters affecting resolution can, in classical HPLC separations at least, lead to the location of a local, rather than the global optimum. For this and other reasons, such as speeding up method development, procedures utilizing the simultaneous variation of all relevant parameters are to be preferred. The application of two of these strategies to the separation of two model compounds was examined. The optimization methods examined were chromatographic response surface mapping and the modified sequential simplex method. The model compounds in questions were oxamniquine (a schistosomicide marketed by Pfizer and Mansil), and UK3883 (a Pfizer compound related to oxamniquine).

15.2 Optimization methods

15.2.1 *Chromatographic response surface mapping*

As previously mentioned, resolution on the AGP CSP is particularly sensitive to changes in eluent pH and percentage of propan-2-ol. The effects of these on the quality of separation can be mathematically expressed by a chromatographic response function (CRF), which describes a chromatogram in terms of resolution and analysis time, as defined by Berridge (1985). By plotting this CRF (on the z-axis) against pH (x) and percentage propan-2-ol (y), the response surface is mapped, the highest point on which equates to the optimal conditions for analysis. The CRF was adapted from that proposed by Berridge (Noctor, 1986) and was as follows:

$$CRF = (Rs/1.5)^2 - 0.25(T_L - 16) - 0.05(7 - T_F)$$

where T_F is the time to first peak and T_L is time to last peak. The factors 0.25 and 0.05 are established empirically.

The data is more easily managed, i.e. the optimum can be more accurately indicated, in the form of an 'iso-CRF' contour plot, in which points representing equal CRF are joined in the form of contours in the plane of pH and percentage propan-2-ol.

The method was applied to the resolution of the enantiomers of oxamniquine. It was assumed that due to the similar chemical nature of the model compounds there would be only minor differences between the response surfaces for oxamniquine and its related compound UK3883.

15.2.2 *Modified sequential simplex approach*

The simplex method of optimization of chromatographic resolution selectively directs the chromatographer through the response surface to an optimum (Berridge, 1982). This is done by firstly evaluating performance with a CRF, as for response surface mapping, for a set of three different starting conditions. These are designated 'worst', 'next to worst' and 'best' according to the value of their CRF. The worst point is rejected, as to proceed in that direction would clearly be to move away from the optimum, and a new set of conditions found by reflecting the worst vertex through the mid-point of a line joining the other two. The resultant point is used to construct a second triangle, and the process is repeated. In this way, the algorithm is selectively guided through the response surface toward the optimum.

The modified sequential simplex (MSS) algorithm differs from the standard method in the addition of the concept of contraction and extension (Berridge,

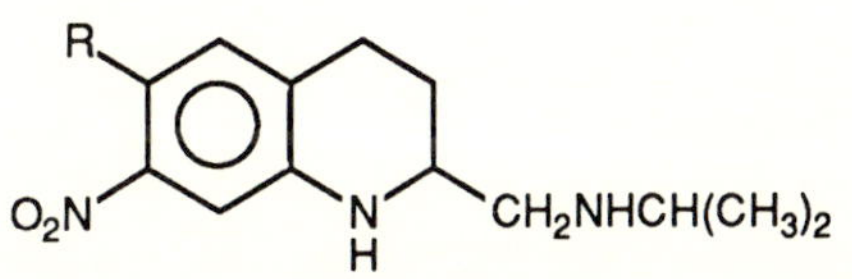

Figure 15.1 Structures of oxamniquine and UK3883. $R = CH_3$, UK3883; $R = CH_2OH$, oxamniquine.

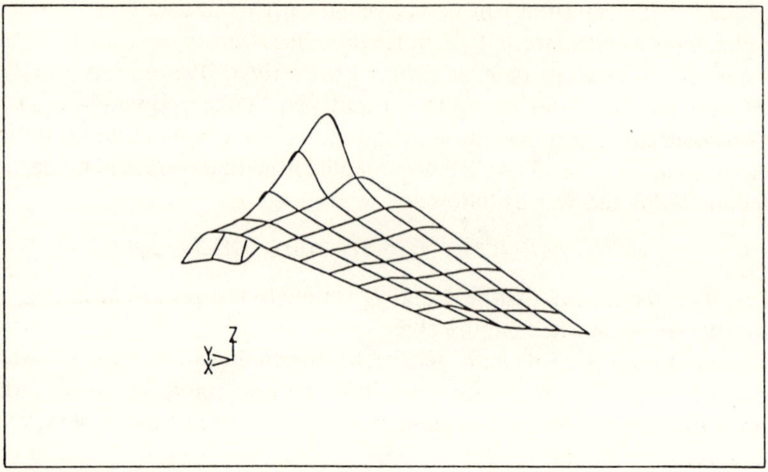

Figure 15.2 Isometric projections of the response surface of oxamniquine. X = pH; Y = percentage of propan-2-ol; Z = CRF.

1985, 1982). The enhancement seeks to direct the algorithm selectively towards the optimum, the zone of data covered becoming progressively focused in the process. The method used was to select three initial eluent mixtures of different pH and percentage propan-2-ol. The CRF figures calculated were then used to generate the next point in the simplex. An off-line microcomputer (BBC-B) was used for this purpose. The method was applied to oxamniquine and the optimum predicted was compared with that previously obtained by CRF mapping. Subsequently, the method was used to optimize the resolution of UK3883, as a test system.

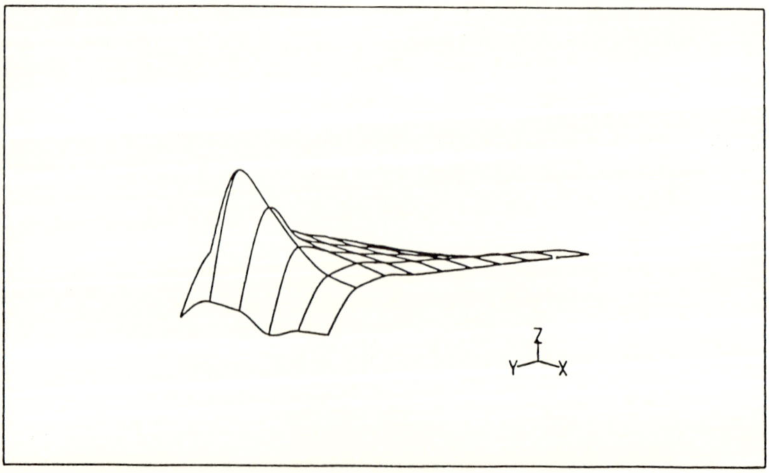

Figure 15.3 'Iso-CRF' contour plot of the oxamniquine response surface.

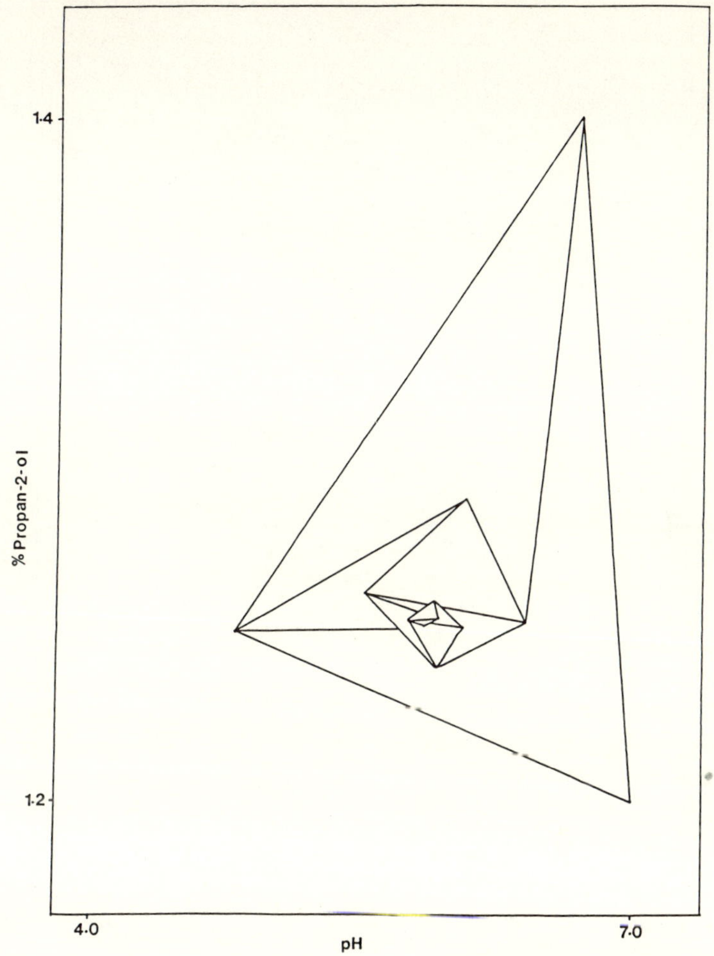

Figure 15.4 Modified sequential simplex diagram for the optimization of the resolution of the enantiomers of oxamniquine.

15.3 Comparison of methods for oxamniquine and its metabolic precursor

Both methods provided useful optimum conditions for the resolution of oxamniquine enantiomers. The response surface plotting method identified an eluent composition of (10mM Na_2HPO_4 + 0.1M NaCl): propan-2-ol (99.5: 0.5% v/v), pH = 5.85.

This yielded k' values for the laevorotatory and the dextrorotatory enantiomers of oxamniquine of 1.22 and 2.44 respectively, the resolution being 2.00 and the selectivity factor (α) being 2.00. The detector response to both isomers was linear in the range 20 ng to 2 μg of injected racemate, the

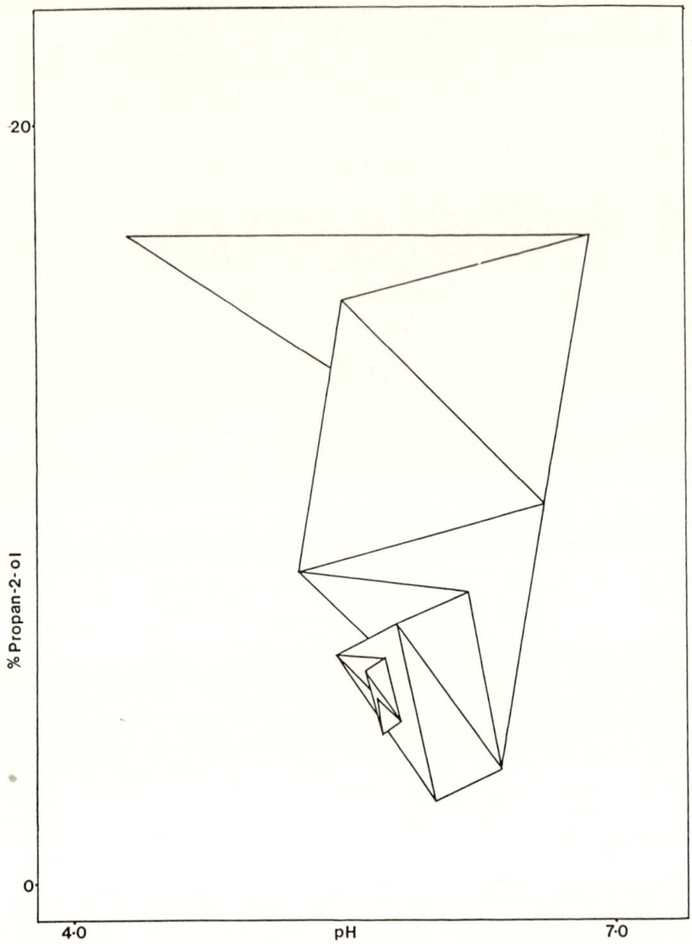

Figure 15.5 Modified sequential simplex diagram for the optimization of the resolution of the enantiomers of UK3883.

detectable limit of the leading enantiomer being < 0.5 ng on column to give a signal-to-noise ratio (SNR) of 2.

The MSS method produced an optimum in 13 operations that was not significantly different to that obtained from the response surface method. The optimum was defined as that zone where the next projected point differed by less than 2% from the previous best. This demonstrated that the simplex algorithm had indeed located the 'true' global optimum despite the difficulty in the fact that the shape of the optimum zone is a long sharp ridge on the response surface. It was assumed that, due to the similar molecular structure of oxamniquine and UK3883, their response surfaces in this system would be

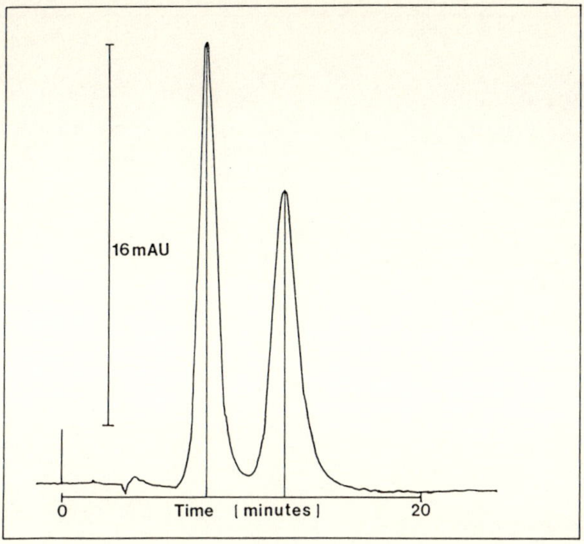

Figure 15.6 Example chromatogram obtained for racemic oxamniquine (25 ng). Conditions as text.

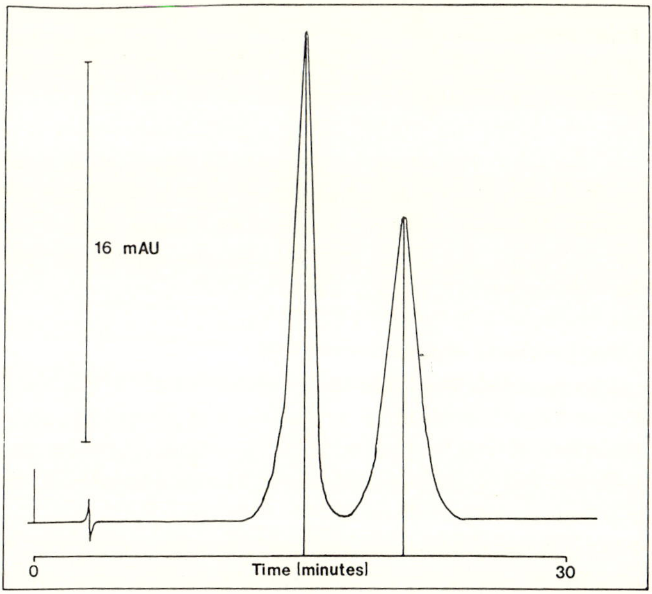

Figure 15.7 Example chromatogram obtained for racemic UK3883 (27 ng), under the optimal conditions given in the text.

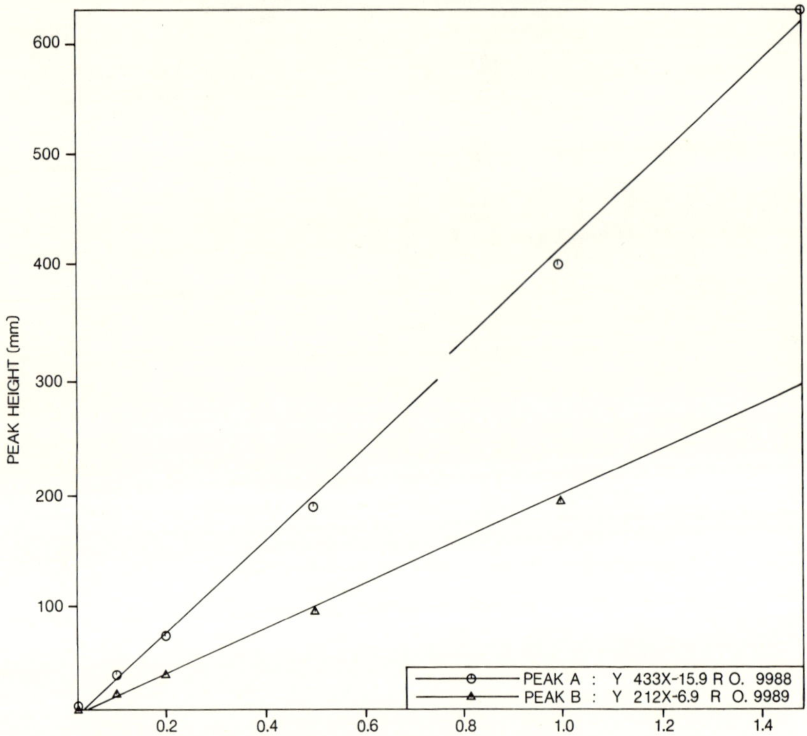

Figure 15.8 Linearity plot for the individual enantiomers of oxamniquine.

broadly similar, with only one optimum present. When the MSS method was applied to the resolution of UK3883, an optimum was located in 17 operations, the final optimized conditions of which were found to be (10 mM NaH_2PO_4 + 0.1 M NaCl): propan-2-ol (95.4: 4.6% v/v), pH = 5.77.

The capacity factors for the first and second peaks were 3.17 and 4.70 respectively, $\alpha = 1.48$, $R_S = 3.50$ and the detection limit for the leading enantiomer was again < 0.5 ng on column (SNR = 2).

Detector response was linear to both enantiomers in the range tested of 20 ng to 2 mg racemate injected on column.

The sensitivity of the AGP column to small changes in analyte molecular structure is perhaps demonstrated by the fact that the enantiomers of one of the compounds could not be resolved in the optimized eluent system found for the other, despite their closely related nature. This is illustrated in the response surface of oxamniquine, which has a very sharply defined optimum zone, as noted above. The response surface (i.e. the variation in resolution *v.* analysis time with % propan-2-ol and pH), was plotted initially for oxamniquine. This is a time-consuming process, but was carried out to ensure there were no local optima in which the modified simplex method subsequently used could become 'trapped'. When the fact that only one optimum was present was verified, the simplex method alone was applied to the closely structurally related (metabolic precursor) UK3883.

15.4 Conclusion

The methods examined have been widely applied to optimization in conventional liquid chromatography. Their extension into the field of chiral HPLC is not trivial, although only two peaks were involved in the present example, because in the complex extracts from, for example, chiral metabolism studies, the presence of other, possibly interfering peaks and of metabolites makes the use of a systematic optimization strategy mandatory. However, it would be dangerous to assume that the behaviour of enantiomers in this type of system is linear, although the response surface map for oxamniquine does show that this is practically the case. The modified sequential simplex method offers advantages in these model separations, particularly in terms of more rapid method development, and in increased certainty that the true optimum is located in the case of complex mixtures.

References

Berridge, J.C. (1982) Unattended optimisation of reversed-phase high-performance liquid chromatographic separations using the modified simplex algorithm. *J. Chromatogr.* **244**, 1–14.

Berridge, J.C. (1985) *Techniques for the Automated Optimizations of HPLC Separations.* Wiley, Chichester, 125–151.

Hermansson, J. (1984) Liquid chromatographic resolution of racemic drugs using a chiral α_1 acid glycoprotein column. *J. Chromatogr.* **298**, 67–78.

Noctor, T.A.G., B.J. Clarke, and A.F. Fell (1986) Chiral separations of drug enantiomers by HPLC. *Anal. Procs.* **23**, 441–443.

Part 5 Future trends and requirements

16 Future trends and requirements

D.R. TAYLOR

16.1 Introduction

As the preceding chapters have shown, the HPLC analysis of a mixture of the enantiomers of an optically-active compound is now usually achievable, either via the preparation of chemically distinct diastereoisomers by reaction with an optically-active reagent, or by direct analysis on a chiral stationary phase (CSP). Although direct analysis on a CSP is the more attractive of these two approaches, in that it avoids the potentially troublesome derivatization step, it can be harder to achieve in practice owing to the rather limited scope of the available chiral phases.

Representatives of the principal classes (Wainer, 1987; Taylor, 1986; Däppen *et al.*, 1986) of CSP are now commercially available; more will undoubtedly emerge as a result of current research aimed at developing phases of higher enantioselectivity but greater generality of application. However, there are those who doubt whether the same success will attend these researches as those which led to the emergence of ODS as the most widely used packing for reversed-phase HPLC. The reason for this pessimistic outlook is the enormous variety of asymmetric structures which require enantioselective analysis, and the relative weakness of the discriminating interactions, recently termed chirality forces (Salem *et al.*, 1987), which give rise to the different retention of enantiomers within the stationary zone.

At a recent conference on chiral HPLC (International Symposium on Chiral Separations, Guildford, UK, September, 1987), it emerged during the discussions that some analysts expected the choice of CSP for a particular separation to become a relatively straightforward matter: either a very generally applicable CSP would be discovered, almost universal in scope, or manufacturers of commercial CSP would be able to define precisely the scope and limitations of their products. These expectations, though understandable, are not very realistic at the present time, because they presuppose a much clearer understanding of the mode of action of CSPs than is currently the case.

The most widely accepted rationale for the recognition of molecular 'handedness' during selective LC retention is that of Dalgliesh (1952), which requires three simultaneous points of interaction between the chiral stationary phase and the analyte molecule. This three-point interaction concept is widely used in the interpretation of separations achieved on CSPs composed of relatively simple molecular structures. Although it has now been shown to have a mathematical basis (Salem *et al.*, 1987), it has limited value in the interpretation of the more complex interaction mechanisms which could arise between analyte molecules and assemblies of molecular structures acting cooperatively, such as one might expect to occur with CSPs based on cellulose or a globular protein.

The designers of bonded HPLC packings capable of discriminating between enantiomers have adopted several different strategies for the provision of the essential molecular interactions. These strategies include the use of

(i) Charge transfer (π–π) interactions
(ii) Dipole–dipole interactions
(iii) H-bonding interactions
(iv) Steric interactions.

Of course, a given chiral packing may use more than one of these processes, and the chromatographic mode employed may be varied between

(i) Ligand exchange LC
(ii) Normal phase (adsorption) LC
(iii) Reverse-phase (partition) LC
(iv) Affinity/inclusion LC
(v) Ion-pairing/ion exchange LC.

Ligand exchange was historically the first to be applied successfully (Davankov, 1980), but the two methods most popular at the present time are undoubtedly adsorption, represented by synthetic multiple interaction phases such as Pirkle's dinitrobenzoyl-*R*-phenylglycine CSP (**1**) and its progeny

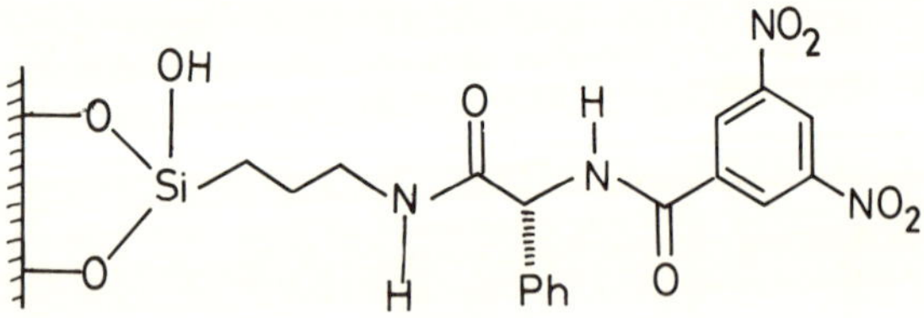

(1)

(Pirkle *et al.*, 1982, 1984; Pirkle and Pochapsky, 1986*a*), and affinity/inclusion, represented on the one hand by the immobilized proteins Enantiopac™ (Hermansson, 1983) and Resolvosil™ (Allenmark *et al.*, 1984) and on the other by the silica-bound cyclodextrins (Cyclobond™) (Armstrong, 1984).

None of these CSPs is ideal. Protein packings are very sensitive in operation, since their globular structure, on which the availability of specific interactions may depend, can vary with pH, pressure, ionic strength and temperature (Wainer *et al.*, 1986*b*; Allenmark, 1986). HPLC analysts are not used to such variability in the behaviour of a single column packing. In contrast, cyclodextrins, being quite rigid structures, are far less susceptible to changes in their environment (Ward and Armstrong, 1986). This means, however, that they can fail to provide the necessary interactions for some analytes, a criticism which applies even more strongly to the relatively simple molecular structures of adsorbant chiral strands such as (**1**).

It is the objective of this concluding chapter to identify trends in chiral separations by HPLC, and to pinpoint possible developments which may make the task of chiral HPLC analysis easier in future. Perhaps an appropriate place to start is with instrumental hardware, since in the author's view there are better prospects for important new developments in this area, which might make our search for a universal CSP seem less urgent.

16.2 Hardware

Hitherto, the vast majority of reported chiral HPLC separations have been achieved on conventional HPLC systems, typically with columns of 25 cm × 4.6 mm i.d. connected to a standard HPLC UV or fluorescence detector. This is a consequence on the one hand of the limited availability of detectors which respond differently to the two enantiomers of an optically active compound, and on the other to the limited demand for CSPs which has restricted the types of columns investigated. However, there are many good reasons why these restrictions should be lifted in future.

16.2.1 *Detectors*

The most obvious choice for a detector to operate in conjunction with a chiral stationary phase is the polarimeter, to be used either on its own or in combination with a UV detector. Unfortunately, commercial analytical polarimeters designed for static sampling, or even for measurements on bulk flowing liquids, offer inadequate detectabilities (*c.* $1.5 \times 10^{-3\circ}$, equivalent to 3×10^{-4} g.mL^{-1} in the light path) at the kind of dilution occurring in a post-column flow-cell and so are unsuitable for analytical HPLC detection without modification. A study of the use of a Perkin-Elmer 241, with a 33-μL cell offering a 10-cm path-length, reported detectabilities of only > 50 μg even at short wavelength (Boehme *et al.*, 1982).

The important requirements for an HPLC polarimeter detector have recently been reviewed by Yeung (1986*b*), who pinpointed the need for a well-optimised polarizer and recommended the use of HeNe laser light to provide moderate but stable light intensity. With such a light source, flow-cells of less than 80 μL have been used with columns of 4.6 mm i.d.; the same detector system even operated satisfactorily with a 25 cm × 1 mm i.d. microbore column (100 000 plates m^{-1}) attached to a 1-μL cell, giving detectability down to 4 ng of a substance with specific rotation of 100° (Bobbitt and Yeung, 1984).

In the UK, plans have been announced recently to manufacture a sensitive polarimeter with microdegree sensitivity, based upon a semiconductor laser diode (2–20 mW at 820 nm), Faraday modulator, photodiode sensor and 1- to 20-μL flow-cell (Goodall and Lloyd, 1988). The high light intensity of this system permits the use of conventional prism polarizers while still affording microgram detectability for carbohydrates in water, for example.

Several advantages would stem from the use of such sensitive polarimetric detectors. The most obvious is that inactive compounds in an inactive mobile phase would be transparent, simplifying the problem of enantiomer resolution on a chiral column, especially if another form of detector was used in a tandem arrangement. Furthermore, the signal produced by an unresolved pair of enantiomers present in unequal concentrations has been shown to be sufficient to determine their proportion; this is achieved by reference to a calibration graph showing the dependence of the ratio of the signals of two detectors, one of which is a polarimeter, upon the enantiomeric excess (the difference between the percentages of the two enantiomers) (Reitsma and Yeung, 1986). Pure standards are of course required. Used in this way, a polarimeter would be valuable even when monitoring the effluent from a conventional achiral column. Another important merit of polarimetric detection in chiral HPLC is that when enantiomers are resolved, their elution order is immediately indicated by the sign of the polarimeter's response. This would be particularly useful when pure enantiomeric standards are difficult to obtain, for example when analysing metabolities of a racemic substance.

Although it is well known that enantiomers rotate the plane of polarized

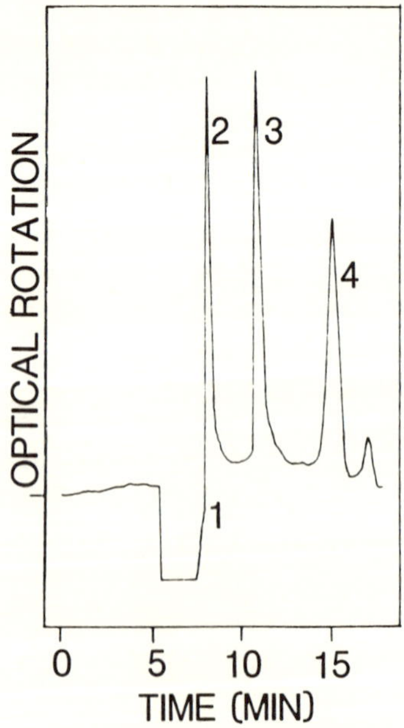

Figure 16.1 Detection by indirect polarimetry: eluent, 95:5 (*R*)-(+)-limonene/isooctane; flow-rate 24 μL min^{-1}; column, 25 cm × 1 mm i.d. 10 μm silica; (1) injection peak; (2) dioctyl phthalate; (3) dibutyl phthalate; (4) diethyl phthalate. From D.R. Bobbitt and E.S. Yeung (1985).

light equally but in opposite directions, it is less commonly appreciated that inactive substances will also rotate the plane of polarized light if an optically active solvent is used: this is termed indirect polarimetry, and may be used in LC detection (cf. indirect UV detection of ionic species). Such a combination of a polarimetric detector with a chiral mobile phase offers an alternative approach to a universal detector, the sensitivity and dynamic reserve of which make it more attractive for microbore systems than, for example, refractive index detection (Bobbitt and Yeung, 1985). A typical chromatogram obtained in a microbore column with *R*-(+)-limonene (**2**) as the mobile phase additive (MPA) is shown in Figure 16.1.

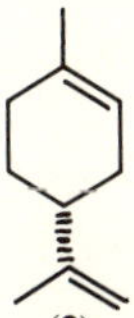

(2)

Circular dichroism, defined as the difference in absorbance of left- and right-circularly polarized light at a specific wavelength, has also been used for detecting chiral molecules (Yeung, 1986*a*). Higher limits of detectability (*c.* 6–20 ng) seem to apply than those reported for the latest generation of polarimetric detectors, however. Laser light is again required (argon ion, 488 nm). It should be noted that circular dichroism absorption does not indicate direction of rotation, and hence it cannot necessarily be deduced which enantiomer emerges under a particular peak envelope. A typical chromatogram is shown in Figure 16.2.

16.2.2 *Miniaturization*

With the advent of microbore (i.d. < 2 mm) and capillary HPLC columns (i.d. < 100 μm), the possibility of using chiral mobile phases (CMPs) as an alternative to CSPs becomes a reality, since flow rates are reduced to less than 50 μL min^{-1} (for a recent review, see Gilbert, 1987). Chirality within a mobile phase may be a consequence of its inherent asymmetry (e.g. butan-2-ol) or of the presence in the mobile phase of an asymmetric modifier or MPA (e.g., β-cyclodextrin, limonene, D-glucose). Neither alternative is attractive owing to the cost involved, unless used in conjunction with miniaturized equipment. Unfortunately, the use of packed capillary or open-tube columns is far from widespread, owing to difficult problems associated with the design of appropriate low-volume injectors, detectors and pumps, and with the avoidance of extra-column dead volumes.

Most work in this area has been carried out in situations where a metal complex or ion pair is formed between the modifier and the analyte (Davankov and Kurganov, 1980). The position of the equilibrium controls the retention, and hence it is essential that the two enantiomers have different formation

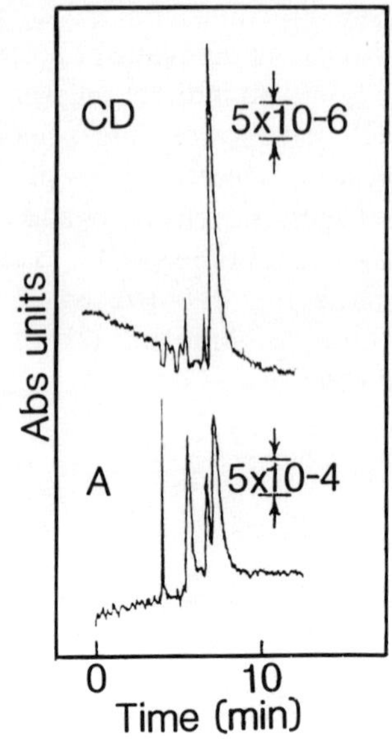

Figure 16.2 HPLC detection of $Co(NH_3)_5\ Cl^{2+}$, $Cr(NH_3)_6^{3+}$, and (+)-$Co(en)_3^{3+}$ in order of elution. CD, circular dichroism. A, absorbance. 10 μL injected, 0.67 mL min^{-1}, 5×10^{-4} M (+)-$Co(en)_3^{3+}$, 500 kHz modulation frequency, 1 s time constant, and a 2-cm pathlength cell. Reproduced with permission from Yeung, E.S. (1986*a*). Copyright © 1986, John Wiley & Sons, Inc.

constants for their complexes. There is to date little evidence that an inherently asymmetric mobile phase produces significant resolutions for a wide range of structural types of racemic analyte on a conventional column packing. Nor have many researchers addressed the question of whether enantioselectivity on CSP is enhanced by asymmetry in the mobile phase, although there are good reasons for supposing that solvent molecules participate to some extent in the formation of diastereoisomeric association complexes. Interestingly, quinine (**3a**), when used as MPA in dichloromethane/acetic acid, appears to enhance the performance of a CSP derived from the closely related acetylquinidine (**3b**) (Pettersson and Gioeli, 1988).

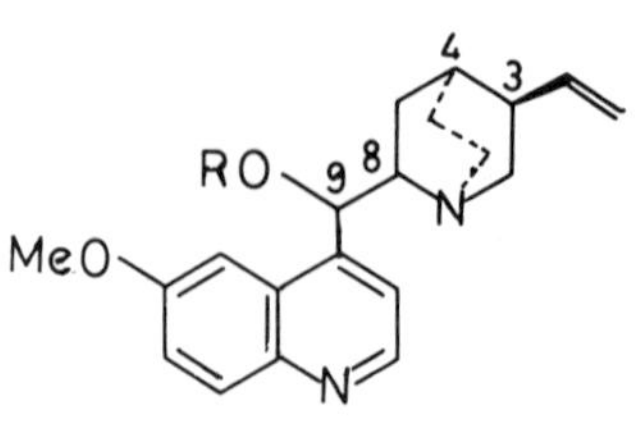

(**3a**) R = H; 3(*R*), 4(*S*), 8(*S*), 9(*R*);
(**3b**) R = Ac; 3(*R*), 4(*S*), 8(*R*), 9(*S*)

16.2.3 *Multicolumn systems*

One potentially interesting development which has not so far been reported is the simultaneous use of more than one type of chiral packing. Probably the most effective way of implementing this idea would be as part of a multi-dimensional HPLC system, in which column switching was fully integrated with an auto-optimization facility. The study of procedures for automatic determination of the optimal conditions for an HPLC analysis has most commonly been conducted on a single packing, the system software and hardware identifying the ideal composition of mobile phase (Berridge, 1985; Schoenmakers, 1986), after a sequence of analyses for which the analyst has provided assessment criteria.

For the analysis of a variety of racemic materials one really needs such a system to have access to a selection of alternative types of CSP, through an appropriate column-switching facility. This would imply perhaps six different chiral columns in parallel, with a corresponding range of mobile phases. Such HPLC systems would eventually approach the complexity of a computerized Expert System (Focus, 1986), because following the changes in resolution for a particular pair of enantiomers as the system switches from one CSP to another, perhaps with substantial change in retention, would present a considerable challenge even with polarimetric detection. Unfortunately, there is no polarimetric equivalent of the photodiode array detector. For such a device to be constructed, for example to acquire circular dichroism measurements at different monitoring wavelengths, each detector element would have to be capable of monitoring simultaneously the light absorbance A and the differential absorbance, $DA = A(+) - A(-)$, so that the dissymmetry ratio or g-value (DA/A) at several wavelengths could be displayed as a function of time. However, there is as yet no evidence that such information could be obtained on-line in HPLC.

16.3 Trends in the development of new chiral packings

There appears to be no downturn in the rate at which new chiral stationary phases are being reported. The various types of CSP have now been classified into broad groups (Wainer, 1987; Taylor, 1986; Däppen *et al.*, 1986), but the number of available commercial CSPs continues to rise faster than our understanding of their underlying mechanism of operation. This may be good for the research scientist but it is confusing for the analyst. We need a more rational approach to the selection of CSPs for specific types of analysis, and this is only likely to emerge when we understand better how they operate.

16.3.1 *Recognition mechanisms*

Attempts to unravel the molecular recognition mechanism by which a CSP resolves a racemic mixture are making most rapid progress for the relatively simple bonded-strand CSP. This is not particularly surprising: such CSPs afford only a limited number of possible interaction sites, and can be spaced out on the silica surface so that interstrand effects are minimized (they are normally ignored!). Discrimination mechanisms based on π-acid/π-base interactions (Wainer, 1986*b*) operating in conjunction with dipole-stacks (Pirkle, 1983) and/or hydrogen-bonding (Dobashi and Hara, 1985) are relatively easy to comprehend and to model. They therefore enable some sort of prediction of their scope and limitations to be made, and the concept of reciprocity to be applied to their design and evaluation (Pirkle *et al.*, 1982, 1987).

It should be possible to apply high-resolution NMR spectroscopy to the study of the interactions which lead to chiral recognition, using model compounds in solution at least. Unfortunately, this valuable parallel technique has not yet been properly exploited for the study of CSPs and their mechanisms. Two exceptional papers point the way: in the first (Shah *et al.*, 1987) the association complexes of the enantiomers of 2, 2, 2-trifluoro-1-(9-anthryl)ethanol (**4**) with butoxycarbonyl-(*R*)-valine (**5**) were studied in this way, and the chemical shifts induced were observed to be greatest for the more retained (*R*)-enantiomer.

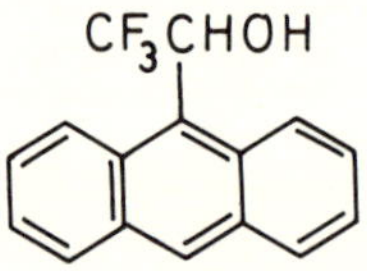

(**4**)

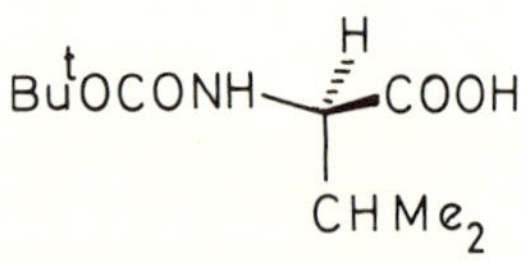

(**5**)

In the second (Pirkle and Pochapsky, 1987), specific intermolecular NOE enhancements were observed for the (*S*, *S*)-diastereoisomeric complex of N-(dinitrobenzoyl)leucine *n*-propylamide (**6**) with N-(2-naphthyl)alanine methyl ester (**7**); these appear to corroborate the chiral recognition model previously proposed for selective retention of the (*S*)-DNB-leucine amide on a CSP derived from naphthylalanine.

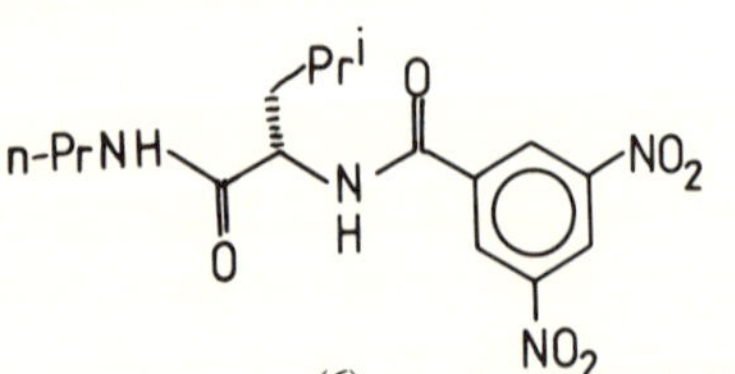

(**6**)

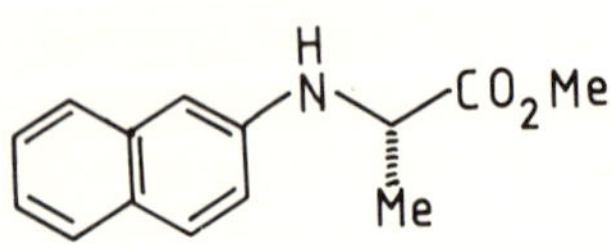

(**7**)

One highly significant observation was that in the latter type of complex, π–π charge transfer is the weakest interaction, i.e., the interaction lost when the less retained enantiomer and the CSP associate. Such studies could be extremely revealing, because the role of the solvent and of conformational variation could be explored; which groups approach most closely when the analyte and CSP associate could also be established.

Somewhat more complex are the inclusion complexes believed to provide the essential discriminations in cyclodextrin CSP (Armstrong *et al.*, 1986), but simple concepts of hydrophobicity, cavity diameter, and hydrogen-bonding with secondary hydroxyls enable a working hypothesis of suitability to be constructed for new analytical problems (Maguire, 1987). The most likely new development in this area is the use of functional group interconversions to transform the H-bonding sites on the rim of the cavity as in (**8**), obtained by derivatization *after* immobilization on silica (Tanaka *et al.*, 1987), and the permethylcyclodextrins (**9**), obtained by methylation *before* surface attachment (Tanaka *et al.*, 1986).

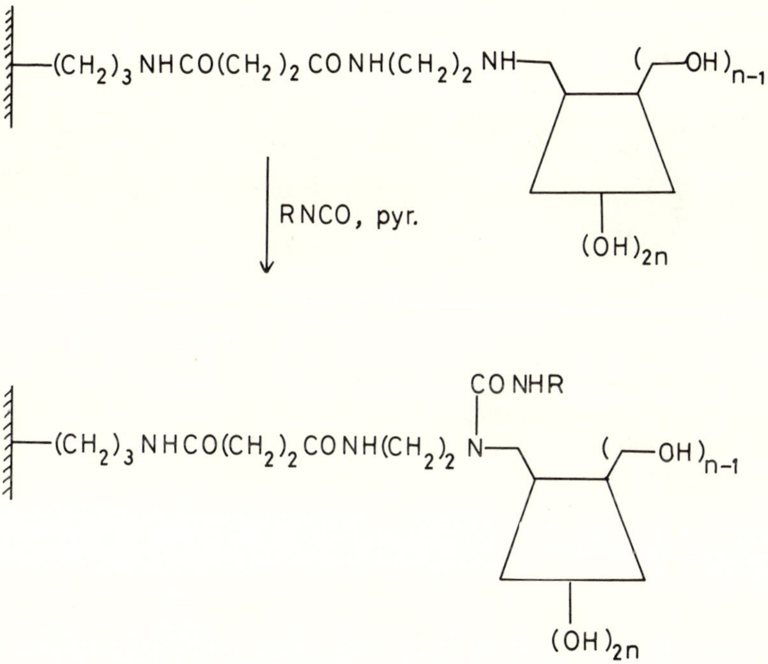

(**8**)

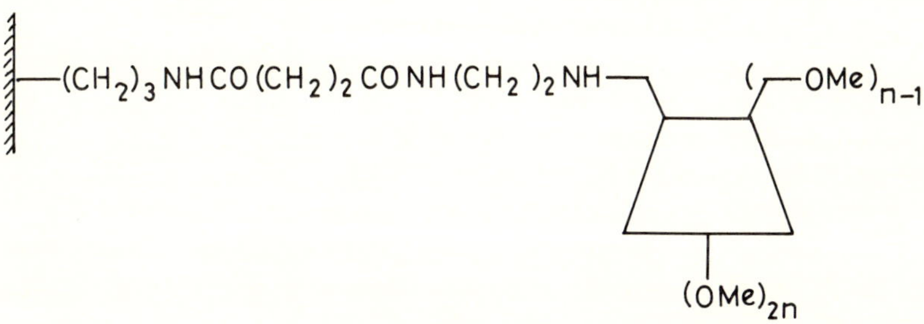

(**9**)

Cyclodextrins have also been used for chiral subcritical fluid chromatography in which modified carbon dioxide was the mobile phase, claimed to be the first example of normal-phase chiral LC separation on a bonded cyclodextrin (Macaudiere *et al*., 1987).

The least well-understood are the polymeric CSPs, which fall into three main classes at present: (i) the cellulose derivatives such as cellulose triesters (**10a**) (Ichida *et al*., 1984) and tricarbamates (**10b**) (Okamoto *et al*., 1984); (ii) the globular proteins such as glycoprotein (Hermansson, 1983) and bovine serum albumin (Allenmark *et al*., 1984); and (iii) the synthetic chiral polymers, of which the most extensively investigated are the polyacrylates (**11**) (Okamoto and Hakada, 1986).

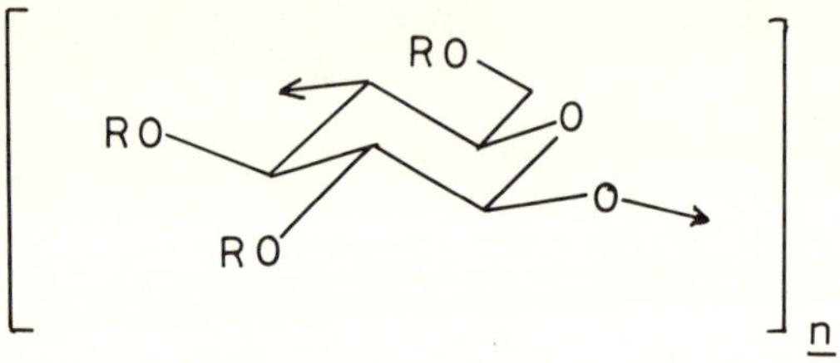

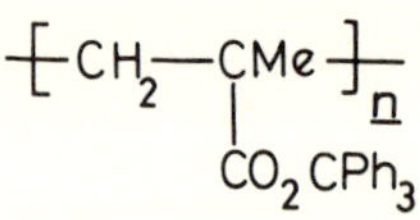

(**11**)

(**10a**) R = COMe, COPh; (**10b**) R = CONHAr

The molecular complexity of such stationary phases militates against any rapid understanding of their mode of action. Attempts are nevertheless being made, for example by bonding to silica the segmental proteins obtained by partial degradation of large globular proteins (Erlandsson and Nilsson, 1987), to identify the sites of strongest interactions. As more and more applications are reported for individual CSPs, their scope and limitations will gradually emerge to provide further clues to their mode of action (Hermansson *et al*., 1987). There is a need for a comprehensive database of separations on such complex CSPs. The task resembles that of probing the effect of cancer-inducing compounds on DNA, or of defining the receptor for a class of drugs, and we need to bear in mind that the use of CSPs is still in its infancy. One important development we should anticipate is for progress on the elucidation of the effects of mobile phase modifiers on these polymeric phases, both during their analytical use (Wainer *et al*., 1987) and as a pre-treatment protocol (Aubel and Rogers, 1987). In the former case, the effects of chain length, bulk and concentration of alcohol mobile-phase modifiers on retention and enantioselectivity provide a convincing demonstration of the presence of solvent molecules at the sites of interaction.

Polymeric CSPs, the asymmetry of which can be due either to the helicity of the polymer or to the presence of pendant asymmetric sites, afford a multiplicity of interaction sites for enantioselective retention. The length and flexibility of the polymer chains will encourage cooperative interactions between adjacent strands, and this no doubt explains why polymeric CSPs enable a wide variety of molecular structures to find a matching, discriminating, segment. However, their molecular complexity poses a number of problems: for example, heterogeneous acetylation of cellulose yields a different crystalline form to homogeneous acetylation, a difference which is reflected both in observed enantioselectivities and their mechanistic explanations (Wainer and Alembik, 1986).

16.3.2 *New types of chiral stationary phases*

At the start of the development of CSPs, the essential prerequisite was that a bondable material was readily available in good optical purity, preferably at moderate cost. Naturally, chemists turned to amino acids and to carbohydrate-related materials such as D-glucose and (*R*, *R*)-tartaric acid; indeed, almost all of the CSPs currently available commercially are chiral by virtue of the presence of amino acid or carbohydrate-related molecular units. The success of this approach has been amply demonstrated by the 'designer' CSPs emerging from Professor Pirkle's group, based on chiral sub-units such as (*R*)-phenylglycine (**1**) and (*S*)-N-(2-naphthyl)valine (**12**).

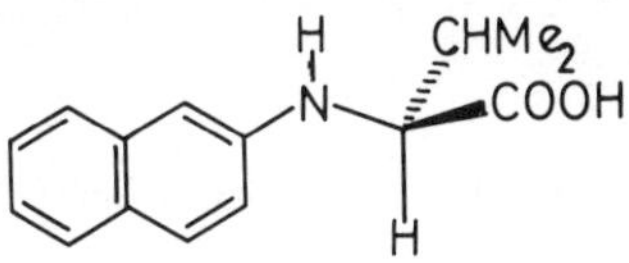

(**12**)

The carbohydrates are particularly popular, and have been attached to silica as polymers (cellulose triesters), as cyclic oligomers (cyclodextrins), as single pyranoses (**13**) (Weller *et al.*, 1987) and in simplest form as (*R*, *R*)-tartaric acid amides (**14**) (Dobashi and Hara, 1985; Bridger, 1987*b*).

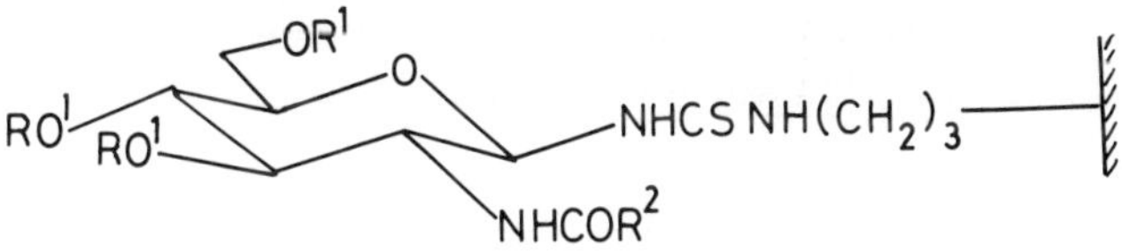

(**13**) $R^1 = Ac$, $R^2 = Me$, $C_6H_3(NO_2)_2$; $R^1 = COPh$, $R^2 = Ph$

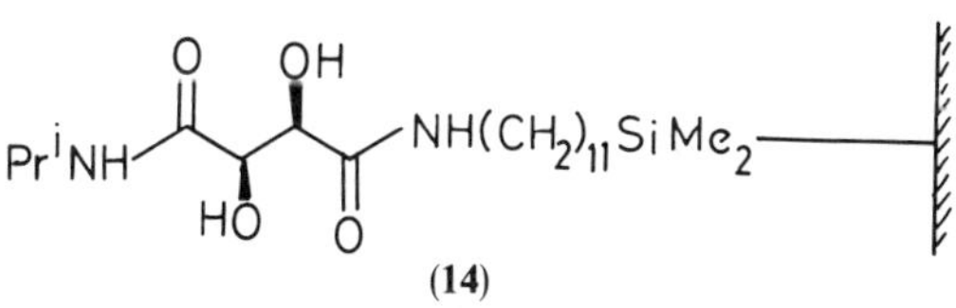

(**14**)

Another group of asymmetric natural products which should prove to be a fruitful source of CSPs is the cinchona alkaloids, such as quinine (**3a**) (Rosini *et al.*, 1985).

One can see that this type of restriction to readily available natural products as starting materials for CSP synthesis will not persist much longer. The literature of organic chemistry currently displays a major preoccupation with asymmetric synthesis (Morrison, 1983–87), fuelled both by the demands of the pharmaceutical industry, and by the elegance of Nature's asymmetric syntheses which provides both a stimulus and a challenge to synthetic chemists. The application of new enantioselective synthetic procedures, combined with an understanding of their mechanism of operation, will make a wider variety of chiral starting materials available on a scale sufficient for CSP manufacture. This may be exemplified by the recent report of the preparation of β-amidoester CSPs (**15**) which were derived from β-lactams (Pirkle, 1987).

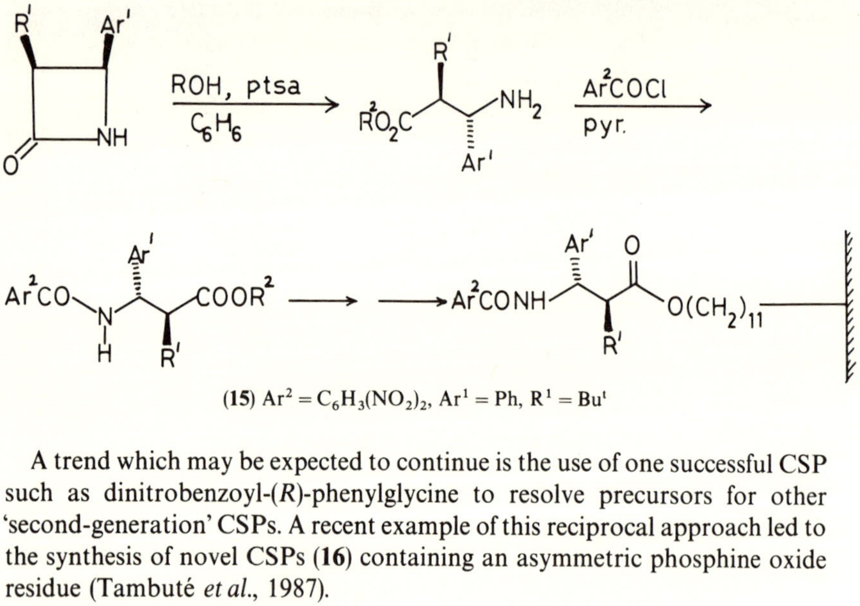

(**15**) $Ar^2 = C_6H_3(NO_2)_2$, $Ar^1 = Ph$, $R^1 = Bu^t$

A trend which may be expected to continue is the use of one successful CSP such as dinitrobenzoyl-(*R*)-phenylglycine to resolve precursors for other 'second-generation' CSPs. A recent example of this reciprocal approach led to the synthesis of novel CSPs (**16**) containing an asymmetric phosphine oxide residue (Tambuté *et al.*, 1987).

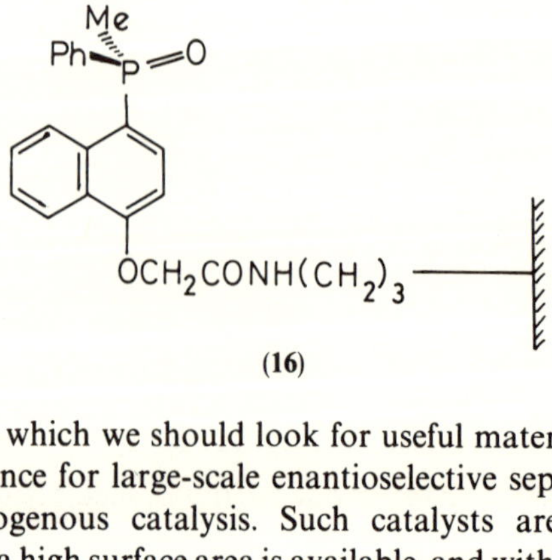

(**16**)

Another area to which we should look for useful materials, likely to be of particular importance for large-scale enantioselective separations, is that of asymmetric heterogenous catalysis. Such catalysts are often composite materials in which a high surface area is available, and within which molecular asymmetry has been immobilized: these are very similar requirements to those for successful CSP in HPLC. These supported catalysts operate by superficially similar mechanisms to CSPs, in that they generally require one or more of the substrates to bind to the immobilized catalytic centre, where asymmetric interactions lead to selectivity in, for example, hydrogenation (Halpern, 1985). The coordination process in such catalytic materials frequently necessitates the presence of a metal ion, but this need not inhibit HPLC separations, as evidenced by the success of ligand exchange CSPs and, in a quite different way, by chiral complexation materials for GC. Typical of the structures reported for such catalysts is polymer-supported DIOP (**17**) (Dumont *et al.*, 1973).

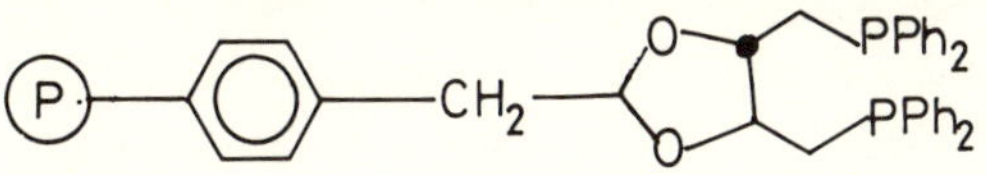

(**17**)

The particular attractiveness of this area as a source of CSPs would be the scale of manufacture which one could eventually envisage for such catalysts, suggesting that they might provide a low-cost source of CSPs for large-scale LC.

Very promising composite materials have recently been reported (Unger *et al.*, 1988). These CSPs were made by free-radical polymerization of chiral acrylamides such as (**18**), pre-adsorbed onto amino- or diol-bonded silica, giving a grafted chiral composite with good mechanical strength under pressure, the main drawback with swollen polystyrene-divinylbenzene resin matrices.

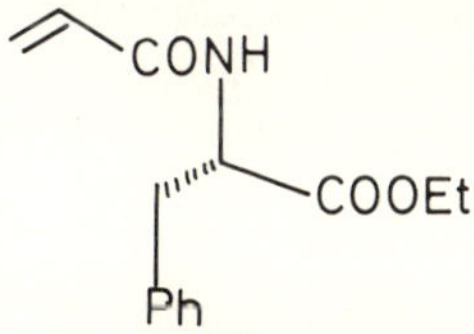

(**18**)

Such composites are also claimed to have good loadability, and, since they could be very cheap if manufactured in bulk, are of great interest for large-scale enantioseparations. Another promising source of materials for such composites would be the chiral polyvinyl recently obtained by template cyclocopolymerization of (**19**) (Wulff *et al.*, 1987).

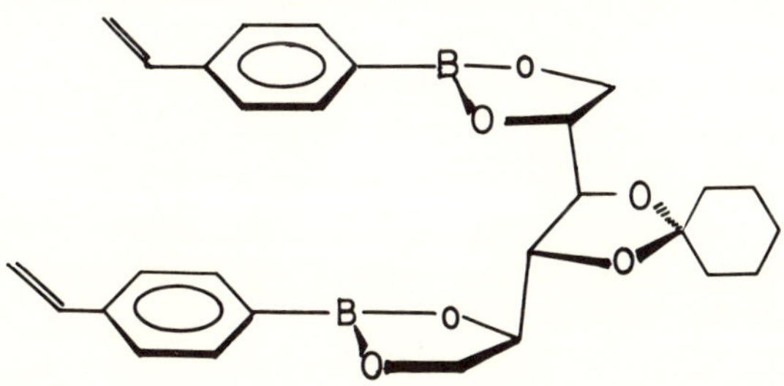

(**19**)

It seems most unlikely that the present very inflated prices for some CSPs can be maintained: manufacturing companies requiring such CSPs for pilot-scale operations will eventually make their own. The development of processes dependent on large-scale LC for chiral separations, although virtually non-existent at present, cannot long be delayed. The principal problem remaining is that of identifying which CSP to select for large-scale production, and this is a mechanism-related problem.

16.3.3 *Use of computer-assisted modelling*

One modern tool for the study of the interaction mechanisms of CSP is computer-assisted molecular modelling (CAMM). Powerful software packages allow, amongst other things, the structures of possible diastereoisomeric complexes to be visualized and optimized. By optimization in this context is meant the computation of bond angles for favoured conformations of the bound strand and the analyte enantiomers. Such knowledge of the conformation is essential if the two alternative association complexes are then to be modelled and compared, in order to determine which complex maximizes the intermolecular attractions between the bound strand and the more-retained enantiomer of the analyte. However, many assumptions must currently be made in undertaking these computations.

For example, such computations typically ignore solvent molecules, although we know perfectly well that the stationary zone is occupied by virtually stagnant solvent, the molecular size and shape of which can affect enantioselectivity (Wainer *et al.*, 1987). In the case of mixed solvents, the problems are even greater, since it would not necessarily be correct to assume that the stagnant zone has a composition identical to the freely moving eluant.

As computing power becomes ever greater and cheaper, CAMM systems will incorporate even more sophisticated packages to predict stable conformations and interaction forces. Consequently their ability to explain observed enantiomeric discriminations will increase: nevertheless, at present we are still far from the day when such a system will indicate the appropriate CSP for a given separation.

Lipkowitz and co-workers (1987*a*) have undertaken a detailed study of one of the most popular asymmetric 'strand' packings, that of N-dinitrobenzoyl-(*R*)-phenylglycine (**1**), the covalent Pirkle CSP. They used W. Clark Still's MODEL program, supported by MOPAC for MNDO calculations and Allinger's MM2 empirical force-field computation to map the potential energy surface, establishing that the favoured conformation is *syn* rather than *anti* about the glycine's C–N bond, as in (**20**). There is about 5 kcal mol^{-1} between the two forms. They have also modelled the association complex between (R)-(**1**) and the enantiomers of trifluoroanthrylethanol (**4**) (Lipkowitz *et al.*, 1987*b*), correctly predicting that the (*S*)-isomer should be more strongly retained. Their predicted association complexes, however, differed significantly from those previously proposed (Pirkle *et al.*, 1980) in that the two enantiomers of

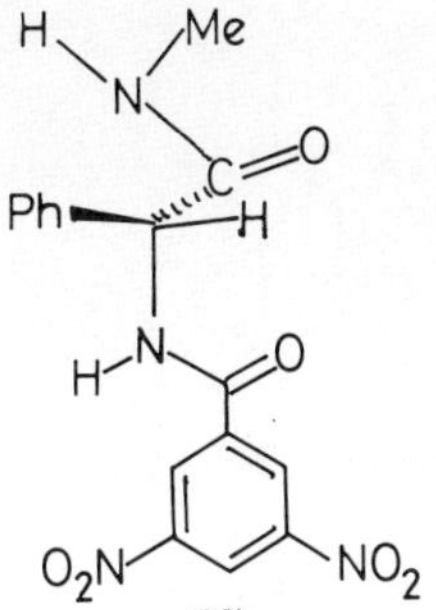

(20)

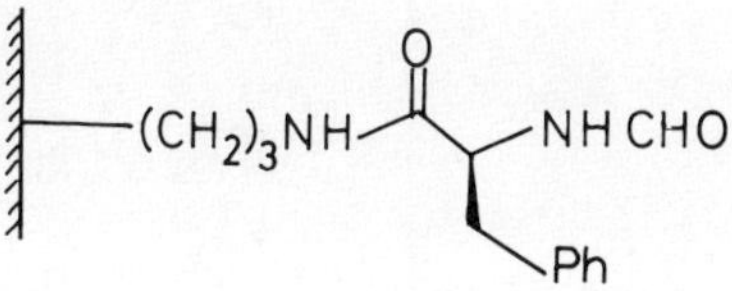

(21)

the analyte adopted different conformations when associating with the CSP.

Bridger *et al.* (1987*a*, *b*) have carried out similar CAMM studies of CSP (**21**) and (**22**) derived respectively from N-formyl-(*S*)-phenylalanine and nitrophenyl-(*R*, *R*)-tartramide, while complexes of β-cyclodextrin with (*R*)- and (*S*)-propranolol have been studied by Armstrong *et al.* (1986), which required molecular structures of greater complexity, but greater rigidity, to be modelled.

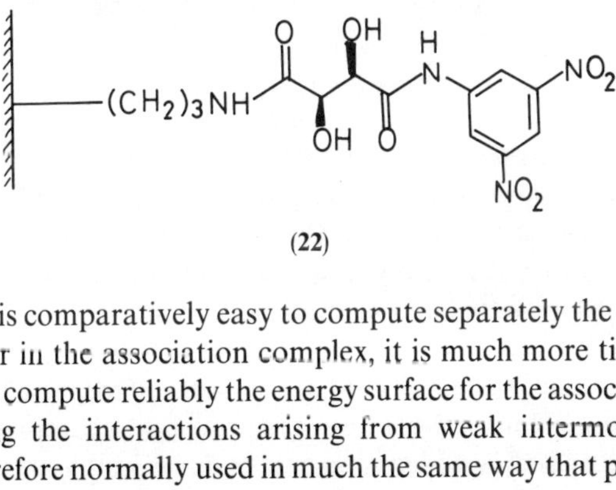

(22)

Although it is comparatively easy to compute separately the conformations of each partner in the association complex, it is much more time-consuming and difficult to compute reliably the energy surface for the association complex itself, including the interactions arising from weak intermolecular forces. CAMM is therefore normally used in much the same way that physical models (e.g. CPK models) are used. The association complexes are compared visually, with interatomic distances being compared and the angles at which hydrogen-bonds might arise being considered. Even with these limitations, alternative mechanisms can often be discerned and placed in order of probability. Such computational procedures will certainly become both faster and more reliable as transputers and other devices of ultra-high computing power become available.

16.3.4 *Conclusions*

HPLC analysts need to recognize that the very specificity which is demanded of an enantioselective CSP militates against its general applicability. To compensate for this inherent drawback, an interesting possibility would be the development of packings which combine several different recognition mech-

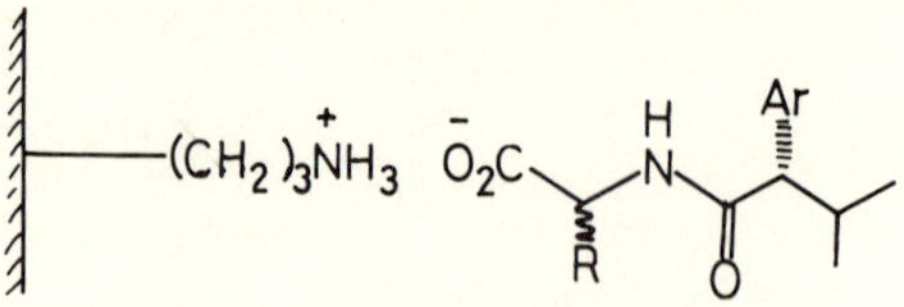

(**23**) R = Ph (*R*, *S*) or *i*-Pr (*S*, *S*)

anisms, although initial attempts in this direction, such as CSP (**23**) (Ôi *et al.*, 1983) were not strikingly successful.

Another disadvantage associated with many current CSPs is the necessity for (achiral) derivatization of the racemic material to enhance the selective interactions. That this need not be a drawback has been demonstrated with amazing force by the use of bis-derivatives such as (**24**) (Pirkle and Pochapsky, 1986*b*) to enhance the interactions leading to enantioselective retention, producing separations between enantiomers with time intervals in hours!

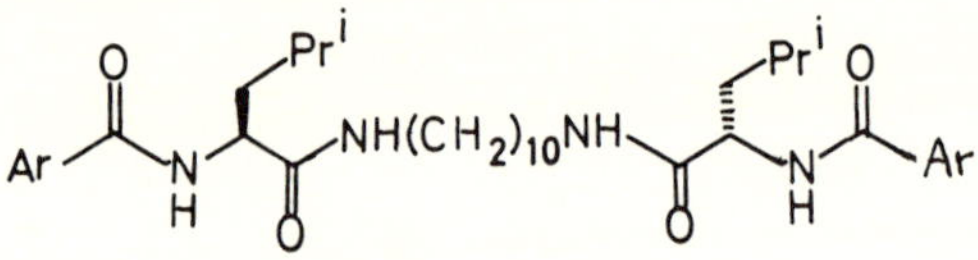

(**24**) Ar = $C_6H_3(NO_2)_2$

Given sufficiently powerful interactions such as these provide, quite short column lengths could be used, possibly in combination with a conventional, achiral, bonded phase which permits the racemic pair of isomers to be resolved from other components. In this context, the use of bonded racemic structures for packing columns to be used in series with effective CSPs has been recommended as a way of adjusting resolutions (Pirkle *et al.*, 1987). The new connectable cartridge systems would be very attractive for such a mixed-phase approach, although in at least one case a single column containing two different chiral phases has been found to perform well for a range of solutes (Kip *et al.*, 1986). Alternatively, column switching devices could provide access to a desired combination of chiral and achiral phases.

16.4 Preparative liquid chromatography

Large-scale LC separations are now receiving very active investigation: three international symposia have already been dedicated to the subject, and columns of 45-cm diameter have been constructed and shown to operate according to the general principles of narrower-diameter LC columns. Worldwide, between 25 and 50 chemical or biotechnology companies are believed to be operating LC columns at the pilot scale or beyond, while semipreparative and preparative HPLC are very widely used in research laboratories, particularly for isomer separations. It is the purpose of this section to consider how such developments are likely to affect enantiomer separations.

Several terms are used to describe LC operations which yield fractions large enough to be worthwhile collecting for subsequent use. A classification will be adopted here which distinguishes (i) semipreparative HPLC in 10–25 mm i.d. columns, useful for obtaining pure calibration standards or samples for spectroscopic analysis; (ii) preparative HPLC in 25–50 mm i.d. columns, useful for quantities adequate for small-scale synthetic work; (iii) pilot-scale LC in 50–110 mm i.d. columns, used to generate 100 g to several kg of material, and (iv) production-scale LC in columns of up to 1 m i.d. as a unit process on fine-chemical plant. High performance can be achieved at all diameters, but only at a severe cost, since 10-μm packings sell at about £1000 kg^{-1}, and solvent costs will be considerable.

16.4.1 *Preparative chiral separations*

In a recent review of chiral preparative separations (Pirkle and Hamper, 1987), approximately 12 different types of CSP were cited as useful for preparative enantiomer separations at scales ranging from 1 mg to 20 g per pass. The largest single separation cited was for ligand-exchange LC on resin-supported hydroxyproline, but this record has since been superseded. According to one report (Pirkle, 1987), the enantiomers of N-(2-naphthyl)valine esters (**25**) were separated in a 125 mm × 120 cm column containing 13 kg of dinitrobenzoyl-(*S*)-leucinamidopropylsilica (**26**). Each 5-h pass separated 125 g of racemate with an α-value of 5.

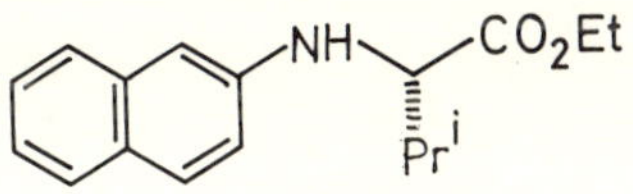

(**25**)

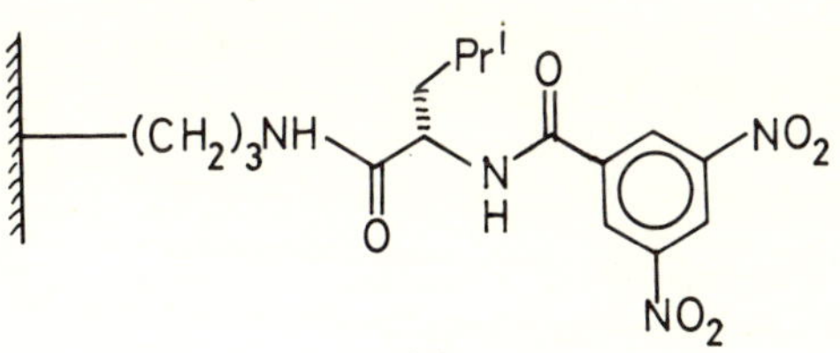

(**26**)

Naturally the scale of separations achievable depends on resolution, which in turn depends upon particle diameter (normally 10–20 μm for chiral separations at this scale), column diameter and the selectivity offered by the CSP. For development work in association with a research programme to establish biological activity, columns of 50 cm × 80–110 mm i.d. are clearly attractive: they require the availability of approximately 5–10 kg of CSP on which the separation of 500 mg–50 g of racemate should be routinely achievable per pass, and in favourable cases loadings could be considerably higher. Solvent costs at this scale of operation are quite modest, since flow-rates are below 500 mL min^{-1}, and runs should normally be completed in less than four hours. Thus, there are no inherent problems, other than cost and availability of bulk quantities of a suitable CSP, to prevent direct enantiomer separations at modest pilot-scale.

16.4.2 *Preparative diastereoisomer separations*

Sometimes, diastereoisomer derivatization and the use of relatively cheap silica will be the preferred approach for preparative-scale LC separations. Recently a route to the enantiomers of gossypol (**27**) was achieved in this way via separation of the corresponding diastereoisomeric Schiff's bases (**28**) derived from (+)-phenylalanine ethyl ester (Tyson, 1988).

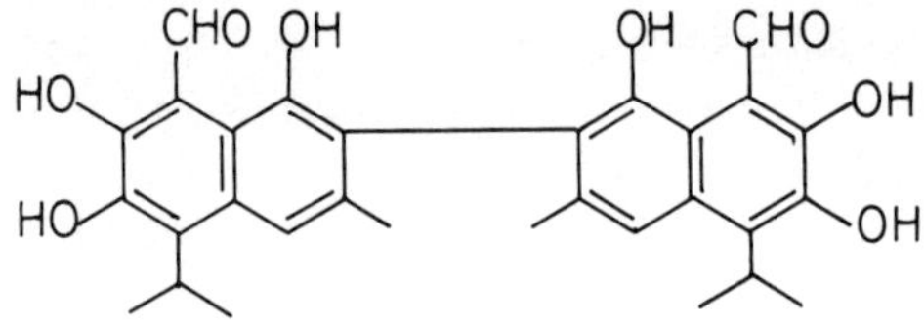

(**27**)

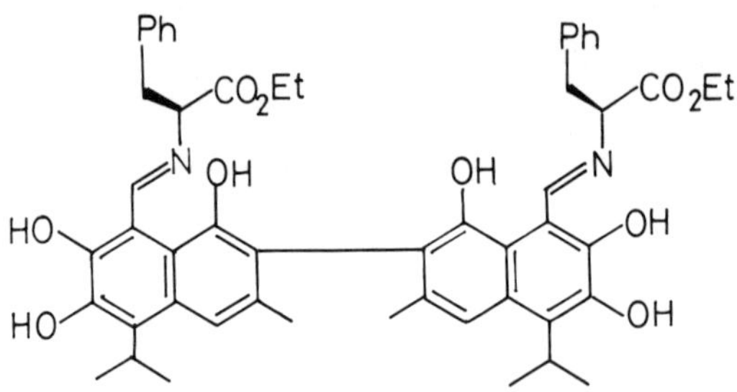

(**28**)

For diastereoisomers such as these, very-large-scale separations should be attainable on the production-scale columns of up to 300 mm which are now available commercially (unpacked, and at a price!).

The potential clearly exists for substantial increases in the scale of chiral separations: on a 300 × 1000 mm column containing about 50 kg packing, separations of up to 1 kg per pass could be envisaged, given adequate α-values. This would indicate that diastereoisomer separations could be achieved at a scale of 10 kg per day (24-h shift operation, 2.4 h per injection). For direct chiral separations, for which it is imperative that column efficiency is not sacrificed by overloading the column, this daily rate would probably have to fall to around 1 kg per day.

16.4.3 *Packings for preparative chiral HPLC*

A recent review has surveyed factors such as loadability, chemical resistance and cost of the packings currently available for general preparative LC (Unger and Janzen, 1986). Apart from affinity chromatography matrices, no CSPs were included! Clearly, the development most needed in this area is that of a low-cost route, of general scope, to a range of CSPs offering a wide range of selectivities.

For specific applications, the easiest current route to a preparative CSP would appear to be via surface modification of silica with a structure designed to manifest complementary interactions with the racemate to be separated. This approach has been used very successfully by Pirkle and co-workers (1984). Neither the globular protein nor the triacetylcellulose packings have yet been demonstrated to have loadabilities suitable for preparative-scale enantiomer separations. A recent paper described the use of a 22-mm i.d. × 500 mm column packed with 20–45 μm silica on to which 8 g bovine serum albumin was deposited (Erlandsson *et al.*, 1986), but suggested that a favourable case would yield a limiting throughput of only 1 mg h^{-1}. Microcrystalline cellulose triacetate has been used semipreparatively for the isolation of β-blocking agents, in two columns of 600 mm × 10 mm i.d. coupled in series (Isaksson and Lamm, 1986), and by the same group for the resolution of racemic alkylidenes (**29**) using columns of 25 mm i.d. × 300 mm (Hussenius *et al.*, 1987). In the latter case, 100 mg injections were possible.

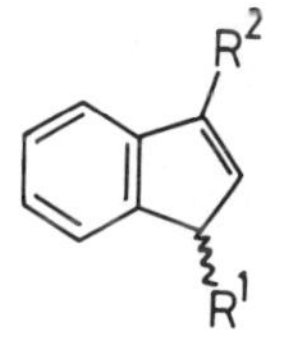

(**29**) R^1 = Me, Et, or *i*-Pr, R^2 = *n*-Pr, *n*-Bu; R^1 = *i*-Pr, R^2 = Me, Et

The costs of silica-supported cellulose or helical polyacrylates are currently prohibitive for up-scale work. Polymeric cyclodextrin has been used for very small-scale preparative separations of alkaloids (Zsadon *et al.*, 1986), but the high-efficiency bonded cyclodextrins only for analytical separations.

The best future prospects for cheap preparative CSPs seem to lie either with on-the-surface polymerization of chiral monomers to yield a chirally modified silica (Unger *et al.*, 1988; Kinkel and Reichert, 1988), or in the development of asymmetric rigid bulk polymers such as polyacrylamides. Quantities of *c.* 250 mg of anticancer 1, 3, 2-oxazaphosphorine enantiomers (**30**) were separated by low-pressure (3-bar) LC on polyacrylamides (**31**) derived from (*S*)-phenylalanine (Blaschke and Maibaum, 1986).

In spite of the limited number of preparative enantiomer separations reported hitherto, it is perfectly clear that pilot-scale separations of this type can and will be undertaken in future for strictly commercial objectives. The

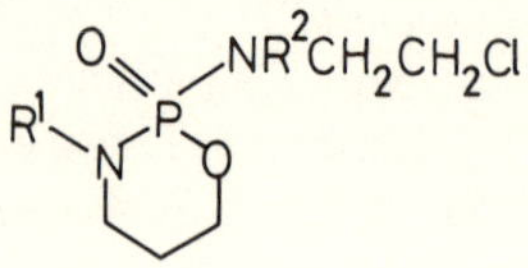

(**30**) $R^1 = CH_2CH_2Cl$, $R^2 = H$ or CH_2CH_2Cl; $R^1 = H$, $R^2 = CH_2CH_2Cl$

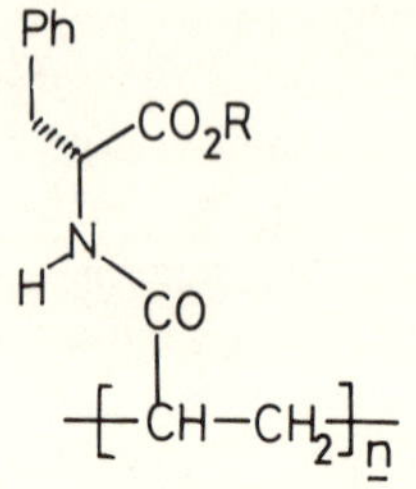

(**31**) R = Me or CH_2Ph

availability of such methodology will in turn affect the way in which development departments design the scaling-up of laboratory synthesis, as well as making available a much greater range of optically pure materials at the multigram scale, the cost of which, while they will remain high as long as CSP manufacturing costs remain high, will bear little relation to the prices currently quoted for pure enantiomers. This in turn has exciting implications for both asymmetric organic synthesis and for the design of new chirals stationary phases for HPLC.

References

Allenmark, S., B. Bomgren and H. Boren (1984) Direct separation of enantiomers on immobilised protein stationary phases 4. Molecular interaction forces and retention behaviour in chromatography on bovine serum albumin as stationary phase. *J. Chromatogr.* **316**, 617–624.

Allenmark, S. (1986) Optical resolution by liquid chromatography on immobilised bovine serum albumin. *J. Liq. Chromatogr.* **9**, 425–442.

Armstrong, D.W. (1984) Chiral stationary phases for the HPLC separation of enantiomers: a mini-review. *J. Liq. Chromatogr.* **7**, 353–376.

Armstrong, D.W., T.J. Ward, R.D. Armstrong and T.E. Beesley (1986) Separation of drug stereoisomers by the formation of β-cyclodextrin inclusion complexes. *Science* **232**, 1132–1135.

Aubel, M.T. and L.B. Rogers (1987) Effects of pretreatment on the enantioselectivity of silica-bound proteins used as HPLC stationary phases. *J. Chromatogr.* **408**, 99–113.

Berridge, J.C. (1985) *Techniques for the Automated Optimization of HPLC Separations.* Wiley, Chichester.

Blaschke, G. and J. Maibaum (1986) Optical resolutions of the racemic anticancer drug ifosfamide and other chiral oxazaphosphorines. *J. Chromatogr.* **366**, 329–334.

Bobbitt, D.R. and E.S. Yeung (1984) Direct and indirect polarimetry for detection in microbore liquid chromatography. *Anal. Chem.* **56**, 1577–1581.

Bobbitt, D.R. and E.S. Yeung (1985) Absorption detection in microcolumn liquid chromatography via indirect polarimetry. *Anal. Chem.* **57**, 271–274.

Boehme, W., G. Wagner, U. Oehme and U. Priesnitz (1982) Spectrophotometric and polarimetric detectors in liquid chromatography for the determination of enantiomers in complex mixtures. *Anal. Chem.* **54**, 709–715.

Bridger, G. (1987*b*) PhD Thesis, University of Manchester Institute of Science and Technology, UK.

Bridger, G., R. Husband, M.J. Tandy and D.R. Taylor (1987*a*) Applications of computerised molecular graphics to enantioseparation in HPLC. *Int. Symp. on Chiral Separations*, Guildford, UK.

Dalgliesh, C.E. (1952) The optical resolution of aromatic aminoacids on paper chromatograms. *J. Chem. Soc.* 3940–3942.

Däppen, R., H. Arm and V.R. Meyer (1986) Applications and limitations of commercially available chiral stationary phases for HPLC. *J. Chromatogr.* **373**, 1–20.

Davankov, V.A. (1980) Resolution of racemates by ligand exchange chromatography. *Adv. in Chromatogr.* **18**, 139–195.

Davankov, V.A. and A.A. Kurganov (1980) Ligand exchange chromatography of racemates 12: HPLC of aminoacids on a polystyrene resin with fixed ligands of the type (R)-N′, N′-dibenzyl-1, 2-propanediamine. *Chromatographia* **13**, 339–341.

Dobashi, Y. and S. Hara (1985) Direct resolution of enantiomers by liquid chromatography with the novel stationary phase derived from (R, R)-tartramide. *Tetrahedron Lett.* **26**, 4217–4220.

Dumont, W., J.-C. Poulin, T.-P. Dang and H.B. Kagan (1973) Asymmetric catalytic reduction with transition metal complexes. II. Asymmetric catalysis by a supported chiral rhodium catalyst. *J. Amer. Chem. Soc*, **95**, 8295–8299.

Erlandsson, P., L. Hansson and R. Isaksson (1986) Direct analytical and preparative resolution of enantiomers using albumin adsorbed to silica as stationary phase. *J. Chromatogr.* **370**, 475–483.

Erlandsson, P. and S. Nilsson (1987) Albumin fragments used as chiral stationary phases. *11th Int. Symp. on Column Liquid Chromatography*, Amsterdam.

Focus, S.A.B. (1986) Expert systems for liquid chromatography. *Anal. Chem.* **58**, 1192A–1200A.

Gilbert, M.T. (1987) *High Performance Liquid Chromatography* Wright, Bristol, Chapter 4.

Goodall, D.M. and D.K. Lloyd (1988) An optical activity detector for HPLC. In *Chiral Separations*, eds. Stephenson, D. and I.D. Wilson, Plenum, London.

Halpern, J. (1985) Asymmetric catalytic hydrogenation: mechanism and origin of enantioselection. In *Asymmetric Synthesis*, ed. Morrison, J.D. Vol. 5, Academic Press, New York, Chapter 3.

Hermansson, J. (1983) Direct liquid chromatographic resolution of racemic drugs using α_1-acid glycoprotein as the chiral stationary phase. *J. Chromatogr.* **269**, 71–80.

Hermansson, J., K. Strom and R. Sandberg (1987) Relationship between enantioselectivity and solute structure on a chiral α_1-acid glycoprotein column. *Chromatographia* **24**, 520–526.

Hussenius, A., R. Isaksson and O. Matsson (1987) Separation of alkyl-substituted indenes by chromatography on microcrystalline triacetylcellulose. *J. Chromatogr.* **405**, 155–162.

Ichida, A., T. Shibata, I. Okamoto, Y. Yuki, H. Namikashi and Y. Toga (1984) Resolution of enantiomers by HPLC on cellulose derivatives. *Chromatographia* **19**, 280–284.

Isaksson, R. and B. Lamm (1986) Semipreparative separation of cyclic carbamates of β-blocking agents on swollen microcrystalline triacetylcellulose. *J. Chromatogr.* **362**, 436–438.

Kinkel, J.N. and K. Reichert (1988) Preparative chromatographic separation of enantiomers on microcrystalline cellulose triacetate and optically active polymeric bonded silica. *2nd Int. Symp. on Preparative and Upscale LC*, Baden-Baden, 1988.

Kip, J., P. Van Haperen and J.C. Kraak (1986) (R)-N-(pentafluorobenzoyl)phenylglycine as chiral stationary phase for separation of enantiomers by HPLC. *J. Chromatogr.* **356**, 423–427.

Lipkowitz, K.B., D.A. Demeter, C.A. Parish, J.M. Landwer and T. Darden (1987*a*) Theoretical studies of a chiral stationary phase used in column chromatography. *J. Comput. Chem.* **8**, 753–760.

Lipkowitz, K.B., D.A. Demeter, C.A. Parish and T. Darden (1987*b*) Enantioselective binding of 2, 2, 2-trifluoro-1-(9-anthryl)ethanol on a chiral stationary phase: a theoretical study. *Anal. Chem.* **59**, 1731–1733.

Macaudiere, P., M. Caude, R. Rosset and A. Tambuté (1987) Resolution of racemic amides and phosphine oxides on a β-cyclodextrin bonded phase by subcritical fluid chromatography. *J. Chromatogr.* **405**, 135–143.

Maguire, J.H. (1987) Some structural requirements for resolution of hydantoin enantiomers with a β-cyclodextrin liquid chromatography column. *J. Chromatogr.* **387**, 453–458.

Morrison, J.D. (ed.) (1983–87) *Asymmetric Synthesis*, Vols. 1–5, Academic Press, New York.

Ôi, N., M. Nagase, Y. Inda and T. Doi (1983) HPLC separation of enantiomers on (S)-2-(4-chlorophenyl)isovaleric acid and its amide derivatives bonded to silica gel. *J. Chromatogr.* **265**, 111–116.

Okamoto, Y., M. Kawashima, K. Yamamoto and K. Hatada (1984) Chromatographic resolution 6. Useful chiral packing materials for HPLC resolution: cellulose triacetate and tribenzoate coated on macroporous silica gel. *Chem. Lett.* 739–742.

Okamoto, Y. and K. Hatada (1986) Resolution of enantiomers by HPLC on optically-active poly-(triphenylmethylmethacrylate). *J. Liq. Chromatogr.* **9**, 369–384.

Pettersson, C. and C. Gioeli (1988) Improved resolution of enantiomers of naproxen by the simultaneous use of a chiral stationary phase and a chiral additive in the mobile phase. *J. Chromatogr.* **435**, 225–228.

Pirkle, W.H. (1983) Stacking of amide dipoles as an alternative to hydrogen bonds in chiral recognition rationales. *Tetrahedron Lett.* **24**, 5707–5708.

Pirkle, W.H. (1987) New developments in chiral stationary phases for HPLC. *11th Int. Symp. on Column Liquid Chromatography*, Amsterdam.

Pirkle W.H. and R. Däppen (1987) Reciprocity in chiral recognition: comparison of several chiral stationary phases. *J. Chromatogr.* **404**, 107–115.

Pirkle, W.H. and B.C. Hamper (1987) The direct preparative resolution of enantiomers by LC on chiral stationary phases. In *Preparative Liquid Chromatography*, ed. Bidlingmeyer, B.A., Elsevier, Amsterdam.

Pirkle, W.H. and T.C. Pochapsky (1986*a*) A new easily accessible reciprocal chiral stationary phase for the chromatographic separation ot enantiomers. *J. Amer. Chem. Soc.* **108**, 352–354.

Pirkle, W.H. and T.C. Pochapsky (1986*b*) Generation of extreme selectivity in chiral recognition. *J. Chromatogr.* **369**, 175–177.

Pirkle, W.H. and T.C. Pochapsky (1987) Chiral molecular recognition in small bimolecular systems: a spectroscopic investigation into the nature of diastereomeric complexes. *J. Amer. Chem. Soc.* **109**, 5975–5982.

Pirkle, W.H., D.W. House and J.M. Finn (1980) Broad spectrum resolution of optical isomers using chiral HPLC bonded phases. *J. Chromatogr.* **192**, 143–158.

Pirkle, W.H., J.M. Finn, B.C. Hamper, J. Schreiner and J.M. Pribush (1982) A useful and conveniently accessible chiral stationary phase for the liquid chromatographic separation of enantiomers. In *Asymmetric Reactions and Processes*, ACS Symp. Ser. **185**, eds. Eliel, E.L. and S. Otsaka, American Chemical Society, Washington DC.

Pirkle, W.H., M. Hyun, A. Tsipouras, B.C. Hamper and B. Banks (1984) A rational approach to the design of highly effective chiral stationary phases for liquid chromatographic separation of enantiomers. *J. Pharm. and Biomed. Anal.* **2**, 173–181.

Pirkle, W.H., R. Däppen and D.S. Reno (1987) Applications for racemic versions of chiral stationary phases. *J. Chromatogr.* **407**, 211–216.

Reitsma, B.H. and E.S. Yeung (1986) HPLC determination of enantiomeric ratios of aminoacids without chiral separation. *J. Chromatogr.* **362**, 353–362.

Rosini, C., P. Altemura, D. Pini, C. Bertucci, G. Zullino and P. Salvadori (1985) Cinchona alkaloids for preparing new easily accessible chiral stationary phases. 2: Resolution of binaphthyl derivatives on silica supported quinine. *J. Chromatogr.* **348**, 79–87.

Salem, L., X. Chapuisat, G. Segal, P.C. Hilberty, C. Minot, C. Leforestier and P. Sautet (1987) Chirality forces. *J. Amer. Chem. Soc.* **109**, 2887–2894.

Schoenmakers, P. (1986) *Optimisation of Chromatographic Selectivity* (*J. Chromatogr.* Library Vol. 35), Elsevier, Amsterdam.

Shah, P., T-B. Hsu and L.B. Rogers (1987) ^{1}H and ^{13}C NMR studies of enantioselective hydrogen bonding of fluoroalkylcarbinol by chiral butoxycarbonyl-D-valine. *J. Chromatogr.* **396**, 31–41.

Tambuté, A., A. Begos, M. Lienne, M. Caude and R. Rosset (1987) New chiral stationary phases containing a phosphorus atom as an asymmetric centre: synthesis and first chromatographic results. *J. Chromatogr.* **396**, 65–81.

Tanaka, M., J. Okasaki, H. Ikeda and T. Shono (1986) Methylated cyclodextrin-bonded stationary phases for liquid chromatography. *J. Chromatogr.* **370**, 293–301.

Tanaka, M., H. Ikeda and T. Shono (1987) Carbamoylated cyclodextrin-bonded stationary phases for liquid chromatography. *J. Chromatogr.* **398**, 165–173.

Taylor, D.R. (1986) Analysis of racemic mixtures. *Laboratory Practice*, January, 49–52.

Tyson, R. (1988) Custom synthesis and process development. *Chem. & Ind.* 15 February, 118–122.

Unger, K.K. and R. Janzen (1986) Packings and stationary phases in preparative column liquid chromatography. *J. Chromatogr.* **373**, 227–264.

Unger, K.K., U. Muller and J.N. Kinkel (1988) Packings and stationary phases in preparative

LC – novel concepts and improved products. *2nd Int. Symp. on Preparative and Upscale LC*, Baden-Baden.

Wainer, I.W. (1987) Proposals for classification of HPLC chiral stationary phases. *Trends in Anal. Chem.* **6**, 125–134.

Wainer, I.W. and M.C. Alembik (1986) Resolution of enantiomeric amides on a cellulose-based CSP: steric and electronic effects. *J. Chromatogr.* **358**, 85–93.

Wainer, I.W., S.A. Barkan and G. Schill (1986*a*) α_1-Acid glycoprotein chiral stationary phase: HPLC application to resolution of enantiomeric drugs. *LC–GC* **4**, 422–430.

Wainer, I.W., T.D. Doyle and F.S. Fry (1986*b*) Chiral recognition model for the resolution of ephedrine and related aminoalcohols as enantiomeric oxazolidine derivatives. *J. Chromatogr.* **355**, 149–156.

Wainer, I.W., M.C. Alembik and E. Smith (1987) Resolution of enantiomeric amides on a cellulose tribenzoate chiral stationary phase: mobile phase modifier effects on retention and stereoselectivity. *J. Chromatogr.* **388**, 65–74.

Ward, T.J. and D.W. Armstrong (1986) Improved cyclodextrin chiral phases, a comparison and review. *J. Liq. Chromatogr.* **9**, 407–423.

Weller, O., J. Schulze and W.A. Konig (1987) Separation of racemic pharmaceuticals by HPLC on silica gel modified with carbohydrate residues. *J. Chromatogr.* **403**, 263–270.

Wulff, G., R. Kemmerer and B. Vogt (1987) Chirality of polyvinyl compounds 5. Optically active polymers with structural chirality in the main chain through asymmetric cyclocopolymerisation. *J. Amer. Chem. Soc.* **109**, 7449–7457.

Yeung, E.S. (1986*a*) Circular dichroism detectors. In *Detectors for Liquid Chromatography*, ed. Yeung, E.S., Wiley-Interscience Chemical Analysis Monographs Vol. 89, Wiley, New York.

Yeung, E.S. (1986*b*) Polarimetric detectors. In *Detectors for Liquid Chromatography*, ed. Yeung, E.S., Wiley-Interscience Chemical Analysis Monographs Vol. 89, Wiley, New York.

Zsadon, B., M. Szilasi, L. Decsei, A. Ujhasy and J. Szejtli (1986) Variation in the selectivity in the resolution of alkaloid enantiomers on cross-linked cyclodextrin polymer stationary phase. *J. Chromatogr.* **356**, 428–432.

Appendix

Examples of commercially-available chiral HPLC columns

The following table lists chiral HPLC columns available from suppliers in the UK and USA, with prices. Prices refer to columns of dimensions 250×4–5mm, unless otherwise stated; most of the column descriptions are registered trademarks; numbers in parentheses refer to the chapter(s) in which the class of CSP is discussed.

UK

J. T. Baker U.K., PO Box 9, Hayes Gate House, 27 Uxbridge Road, Hayes, Middlesex, UB4 0JD.

DNBG, ionic, 5μm; £350 (6)
DNBG, covalent, 5μm; £400 (6)
DNB Leu, covalent, 5μm; £400 (6)
CHIRALPAK OT (+) £855 (10)
CHIRALPAK OP (+) £855 (10)
CHIRALCEL OA £855 (9)
CHIRALCEL OB £855 (9)
CHIRALCEL OC £855 (9)
CHIRALCEL OK £855 (9)
CHIRALPAK WH £855 (5)
CHIRALPAK WM £855 (5)
CHIRALPAK WE £855 (5)
CROWNPAK CR £855 (12)
CHIRALCEL OC £855 (8)
CHIRALCEL OF £855 (8)
CHIRALCEL OG £855 (8)
CHIRALCEL OJ £855 (8)

Technicol Ltd., Brook Street, Higher Hillgate, Stockport, Cheshire, SK1 3HS
as per J. T. Baker U.K. plus:

CYCLOBOND I; £260 (8)
CYCLOBOND I, acetylated; £290 (8)
CYCLOBOND II; £260 (8)
CYCLOBOND III £260 (8)
CYCLOBOND III acetylated; £290 (8)
RESOLVOSIL (150mm×4mm); £310 (7)

Pharmacia Ltd., Pharmacia LKB Biotechnology, Pharmacia House, 351 Midsummer Boulevard, Central Milton Keynes, MK9 3YY.

ENANTIOPAC (100mm×4mm) £404 (7)

Other U.K. suppliers

Field Analytical, Weybridge, Surrey (5, 6, 7)
HPLC Technology, Macclesfield, Cheshire (5, 6, 7)
Supelchem, Sawbridgeworth, Hertfordshire (6)
Camlab, Cambridge, Cambridgeshire (5, 6)
Hichrom, Reading, Berkshire (6)
Jones Chromatography, Hengoed, Mid-Glamorgan (6)
Phase Sep, Deeside, Clwyd (6)

USA

Regis Chemical Company, 8210 Austin Avenue, PO Box 519, Morton Grove, Illinois 60053, USA.

REGIS, ionic, D-phenylglycine; $400 (6)
REGIS, ionic, L-leucine; $420 (6)
REGIS, covalent, D-phenylglycine; $500 (6)
REGIS, covalent, L-phenylglycine; $525 (6)
REGIS, covalent, L-leucine; $525 (6)
REGIS, covalent, D-leucine; $625 (6)
SUMITOMO, covalent, OA-3100 $1150 (6)
REGIS, covalent, D-naphthylalanine; $600 (6)
REGIS, covalent, D-naphthylalanine; $600 (6)
SUMITOMO, covalent, OA-1000 (150mm×4mm); $750 (6)
SUMITOMO, covalent, OA-1100 (150mm×4mm); $750 (6)
SUMITOMO, covalent, OA-2200 (150mm×4mm); $750 (6)
SUMITOMO, covalent, OA-4000 (150mm×4mm); $1150 (6)
SUMITOMO, covalent, OA-4100 (150mm×4mm); $1150 (6)
SUMITOMO, covalent, OA-3000 (150mm×4mm); $750 (6)

Supelco Inc., Supelco Park, Belle Fonte, Philadelphia 16823-0048.

SUPELCOSIL LC-(R)-UREA; $300 (6)

Bodman Chemicals, PO Box 500, Media, Philadelphia 1906.

Regis Pirkle columns plus:
TSK GEL ENANTIO P1; $995 (6)
TSK GEL ENANTIO L1; $995 (5)
TSK GEL ENANTIO L2; $995 (5)

Other US suppliers

Alltech Associates, Dearfield, Il (5, 6, 7)
Rainin, Emeryville, CA (7, 8)
J T Baker, Phillipsbury, NJ (5, 6, 9)
ASTEC, Whippany, NJ (8)

European suppliers

Hewlett-Packard, Waldron, West Germany (10)
Merck, Darmstadt, West Germany (10)
Macherey-Nagel, Düren, West Germany (5, 6, 7)

Index